Springer Theses

Recognizing Outstanding Ph.D. Research

Aims and Scope

The series "Springer Theses" brings together a selection of the very best Ph.D. theses from around the world and across the physical sciences. Nominated and endorsed by two recognized specialists, each published volume has been selected for its scientific excellence and the high impact of its contents for the pertinent field of research. For greater accessibility to non-specialists, the published versions include an extended introduction, as well as a foreword by the student's supervisor explaining the special relevance of the work for the field. As a whole, the series will provide a valuable resource both for newcomers to the research fields described, and for other scientists seeking detailed background information on special questions. Finally, it provides an accredited documentation of the valuable contributions made by today's younger generation of scientists.

Theses are accepted into the series by invited nomination only and must fulfill all of the following criteria

- They must be written in good English.
- The topic should fall within the confines of Chemistry, Physics, Earth Sciences, Engineering and related interdisciplinary fields such as Materials, Nanoscience, Chemical Engineering, Complex Systems and Biophysics.
- The work reported in the thesis must represent a significant scientific advance.
- If the thesis includes previously published material, permission to reproduce this must be gained from the respective copyright holder.
- They must have been examined and passed during the 12 months prior to nomination.
- Each thesis should include a foreword by the supervisor outlining the significance of its content.
- The theses should have a clearly defined structure including an introduction accessible to scientists not expert in that particular field.

More information about this series at http://www.springer.com/series/8790

Lars Von der Wense

On the Direct Detection of ^{229m}Th

Doctoral Thesis accepted by
Ludwig-Maximilians-University, Garching, Germany

Springer

Author
Dr. Lars Von der Wense
Department of Medical Physics
Ludwig-Maximilians-University
Garching, Bavaria
Germany

Supervisor
Prof. Peter G. Thirolf
Ludwig-Maximilians-University
Garching, Bavaria
Germany

ISSN 2190-5053 ISSN 2190-5061 (electronic)
Springer Theses
ISBN 978-3-319-70460-9 ISBN 978-3-319-70461-6 (eBook)
https://doi.org/10.1007/978-3-319-70461-6

Library of Congress Control Number: 2017957623

Printed on acid-free paper

This Springer imprint is published by Springer Nature
The registered company is Springer International Publishing AG
The registered company address is: Gewerbestrasse 11, 6330 Cham, Switzerland

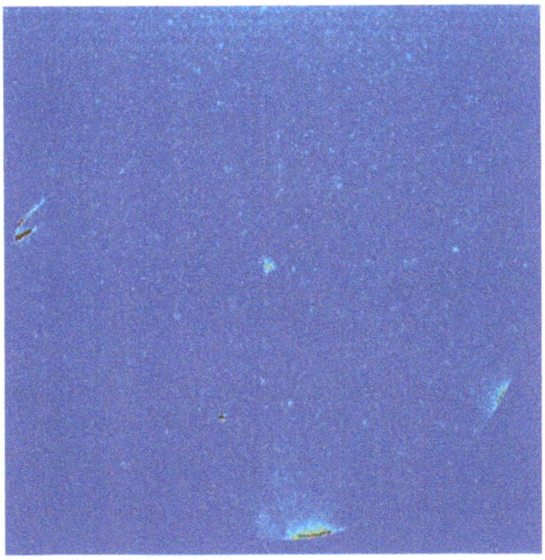

First indication of the decay of ^{229m}Th (central spot) as observed in the raw data on October 15, 2014. The signal was obtained when collecting $^{229}Th^{3+}$ ions, produced in the α decay of ^{233}U, with low kinetic energy directly on the surface of a microchannel-plate detector connected to a phosphor screen. The picture shows a CCD camera image of the phosphor screen with a field-of-view of about 75 $\times$ 75 mm^2 and 400 s exposure time. Intense signals at the edge of the detector originate from field emissions. After a multitude of exclusion measurements, the observed signal could be unambiguously attributed to the thorium isomeric decay.

I dedicate this work to my grandparents
Erna and Walter Liedtke
as well as
Eva-Maria and Herbert Wendorf

Supervisor's Foreword

The 'flow of time' affects our life in all its aspects, from surrounding nature to structuring our daily activities or to the global memory of human history. Presently the precision of our 'official' timekeeping is provided by atomic clocks, where electronic transitions in the atomic shell serve as a frequency standard. Optical atomic clocks already achieve a superb precision of 1 second within 10 times the age of the universe, hardly conceivable to be further improved. Nevertheless, there is the intriguing perspective of a nuclear clock outperforming even the best optical atomic clocks. Since atomic nuclei are about 5 orders of magnitude smaller than atoms with their spacious electron shells, they are much less sensitive to external perturbations like electric or magnetic fields. Therefore a considerably improved stability of a nuclear frequency standard can be expected, based on a suitable transition in the nuclear level scheme. And here comes the problem: amongst all the presently ca. 3400 known atomic nuclei with their currently known ca. 178,000 excited energy levels, there is only one candidate that qualifies for a nuclear clock transition: the lowest excited state in the whole nuclear landscape, namely the isomeric first excited state of the actinide isotope ^{229}Thorium with a presently assumed excitation energy of 7.8(5) eV (corresponding to a transition wavelength of about 160 nm) and an expected half-life of a few thousand seconds. Thus an extremely narrow relative linewidth of the isomeric ground-state transition of $\Delta E/E \sim 10^{-20}$ can be expected, which would be ideally suited to serve as an ultra-precise nuclear frequency standard. This so-called 'thorium isomer' was known from indirect measurements since about 40 years, however, and despite intense world-wide efforts, could never be directly detected. Thus the nuclear clock stayed purely hypothetical. The long-searched direct detection of the isomeric deexcitation to the ground state was now for the first time achieved by Lars von der Wense in the framework of his Ph.D. thesis, paving the road towards a nuclear clock with far-reaching consequences in many fields, just mentioning here considerably improved satellite-based navigation (GPS), relativistic geodesy or the search for temporal variations of fundamental constants like the fine-structure constant.

Receiving his scientific education in physics and mathematics at the University of Hamburg, Lars earned his diploma in physics with a theoretical topic in string theory. After having moved to the University of Cambridge in the UK he passed with honors as a Master of Advanced Study in Applied Mathematics. Despite this strong theoretical background he bravely embarked on a new challenge when taking over the strongly technology-driven experimental task for his Ph.D. project in Munich to search for the direct decay of the thorium isomer. Using a technologically highly complex and multi-stage experimental setup he succeeded to generate a pure particle beam of the thorium isomer and could unambiguously detect a signal of the energetic de-excitation of this exotic excited nuclear state. It should be noted that final success was not achieved in the photonic decay branch, which is investigated in all competing projects worldwide (and also in the first phase of Lars' thesis work), but he draw the right conclusions from the non-observation of decay photons and re-directed the experimental focus towards the internal conversion decay channel, searching for conversion electrons rather than ultraviolet photons, which finally led to a highly visible breakthrough. His thesis describes the impressive scope of experimental work (and theoretical foundations) that were required to reach this goal. This achievement gained widespread attention worldwide and Lars von der Wense published this pioneering result in May 2016 in 'Nature', followed by various media coverages and conference invitations. At the end of 2016, these results were chosen by the editors of 'Physics World' into the Top-10 list of most important breakthroughs in physics of 2016.

I am very pleased that Springer publishes Lars von der Wense's excellent thesis and I am convinced that readers will also appreciate this experimental 'piece of art'.

Garching, Germany
August 2017

PD Dr. Peter G. Thirolf

Abstract

The measurement of time has always been an important tool in science and society. Today's most precise time and frequency measurements are performed with optical atomic clocks. However, these clocks could potentially be outperformed by a 'nuclear clock,' which employs a nuclear transition instead of an atomic shell transition for time measurement. Among the 176,000 known nuclear excited states, there is only one nuclear state that would allow for the development of a nuclear clock using currently available technology. This is the isomeric first excited state of ^{229}Th, denoted as ^{229m}Th. Despite 40 years of past research, no direct decay detection of this nuclear state has so far been achieved.

In this thesis, measurements are described that have led to the first direct detection of the ground-state decay of ^{229m}Th. Two decay channels (radiative decay and internal conversion) are experimentally investigated. Only the investigation of the internal conversion decay channel has led to the successful observation of the first excited isomeric nuclear state of ^{229}Th. Based on this direct detection, a new nuclear laser excitation scheme for ^{229m}Th is proposed. This excitation scheme circumvents the general assumed requirement of a better knowledge of the isomeric energy value, thereby paving the way for nuclear laser spectroscopy of ^{229m}Th.

Many of the presented results have so far been unpublished. This includes results of the investigation of a potential radiative decay channel of ^{229m}Th, a negative result in the search for an isomeric decay during extraction of $^{229}Th^{1+}$ as well as the, investigation of the isomeric decay in thorium molecules and on an MgF_2-coated surface.

Acknowledgements

The presented experimental results are the outcome of joint work to which many people have contributed in a direct or indirect way. Only the supportive and encouraging atmosphere within our own research group as well as strong external collaborations have made the direct detection of ^{229m}Th become possible. In the following, I will try to give credits and say thank you to all people who have contributed to this work and made the time of my Ph.D. thesis the positive and successful experience that it has been.

For offering me the opportunity to work on this very fruitful topic, his unsurpassed support during the whole work of my Ph.D. thesis, his encouragement and also for giving me the freedom to explore new paths, I am extremely grateful to my supervisor PD Dr. Peter G. Thirolf. Without his support, this work would have been impossible.

I am especially grateful to Benedict Seiferle, my longtime companion in all experimental work. Without his refreshing humor during the uncounted joint hours in front of the experiment, the fruitful and enlightening discussions, and his deep physical insights, the time of my Ph.D. would not have been half as rich and delightful as it was.

Special thanks go to my fellow labmates for all the joint time and shared experiences: Dr. Saad Aldawood, Dr. Christian Lang, Dr. Christine Weber, Silvia Liprandi, Michael Mayerhofer, Tim Binder, Dr. Lorant Csige, Dr. Sergey Gasilov, Stefan Gärtner as well as my former roommate Peter Müller. Further thanks go to Prof. Katia Parodi and the whole chair for medical physics.

I would also like to thank Dr. Mustapha Laatiaoui for fruitful discussions, preparatory measurements, and support in performing the experiments, as well as Dr. Jürgen B. Neumayr for the development and construction of the buffer-gas stopping cell, which provides the basis for all our experimental investigations.

For many discussions and encouragements, I am grateful to Prof. Dieter Habs, Prof. Jörg Schreiber, Peter Hilz, Tobias Ostermayr, Daniel Haffa, Klaus Allinger, Florian Lindner, Christian Kreuzer, and Prof. Paul Bolton.

For technical support and support regarding radiation protection as well as source production, I am grateful to Dr. Hans-Friedrich Wirth, Hans-Ulrich Friebel, Dr. Hans-Jörg Maier, Simon Stork, Dr. Ludwig Beck, Dr. Peter Hartung as well as the whole operator team and workshop team of the Maier-Leibnitz Laboratory Tandem accelerator facility. Further thanks go to Rolf Oehm and his whole workshop team, Dr. Johannes Wulz and the construction office, as well as to Johann Krapfl and the electronic workshop.

During the time of my Ph.D. thesis, our group has been working in close collaboration with the University of Vienna. I am very grateful to Dr. Simon Stellmer and Prof. Thorsten Schumm for initiating the European Union's Horizon 2020 Project 'nuClock.' This project has not only led to a big financial support of our group, including the possibility to continue my own research, but also triggered a very active communication and sharing of experiences within the community. I am especially grateful to Dr. Simon Stellmer for many fruitful discussions and his detailed support and corrections during the process of our publications, which have significantly improved the presentation and outreach of our work. Further, I would like to acknowledge the temporary loan of equipment, which has opened the door for some of the recent experiments.

I would also like to thank our collaborators from the Institute for Nuclear Chemistry at the University of Mainz, Prof. Christoph E. Düllmann, Christoph Mokry, Dr. Jörg Runke, Dr. Klaus Eberhardt as well as Prof. Norbert G. Trautmann. Without their support in production of new uranium sources, most of the experiments could not have been conducted.

Further thanks go to our close collaborators at the Max Planck Institute for quantum optics: Prof. Thomas Udem for many enlightening discussions, his strong interest in the topic, and not least for acting as a second referee for this thesis; Dr. Johannes Weitenberg, Dr. Tobias Lamour, Dr. Akira Ozawa, and Dr. Elisabeth Peters for support and discussions as well as Prof. Theodor W. Hänsch for encouragement and financial support during the work of this thesis.

Only recently the group of Dr. Ekkehard Peik from the PTB in Braunschweig has started experimental investigations of the ^{229m}Th-shell hyperfine structure in collaboration with us. For great experimental collaboration and fruitful discussions, I would like to thank Johannes Thielking, Dr. Przemyslaw Glowacki, Dr. Maxim Okhapkin, Dr. David-Marcel Maier as well as Dr. Ekkehard Peik.

For ongoing collaborative support in the development and construction of a cryogenic Paul trap, I would like to thank Dr. Lisa Schmöger, Peter Micke, Dr. Maria Schwarz as well as Dr. José Crespo and Dr. Piet O. Schmidt from the Max-Planck-Institute for Nuclear Physics in Heidelberg and the PTB in Braunschweig.

For support in the development and construction of a high-resolution electron spectrometer, I would like to thank Dr. Francesco Allegretti and Prof. Peter Feulner from the Technical University of Munich.

For support in the development of a mass spectrometer, discussions, and shared information, I am grateful to Dr. Wolfgang Plaß, Dr. Moritz P. Reiter, Christine Hornung, Ann-Katrin Rink, and Dr. Emma Hättner.

Further, I would like to thank all collaborators of the 'nuClock' consortium for fruitful discussions, support, and encouragements. Besides the already mentioned, this includes Dr. Adriana Pálffy from the Max-Planck-Institute for Nuclear Physics in Heidelberg, who I would also like to thank for her kind offer to act as a committee member for this thesis, Pavlo Bilous, Dr. Georgy Kazakov, Prof. Iain Moore, Dr. Volker Sonnenschein, Ilkka Pohjalainen, Sarina Geldhof, Prof. Christian Enss, Dr. Andreas Fleischmann, and Dr. Jürgen Stuhler.

For sharing important information and enlightening discussions, I am grateful to Dr. Jason T. Burke, Dr. Markus Sondermann, Prof. Piet van Duppen, Prof. Feodor F. Karpeshin as well as Prof. Evgeny V. Tkalya.

For support and encouragements during my Ph.D. thesis as well as my studies, I would like to thank my friends Felix Tennie, Jasmin Meinecke, Mandy Schlimm, Sarah Albrecht, Annika Reiss, Rosemarie Hackenberg as well as Steffen Volgmann.

I am extremely grateful for the unrestricted support that I have received from my girlfriend Johanna Soll. Due to her infinite encouragements and patience, she has made the time of Ph.D. thesis great again.

For unfailing support and encouragements during my whole Ph.D. thesis, my studies as well as my entire life, I would like to thank my parents Heidrun and Jens von der Wense as well as Thomas Liedtke and Anja Wietzke. Without their support, this work would have been impossible. I also want to thank my sister Johanna von der Wense for encouragements and refreshing discussions.

I would like to express my thanks and gratitude to my aunt Waltraut Oberhellmann for her generosity and financial support throughout my studies. I am grateful to Michael and Renate Gude for the longtime financial support and the marvelous time in Cologne within the frame of the Gude scholarship.

The presented work was supported by DFG (Th956/3-1), by the European Union's Horizon 2020 research and innovation programme under grant agreement 664732 'nuClock' and by the LMU department for Medical Physics via the Maier-Leibnitz Laboratory.

Contents

Chapter 1
Introduction

^{229m}Th is a metastable nuclear excited state, exhibiting a special position in the nuclear landscape as it possesses the lowest nuclear excitation energy of all about 176,000 known nuclear levels [1]. The energy has been constrained to 7.8 ± 0.5 eV and is therefore at least four orders of magnitude lower than the energy range of typical nuclear transitions [2, 3]. The special standing of ^{229m}Th compared to all other known nuclear isomeric states is visualized in Fig. 1.1, where all known nuclear isomeric states (blue circles) are shown in an energy-half-life diagram. Also selected atomic transitions used for optical clocks are shown for comparison (red circles). The low nuclear excitation energy of ^{229m}Th has led to a multitude of potential applications of which the most promising is the development of a nuclear clock that may outperform all existing atomic clock technology [4, 5]. In the following, a short history of time measurement is provided, the basic principles of clocks are discussed and the concept of a nuclear clock is introduced.

1.1 The History of Time Measurement

During the evolution of time keeping devices, clocks were developed which technologically differ from each other in the maximum imaginable way. However, despite these differences, the abstract underlying principle remained unchanged. What is required for all clocks is a physical package that exhibits oscillatory or continuous changes and acts as a time giving device. Further, some mechanism is used which counts the number of oscillations or measures the changes and translates them into a time value.

Already in historic cultures, the measurement of time has played an important role for society. Early evidence for time measurement dates back to the Sumerians, Babylonians, Egyptians and ancient Chinese at about 3000 BC [7]. At that time celestial bodies were used to develop a calendar that subdivides the year into months. To keep the daytime, mostly the position of the sun was used, measured with the help of obelisks, shadow clocks, sundials and gnomons. The time-giving device in

L. Von der Wense, *On the Direct Detection of ^{229m}Th*,
Springer Theses, https://doi.org/10.1007/978-3-319-70461-6_1

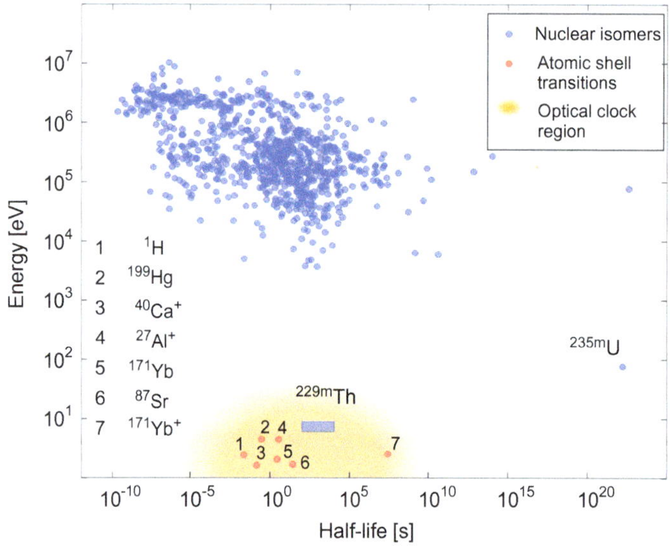

Fig. 1.1 Energy-half-life distribution of known nuclear isomeric states (blue circles) [6]. The unique standing of ^{229m}Th (expected region shown as a blue box) is visualized. Besides ^{229m}Th, also ^{235m}U possesses a low excitation energy (about 76 eV), however, at a significantly longer half-life (see Sect. 2.3.4). Purely radiative half-lives are shown for ^{229m}Th and ^{235m}U, this being the relevant parameter for the development of a nuclear clock. Also several selected atomic shell transitions are shown, used for metrology in the optical region (red circles). With kind permission of Nature research [44]

this case is the earth's rotation, which is very accurate. The length of the day varies only by a few milliseconds around its average value on short time scales [8]. On long time scales the earth rotation is slowing down due to tidal effects leading to a prolongation of the day by about 1 s in 60,000 years [9]. However, the read-out precision of sundials is significantly lower and amounts to only a few minutes. For this reason, a sundial, if properly oriented, can make a very accurate, but at the same time a not very precise clock.

In parallel to sundials, also water clocks played an important role for time measurement [7]. In these devices the flow velocity of water is used for time keeping. As the viscosity of water significantly depends on its temperature, which might have been subject to large variations over the period of a day in water clocks at ancient times, the accuracy of these clocks was rather modest compared to that of sundials and a day by day calibration was required. On short timescales, however, water clocks can be read out to higher precision and more easily operated, thereby providing advantages that justified their use as the most precise and commonly used clocks until the development of the pendulum clock in the 17th century. Also hourglasses, fire clocks and candle clocks became practical instruments for daily time measurement.

First mechanical clocks that were driven by falling weights appeared in the end of the 13th century in churches [10]. One of the oldest clocks of this type still in working condition is the Salisbury cathedral clock. These early mechanical clocks used a foliot as the time-giving device, a sort of horizontal pendulum which used adjustable weights for an increased moment of inertia. In combination with a verge escapement the foliot reduces the speed of a wheel which is driven by a falling weight. These clocks were not very accurate, and deviations from the time as defined by the rotation of the earth of 15 min or more per day may have occurred [11].

The first accurate mechanical clocks were pendulum clocks as developed by Christian Huygens in 1656, providing an error of less than 1 min per day and later less than 10 s per day [12, 13]. In 1675 Huygens also developed the first portable pocket watches that were based on the spiral spring mechanism. For centuries, pendulum clocks remained the most accurate clocks and were constantly developed. Important improvements were implemented by John Harrison, who invented a pendulum clock achieving 1 s accuracy in 100 days and who also solved the longitude problem by developing the most accurate marine chronometer [14]. The most accurate pendulum clock ever built is the Shortt-Synchronome free pendulum clock developed in 1921 [15]. It achieves an accuracy of about 1 s per year and was the first clock being more accurate than the earth rotation itself. Between the 1920s and 1940s, the Shortt clocks were operated in many countries as the most precise timekeeping devices.

Pendulum clocks were outperformed in the 1940s by quartz clocks. The first quartz clocks, built already in 1927 by Joseph Horton and Warren Marrison, provided a precision of 10^{-7}, corresponding to about 1 s in 4 month and therefore significantly less than the Shortt pendulum clock [16]. The oscillation of a quartz clock is based on the piezo-electric effect and no macroscopic mechanical motion is required for timekeeping, which is a technological advantage over mechanical pendulum clocks and led to accuracies approaching 10^{-9} in the 1940s [16]. This precision was much better than even the best mechanical pendulum clocks that were for this reason mostly substituted by quartz crystals.

The age of the atomic clocks began in 1949, when Harold Lyons built the first ammonia maser at the National Bureau of Standards in the US [17]. Developing an atomic clock had already been suggested 4 years earlier by Isidor Rabi [18]. In an atomic clock the frequency of light that corresponds to a certain energy difference between atomic shell states is used for time measurement. The first atomic clock was a proof-of-principle device and less accurate than the quartz clocks at that time. However, improvements led to the development of an accurate cesium standard by Louis Essen and Jack Parry in 1955 at the National Physical Laboratory in the UK, providing an accuracy of 10^{-9} [19]. Due to technical improvements, the accuracy of atomic clocks improved significantly in the following years, approaching 10^{-13}, corresponding to 1 s in 300,000 years in the mid 1960s [20]. This led in 1967 to the decision to redefine the second to be 9,192,631,770 times the period of the ^{133}Cs ground-state hyperfine transition. Previously the second was still defined based on the earth rotation. A further important improvement in time measurement was made in the end of the 1980s, when laser cooling allowed for longer interrogation cycles of the atomic beams, leading to the development of atomic fountain clocks with accuracies in the 10^{-16} range [21].

The next technological step was largely driven by the development of the frequency comb in 1998, that allows for direct measurement of frequencies in the optical range, five orders of magnitude larger than previously possible [22]. This improvement was immediately applied for the development of single-ion optical clocks. The first clock of this type was presented in 2001 and was a single-ion $^{199}Hg^{+}$ clock, providing a fractional frequency instability of $7 \cdot 10^{-15}$ measured in 1 s of averaging [23]. In the meantime a multitude of single-ion optical clocks were developed and are operated in metrology institutes around the world. A 2008 comparison of two single-ion optical clocks revealed a fractional uncertainty of $5.2 \cdot 10^{-17}$ [24] and in 2016 the operation of a single ion Yb^{+} optical clock at $3 \cdot 10^{-18}$ uncertainty was reported [25]. In parallel, also optical lattice clocks were developed. These clocks are based on neutral atoms instead of ions, providing the advantage that no Coulomb interactions between individual particles are present, which results in high achievable densities when atoms are stored in an optical lattice, thereby significantly increasing statistics. By today optical lattice clocks are the most precise clocks in the world, approaching a total frequency uncertainty of 10^{-18} [26, 27].

A schematic of the operational principle of modern optical atomic clocks is shown in Fig. 1.2. A physics package in form of a laser-cooled single ion in a Paul trap or several thousands of neutral atoms confined in an optical lattice is irradiated by a spectroscopy laser, locked to an ultra-stable reference cavity. The resonance line of the physics package is detected and the spectroscopy laser is accordingly tuned to the resonance center by a feedback loop. Counting of the frequency in the optical range is performed with the help of a frequency comb, which then provides the time information.

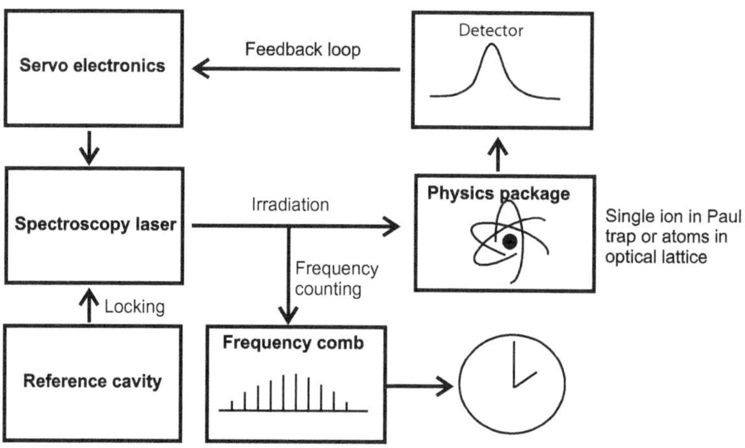

Fig. 1.2 Schematic concept of an optical atomic clock. A spectroscopy laser is tuned to the resonance line of a physics package. The laser frequency can thus be considered as constant and is counted with the help of a frequency comb. After a certain (predefined) number of periods of the light wave generated by the spectroscopy laser has passed one second has elapsed [28–30]

1.2 Quality of Clocks

Independent of the exact process of time measurement, there must be way how to compare the quality of different clocks. The quality of a clock will depend on various factors, however, there are two quantities which play the most important role when discussing the quality of clocks: Accuracy and stability [30]. Quite generally, the accuracy describes how much a measured value differs from the correct value. In the context of time measurement, two different accuracies have to be distinguished: The accuracy of the actual time measurement and the frequency accuracy of the clock. The time accuracy (usually just referred to as "accuracy") expresses how much a clock deviates from the actual time and is given in relative units, e.g. if a clock deviates by 1 s from the correct time, 24 h after it was synchronized with the correct time, the clock has an accuracy of 1 s per day, or $1.1 \cdot 10^{-5}$. The definition of accuracy requires the comparison with a time value that is said to be "correct". In contrast, the frequency accuracy (usually referred to as the "systematic frequency shift") is a measure for the frequency deviation of a real oscillator to the ideal value. If the ideal frequency of an oscillator is given as ω_0, the real oscillator will most likely exhibit a systematic shift leading to a real frequency $\omega_{\text{real}} = \omega_0(1 + \epsilon)$. Here ϵ denotes the systematic frequency shift.

If many clocks of the same type were compared, each clock can be expected to possess a slightly different systematic frequency shift, leading to a distribution of frequency shifts. The size of this distribution will be the larger, the more sensitive the clock reacts to external perturbations. For this reason, the uncertainty in ϵ, usually referred to as the "systematic frequency uncertainty" is an important parameter for the quality of a clock. In case that the ideal oscillator value ω_0 would be set as a standard used to define the correct time, the systematic frequency uncertainty would be a measure for the time accuracy that could on average be achieved with the type of oscillator. This is the case, because one can correct for the systematic frequency shift ϵ up to its uncertainty. The remaining value will then lead to an unavoidable deviation of the frequency used for time measurement to the ideal frequency and thus to a continuously growing error in time measurement, which equals the time accuracy. For this reason, the total systematic frequency uncertainty is often referred to as the clock's accuracy.

The stability of a clock is a measure for statistical fluctuations that occur during the frequency measurement. If the ideal frequency is denoted with ω_0, the measured frequency will in general be time dependent in accordance with $\omega_{\text{meas}}(t) = \omega_0(1 + \epsilon + y(t))$. Here $y(t)$ is called the frequency noise and the variance of $y(t)$ is a measure for the stability of a clock. The variance is given as the Allan variance $\sigma_y^2(\tau)$, which is defined as the comparison of two successive frequency deviations for a given sampling period τ. More exactly, we have $\sigma_y^2(\tau) = 1/2\langle(\bar{y}_{n+1} - \bar{y}_n)^2\rangle$, with $\bar{y}_n$ as the nth fractional frequency average over the sampling period τ. For current atomic clocks, the best Allan deviations $\sigma_y(\tau)$ that can be achieved are limited by the quantum projection noise (QPN) and given by [31]

$$\sigma_y(\tau) \approx \frac{1}{Q}\sqrt{\frac{T}{N\tau}}. \quad (1.1)$$

Here $Q = \omega/\Delta\omega$ is the quality factor of the resonance, T the time of the interrogation cycle (which cannot exceed $1/\Delta\omega$), N the number of irradiated atoms and τ the sampling period. For the most stable atomic lattice clock, an Allan deviation of $2.2 \cdot 10^{-16}/\sqrt{\tau}$ for 1 s interrogation cycle and thus close to the QPN limit was reported [27]. A total frequency uncertainty of $2.1 \cdot 10^{-18}$ was achieved. The clock parameters for this strontium lattice clock are $Q = 2 \cdot 10^{14}$ and $N \approx 2000$. The best single-ion clock currently in operation ($^{171}Yb^{1+}$ E3) provides a total frequency uncertainty of $3 \cdot 10^{-18}$ and an Allan deviation of $5 \cdot 10^{-15}/\sqrt{\tau}$ ($T = 1$ s, $Q \approx 10^{23}$, $N = 1$) [25]. The calculated QPN limit is 10^{-23}, significantly below the obtained value. The difference originates from an extremely small natural linewidth of the transition in the nHz region, which is drastically smaller than the bandwidth of any laser available for excitation. For this reason the stability is not limited by the transition linewidth, but instead by the bandwidth of the laser light and only limited advantage is gained from the extremely small natural linewidth of the transition.

1.3 Advantages of a Nuclear Clock

Although the accuracy that is achieved by the best atomic clocks today is already stunning, approaching 10^{-18} corresponding to an error of 1 s after $3 \cdot 10^{10}$ years, significantly longer than the age of the universe, it is reasonable to ask if it is possible to push the limits further. The performance of optical lattice clocks is by today mostly limited due to the influence of external perturbations like electric and magnetic fields. A natural way of improving clock performance would be to use nuclear transitions instead of atomic shell transitions for time measurement. Conceptually, nuclear transitions provide three advantages (see also Ref. [32]): 1. The atomic nucleus is about 5 orders of magnitude smaller than the atomic shell, which leads to significantly reduced magnetic dipole and electric quadrupole moments and therefore to a higher stability against external influences, resulting in an expected improved accuracy and stability of the clock. 2. The transition frequencies are larger, thereby in principle allowing for small Allan deviations. 3. The nucleus is largely unaffected by the atomic shell, for this reason it is intriguing to develop a solid-state nuclear clock based on Mössbauer spectroscopy. Such a solid-state clock could contain a high density of nuclei of about 10^{14} mm^{-3}, thus leading to much improved statistical uncertainties when compared to atomic lattice clocks providing about 10^4 atoms.

Generally, two different classes of nuclear clocks can be considered: (1) Solid-state nuclear clocks that make use of laser-based Mössbauer spectroscopy and would have the advantage of a high density of nuclei, leading to high statistics, but may eventually suffer from line broadening effects and (2) single-ion nuclear clocks, which provide drastically lower statistics, but for which the environmental conditions could be by far better controlled. The operational principles very much resemble those of atomic optical lattice clocks and atomic single-ion clocks (see Fig. 1.2), except that the atomic transition is replaced by a nuclear transition.

1.4 Gamma Decay Versus Internal Conversion

There is one difference which slightly complicates the discussion of nuclear clocks compared to atomic clocks: While for atomic transitions only one decay channel of the excited states enters the discussion, namely radiative decay via emission of a photon, nuclear transitions may exhibit more relevant decay channels. The most important ones in this context are radiative decay (γ decay) and internal conversion (IC). During the internal conversion process the nucleus couples to the atomic shell, transferring the excited state's energy to an electron, which is subsequently emitted into the continuum. The fraction of IC compared to γ decay is called the internal conversion coefficient and usually denoted as α_{ic}. Typically, IC is even the dominant decay channel of excited nuclear states with low energy, long lifetime and large atomic number. The possibility of IC leads to a shortening of the excited state's lifetime and thereby to a line broadening, which is unwanted when considering clock performance. Also care has to be taken when calculating nuclear excitation probabilities, as only the linewidth corresponding to the radiative decay rate enters the calculation. Thus nuclear states that appear to have a rather short lifetime might still have a negligible excitation probability if the internal conversion coefficient is large.

One possibility to reduce the IC decay branch of a nuclear transition is to increase the charge state of the considered ion. For nuclear excited states of low energy, there exists a certain charge state for which the ionization potential of the ion starts to exceed the energy of the nuclear state. If this is the case, the excited state's decay does not provide sufficient energy to lead to further ionization and the IC decay is energetically forbidden. Note, that this will lead to an increased lifetime of the nuclear state and, correspondingly, to a smaller linewidth of the transition. It will, however, not result in an enhanced excitation probability of the nuclear state, as this value is connected to the radiative decay rate, which remains unaffected from a reduced IC decay rate. Even when considering low-energy nuclear states, the suppression of the IC decay channel would typically require the use of highly-charged ions. One example is the low-energy state $^{205\text{m}}$Pb (2.3 keV, 24.2 μs, E2), where the lifetime of 24.2 μs is rather low considering a potential application for a nuclear clock. The IC coefficient is however large, with a value of $\alpha_{\text{ic}} \approx 4.4 \cdot 10^8$ and from Pb^{46+} onwards the IC decay channel can be expected to be energetically suppressed, leading to a drastic prolongation of the lifetime to up to 10^4 s, thus in principle providing a good candidate for a highly-charged-ion nuclear clock. For solid-state nuclear clocks, the ionization potential has to be replaced by the material's band gap, however, there is practically no overlap of the energy range of band gaps with the nuclear energy scale. For this reason there is no hope for the suppression of the IC decay channel in the solid-state nuclear clock concept (^{229m}Th provides an exception in this context).

1.5 What Makes ^{229m}Th Unique

By today, there is a major obstacle that prevents the development of a broad variety of nuclear clocks. This is that even energies of several keV, corresponding to wavelengths of 1 nm and below, are considered as low-energy transitions on nuclear energy scales and by today the frequency-comb technology has not been sufficiently extended into the x-ray region in order to drive most of the nuclear transitions. As the frequency comb technology is a young and fast improving field of technology, one might expect that the availability of frequency combs in the energy range of a few keV with sufficient intensities to drive nuclear transitions and thus to allow for high-precision Mössbauer spectroscopy is just a matter of time. In the meantime, however, the research on nuclear clocks has to focus on a single nuclear state providing an extraordinary low energy of only about 7.8 eV above the nuclear ground state [2, 3]. This is the first excited state of ^{229}Th, denoted as ^{229m}Th, where the superscript m stands for metastable and expresses that a comparatively long lifetime for this state is expected. To give an impression on the scarcity of nuclear states of such low energy: There is only one further nuclear state known with an energy of less than 1 keV above its nuclear ground state, which is ^{235m}U, providing an energy of about 76 eV (see Fig. 1.1). This isomeric state, however, does not allow for the development of a nuclear clock, due to its extremely long radiative lifetime of about $3.8 \cdot 10^{22}$ s, which is 100,000 times the age of the universe, not allowing for any significant laser coupling. For this reason, ^{229m}Th is by today the only realistic candidate for the development of a nuclear frequency standard. In this thesis only ^{229m}Th will be considered.

1.6 The ^{229}Th Nuclear Clock Proposal

The observation of a nuclear excited state of only a few eV above the ground state [33] attracted some attention from the scientific community. Already in 1991 a theory paper appeared, predicting an increasing interest of physicists from other disciplines like *"optics, solid-state physics, lasers, plasma and others"*, different decay channels and lifetimes were discussed in dependence on the transition energy [34]. In a 1996 paper of E.V. Tkalya et al. a list of potential applications is provided, which also contains the potential for the *"development of a high stability nuclear source of light for metrology"* [35]. While this proposal does not include the development of a nuclear clock based on ^{229}Th, it contains a hint pointing at an expected high stability. A few years later also the decay of ^{229m}Th in solid state materials was discussed [36].

On this basis, a ^{229m}Th nuclear clock was proposed in 2003 by E. Peik and C. Tamm, shortly after the development of the frequency-comb technology, which for the first time allowed precision laser spectroscopy in the optical region [4]. This paper contains both nuclear-clock concepts: A clock based on Th ions as well as a solid state nuclear clock.

^{229m}Th, besides being an excited nuclear state of extraordinary low energy, possesses a lifetime corresponding to its radiative decay channel of expectedly about 10^4 s [37, 38]. This leads to a large quality factor of the corresponding resonator of $\omega/\Delta\omega \approx 10^{20}$ and a small absolute bandwidth of 10^{-4} Hz in case that IC is suppressed. The ground and isomeric states have spin and parity values of $5/2^+$ and $3/2^+$, respectively, leading to a multipolarity of M1 of the isomer-to-ground-state transition, which results in a sufficiently large cross section to allow for direct laser excitation. Further, the ground state of ^{229}Th has a half-life of 7932 years, which makes it relatively easy to handle moderate quantities of the material. All properties are requirements for the development of a ^{229}Th-based nuclear clock.

In their 2003 nuclear clock concept Peik and Tamm proposed to perform nuclear laser spectroscopy with $^{229}Th^{3+}$ [4]. The 3+ charge state was proposed because it possesses a favorable electronic configuration with a radon-like core and one valence electron, exhibiting a closed 3-level Λ system and a closed 2-level system, which can be employed for laser cooling. Due to the small magnetic dipole and electric quadrupole moments of the nucleus, the direct coupling of external perturbing fields to the nucleus are negligible. However, shell-nucleus coupling via the hyperfine interaction still has to be considered as a potential source of perturbations. One of the central ideas of the 2003 nuclear-clock concept is that one could choose an excited state of the electronic shell in a way that the combined quantum numbers of shell plus nucleus are "good" in a sense that the entire system provides lowest sensitivity to external perturbations [32]. It was shown that the metastable $7s^2S_{1/2}$ shell state in Th^{3+} with 1 s lifetime would be an appropriate choice. In this way, perturbing effects due to the linear Zeeman effect, the tensor part of the quadratic Stark effect and atomic quadrupole interactions could be avoided. Further, as no shifts can play a role which are entirely dependent on the electronic quantum numbers, no shifts from static electric fields, electromagnetic radiation or collisions have to be considered, leading to the proposal of a highly stable nuclear clock. The population of the nuclear excited state could be monitored by making use of the change of the nuclear spin state, resulting in a different hyperfine structure of the cooling transitions (double-resonance method). It was also suggested that even a solid-state nuclear clock could be realized, if ^{229}Th was embedded into a material providing a band gap larger than the nuclear isomer's energy, for which (3.5 ± 1) eV was the assumed value at that time [4, 32].

This pioneering nuclear clock concept, besides promising to lead to an extraordinary stable clock, has two disadvantages [5]: (1) The quadratic Zeeman effect cannot be suppressed and is estimated to be 1 kHz at 0.1 mT, which is comparable to usual atomic clocks. (2) The concept requires to excite the $^{229}Th^{3+}$ ions into the $7s^2S_{1/2}$ atomic shell state, which possesses 1 s lifetime (instead of the expectedly $\sim 10^4$ s lifetime of the nuclear isomer) and will thus lead to a reduced quality factor of the resonance.

A more recent proposal by C. Campbell et al. aims at a solution of these problems [5]. Here the idea is to use a pair of stretched nuclear hyperfine states for the clock transition, while $^{229}Th^{3+}$ remains in its $5f^2F_{5/2}$ electronic ground state. A careful analysis of the expected systematic uncertainties of a corresponding $^{229}Th^{3+}$ single-ion nuclear clock was performed, resulting in an expected total uncertainty approaching 10^{-19}, thereby potentially outperforming all existing atomic-clock technology [5]. The development of a nuclear clock based on ^{229}Th ions is currently envisaged by three groups in the world, with more eventually to follow [39–41].

The concept of a solid-state nuclear clock based on Mössbauer spectroscopy of ^{229m}Th, as first proposed in Ref. [4], was further developed by two groups [42, 43]. A long-term fractional frequency accuracy of $2 \cdot 10^{-16}$ has been predicted for this device [42], with a fractional instability approaching 10^{-19} [43].

1.7 Constraining the Transition Energy

The development of a ^{229}Th-based nuclear clock requires direct laser excitation of the isomeric nuclear state. However, due to the narrow radiative bandwidth of the nuclear transition of expectedly only 10^{-4} Hz, the laser-nucleus interaction is weak and improved constraints on the transition energy are required as a prerequisite for the nuclear laser excitation of ^{229}Th ions in a Paul trap. Up to today it was not possible to constrain the isomeric energy value sufficiently in order to allow for the development of a nuclear clock. For this reason major experimental investigations are continued, aiming for a reduction of the uncertainty of the ^{229m}Th energy value. Except for a few experiments, most of the experimental concepts were aiming for a direct decay detection of the isomer-to-ground-state transition in order to constrain the transition energy. While previous experiments have failed to observe a direct decay signal of ^{229m}Th, this thesis presents the first time direct detection of the isomeric state [44]. This observation paves the way for a precise energy determination and therefore for the development of a ^{229}Th-based nuclear clock. This is possible now in two ways: (1) By performing conversion-electron spectroscopy [45] and (2) by using a direct nuclear laser excitation scheme [46]. A detailed review of the experimental efforts that were aiming for a direct detection and energy determination of ^{229m}Th is provided in Chap. 3.

1.8 Prospects for a ^{229}Th Frequency Standard

A ^{229}Th nuclear clock is expected to outperform even the best atomic-shell based clocks due to conceptual advantages [4, 5]. An estimate of the total uncertainty budget results in an expected value approaching 10^{-19}, which is by about an order of

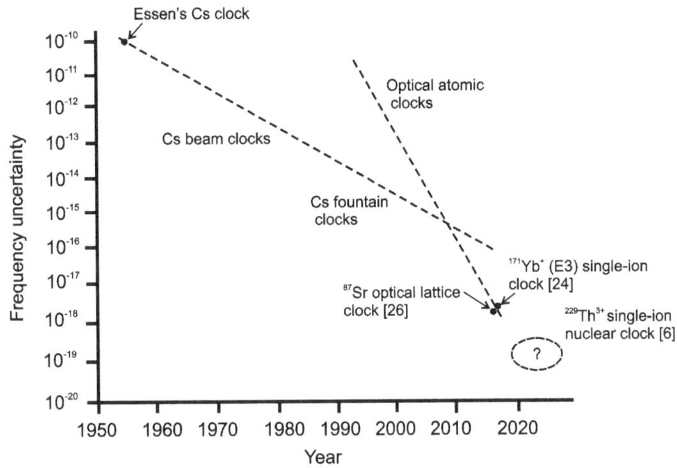

Fig. 1.3 Evolution of the frequency uncertainty of atomic clocks [29]. The frequency uncertainty of a single-ion nuclear clock based on ^{229}Th^{3+} is expected to approach 10^{-19} [5]

magnitude better than the best atomic clocks currently in operation. The expected performance of a single-ion ^{229}Th^{3+} nuclear clock compared to optical atomic clocks is visualized in Fig. 1.3. Here the development of the frequency uncertainty of atomic clocks is shown as a function of time. Further, a solid-state nuclear clock, although being less accurate, could provide a very compact and robust device for time measurement and thus being of significant practical use [4, 42, 43].

Due to its high expected accuracy and precision, a ^{229m}Th-based nuclear clock may not only provide an alternative to existing atomic-clock technology [27], which has a wide range of applications (e.g. as frequency standards, in satellite-based navigation and data transfer), but may also open a new field of utilization, like in geodesy and earth-quake detection [47]. Even further applications for a nuclear frequency standard may be found in fundamental physics, for example the test of general relativity [27], dark matter search [48], gravitational wave detection [49] and the probe for time variations of fundamental constants [50]. Besides a nuclear clock, also other ideas based on the special properties of ^{229m}Th have been proposed. These include the first development of a nuclear-based laser [51], nuclear quantum optics [52] and a highly stable qubit for quantum computing [53].

Of course there is the hope that ^{229}Th will just be a first candidate for an ultra-precise nuclear clock and that, with ever evolving frequency-comb technology, a whole new generation of clocks based on nuclear transitions could be developed [54].

References

1. NNDC Interactive chart of nuclides, Brookhaven National Laboratory, Brookhaven. http://www.nndc.bnl.gov/chart Accessed 22 Mar 2016
2. Beck BR et al (2007) Energy splitting of the ground-state doublet in the nucleus ^{229}Th. Phys Rev Lett 109:142501
3. Beck BR et al (2009) Improved value for the energy splitting of the ground-state doublet in the nucleus ^{229m}Th. LLNL-PROC-415170
4. Peik E, Tamm C (2003) Nuclear laser spectroscopy of the 3.5 eV transition in ^{229}Th. Eur Phys Lett 61: 181
5. Campbell CJ et al (2012) Single-Ion nuclear clock for metrology at the 19th decimal place. Phys Rev Lett 108: 120802
6. Tuli JK (2011) Nuclear wallet cards, 8th edn. National Nuclear Data Center, Brookhaven National Laboratory, New York
7. Higgins K, Miner D, Smith CN, Sullivan DB (2004) A Walk Through Time (version 1.2.1), National Institute of Standards and Technology, Gaithersburg, MD. http://physics.nist.gov/time Accessed 16 Sept 2016
8. Hide R, Dickey JO (1991) Earth's variable rotation. Science 253(5020):629–637
9. Stephenson FR (1997) Historical eclipses and earth's rotation. Cambridge University Press, Cambridge
10. Usher AP (1929) History of mechanical inventions. Oxford University Press, Oxford
11. Lienhard JH The first mechanical clocks, Engines of Our Ingenuity No. 1506. http://www.uh.edu/engines/epi1506.htm Accessed 16 Sept 2016
12. Bennet M et al (2002) Huygens'clocks. Proc R Soc Lond A 458:563–579
13. Sorge F, Cammalleri M, Genchi G (2016) On the birth and growth of pendulum clocks in the early modern era. Essays on the history of mechanical engineering. Springer, Berlin, pp 273–290
14. Gould RT (1923) The marine chronometer: its history and development. J.D. Potter, London
15. Bosschieter JE Shortt's free pendulum, A History of the Evolution of Electric Clocks. http://www.electric-clocks.eu/clocks/en/page10.htm Accessed 3rd Sept 2017
16. Marrison WA (1948) The evolution of the quartz crystal clock. Bell Syst Tech J 27(3):510–588
17. Lyons H (1949) Instruments. At Clock 22:133–135
18. Forman P (1998) Atomichron: the atomic clock from concept to commercial product. In: IEEE ultrasonics ferroelectrics and frequency control society
19. Essen L, Parry JVL (1955) An atomic standard of frequency and time interval: a caesium resonator. Nature 176:280–282
20. Ramsey NF (1983) History of atomic clocks. J Res Natl Bur Stand 88:301–318
21. Wynands R, Weyers S (2005) Metrologica. At Fountain Clocks 42:64–79
22. Udem Th, Holzwarth R, Hänsch TW (2002) Optical frequency metrology. Nature 416:233–237
23. Diddams SA et al (2001) An optical clock based on a single trapped $^{199}Hg^{+}$ ion. Science 293:825–828
24. Rosenband T et al (2008) Frequency Ratio of Al^{+} and Hg^{+} Single-ion optical clocks. Metrology at the 17th decimal place. Science 319:1808–1811
25. Huntemann N et al (2016) Single-ion atomic clock with $3 \cdot 10^{-18}$ systematic uncertainty. Phys Rev Lett 116:063001
26. Bloom BJ et al (2014) An optical lattice clock with accuracy and stability at the 10^{-18} level. Nature 506:71–75
27. Nicholson TL et al (2015) Systematic evaluation of an atomic clock at $2{\cdot}10^{-18}$ total uncertainty. Nat Commun
28. Campbell GK, Phillpis WD (2011) Ultracold atoms and precise time standards. Philos Trans R Soc A 369:4078–4089
29. Gill P (2011) When should we change the definition of the second? Philos Trans R Soc A 369:4109–4130

30. Riehle F (2017) Optical clock networks. Nat Photonics 11:25–31
31. Ludlow AD et al (2015) Optical atomic clocks. Rev Mod Phys 87:637–699
32. Peik E, Okhapkin M (2015) Nuclear clocks based on resonant excitation of γ-transitions. C R Phys 16:516–523
33. Reich CW, Helmer R (1990) Energy separation of the doublet of intrinsic states at the ground state of ^{229}Th. Phys Rev Lett 64:271
34. Strizhov VF, Tkalya EV (1991) Decay channel of low-lying isomer state of the ^{229}Th nucleus. Possibilities of experimental investigation. Sov Phys JETP 72:387
35. Tkalya EV, Varlamov VO, Lomonosov VV, Nikulin SA (1996) Processes of the nuclear isomer ^{229m}Th($3/2^+$, 3.5 ± 1.0 eV) resonant excitation by optical photons. Phys Scripta 53:296–299
36. Tkalya EV, Zherikin AN, Zhudov VI (2000) Decay of the low-energy nuclear isomer ^{229}Th^m($3/2^+$,3.5 ± 1.0 eV) in solids (dielectrics and metals): a new scheme of experimental research. Phys Rev C 61:064308
37. Minkov N, Pálffy A (2017) Reduced transition probabilities for the gamma decay of the 7.8 eV isomer in ^{229}Th. Phys Rev Lett 118:212501
38. Tkalya EV, Schneider C, Jeet J, Hudson ER (2015) Radiative lifetime and energy of the low-energy isomeric level in ^{229}Th. Phys Rev C 92:054324
39. Campbell CJ et al (2011) Wigner crystals of ^{229}Th for optical excitation of the nuclear isomer. Phys Rev Lett 106:223001
40. Zimmermann K (2010) Experiments towards optical nuclear spectroscopy with thorium-229. Ph.D. thesis, University of Hannover, Germany
41. Borisyuk PV et al (2017) Trapping, retention and laser cooling of Th^{3+} ions in a multisection linear quadrupole trap. Quant Electron 47:406–411
42. Rellergert WG et al (2010) Constraining the evolution of the fundamental constants with a solid-state optical frequency reference based on the ^{229}Th nucleus. Phys Rev Lett 104:200802
43. Kazakov GA et al (2012) Performance of a ^{229}Thorium solid-state nuclear clock. New J Phys 14:083019
44. von der Wense L et al (2016) Direct detection of the ^{229}Th nuclear clock transition. Nature 533:47–51
45. Seiferle B, von der Wense L, Thirolf PG (2017) Feasibility study of internal conversion electron spectroscopy of ^{229m}Th. Eur Phys J A 53:108
46. von der Wense L et al (2017) A laser excitation scheme for ^{229m}Th. Phys Rev Lett 119:132503
47. Safronova M (2016) Elusive transition spotted in thorium. Nature 533:44–45
48. Derevianko A, Pospelov M (2014) Hunting for topological dark matter with atomic clocks. Nat Phys 10:933–936
49. Kolkowitz S et al (2016) Gravitational wave detection with optical lattice atomic clocks. arXiv:1606.01859
50. Flambaum VV (2006) Enhanced effect of temporal variation of the fine structure constant and the strong interaction in ^{229}Th. Phys Rev Lett 97:092502
51. Tkalya EV (2011) Proposal for a nuclear gamma-ray laser of optical range. Phys Rev Lett 106:162501
52. Bürvenich TJ et al (2006) Nuclear quantum optics with X-ray laser pulses. Phys Rev Lett 96:142501
53. Raeder S et al (2011) Resonance ionization spectroscopy of thorium isotopes - towards a laser spectroscopic identification of the low-lying 7.6 eV isomer of ^{229}Th. J Phys B 44:165005
54. von der Wense L, Seiferle B, Thirolf, PG (2017) Towards a ^{229}Th-based nuclear clock. In: Measurement techniques, Proceedings of the conference held for the jubilee celebrating 175 years of the D.I. Mendeleev Institute for Metrology (VNIIM) and National Measurement System in St. Petersburg, in preparation

Chapter 2
Theoretical Background

In this chapter some of the fundamental theoretical tools, required to characterize the first excited isomeric state of ^{229}Th, are developed. The first section provides a short introduction to nuclear shell models with a focus on the Nilsson single-particle shell model and rotational bands. The second section describes the fundamental theories of nuclear γ decay and internal conversion, together with their applications to $^{229\mathrm{m}}$Th. The last section introduces the theoretical background for direct nuclear laser coupling and gives examples for $^{229\mathrm{m}}$Th. Also $^{235\mathrm{m}}$U is discussed for comparison, being the isomeric state of second lowest excitation energy.

2.1 Nuclear Shell Models

The ultimate way to describe nuclear transitions would be within a quantum chromodynamical framework. Unfortunately, this theory has not yet been developed to a point which would allow for precise predictions of level schemes. Especially complex nuclei and transitions of lowest energies, as is the case for the first excited state of ^{229}Th, is far beyond the scope of quantum chromodynamics at the current time. Therefore, all possible predictions are within the framework of different collective models, of which the most important one is the nuclear shell model. This model, even with all possible extensions, still does not allow for a precise prediction of nuclear transitions in the eV-range. Instead, it allows for rough estimations, which can then be compared to experimental data. In the following, the nuclear shell model will be discussed together with its extensions as far as of interest in the context of this thesis. The discussion is based on Ref. [1].

L. Von der Wense, *On the Direct Detection of ^{229m}Th*,
Springer Theses, https://doi.org/10.1007/978-3-319-70461-6_2

2.1.1 Nuclear Single Particle Shell Model

Similar to the electrons in an atom, the nucleons can be described as solutions of the Schrödinger equation

$$\hat{H}\psi = E\psi \tag{2.1}$$

for a given Hamiltonian $\hat{H}$. Assuming that a potential is found, in which each nucleon moves independently of all others, the total Hamiltonian can be written as a sum over individual particle Hamiltonians [1]

$$\hat{H} = \sum_i \hat{H}_i = \sum_i \left[-\frac{\hbar^2}{2m}\Delta_i + V(r_i) \right], \tag{2.2}$$

where $V(r)$ is the potential of interest. The Schrödinger equation must then hold for each individual particle

$$\hat{H}_i\psi_i = \left[-\frac{\hbar^2}{2m}\Delta_i + V(r_i) \right] \psi_i = E_i\psi_i. \tag{2.3}$$

The potential in which one nucleon is moving is given by the mean field that is generated by all other nucleons and therefore corresponds to the nucleon density, which is described by Fermi statistics. The basic potential approach is therefore the Woods–Saxon potential

$$V_{\text{Woods–Saxon}}(r) = \frac{-V_0}{1 + e^{(r-R)/a}}, \tag{2.4}$$

where V_0 corresponds to the potential depth, R is the radius at which the potential depth is half of V_0 and a is a parameter corresponding to the steepness of the potential. V_0, R and a have to be experimentally identified. A comparison of theoretical predictions from this model and measured level schemes reveals that this simple approach does not lead to satisfying predictions. Instead also the spin-orbit interaction has to be considered. This is done via an extra term in the potential, which then becomes

$$V(r) = \frac{-V_0}{1 + e^{(r-R)/a}} + V_{ls}(r)\frac{< \hat{l}\hat{s} >}{\hbar^2}. \tag{2.5}$$

Here V_{ls} is a scaling potential and $\hat{l}$ and $\hat{s}$ are the orbital momentum and spin operator, respectively. The operators are coupled as $\hat{j} = \hat{l} + \hat{s}$ which, by applying the commutativity of $\hat{l}$ and $\hat{s}$, gives

$$\hat{l}\hat{s} = \frac{1}{2}\left(\hat{j}^2 - \hat{l}^2 - \hat{s}^2\right). \tag{2.6}$$

As all components of $\hat{j}$ commute with the Hamiltonian, one can find a basis of states $|slj\rangle$, for which the eigenvalue relations of the angular momentum operator hold

$$\begin{aligned}\hat{s}^2|slj\rangle &= \hbar^2 s(s+1)|slj\rangle,\\ \hat{l}^2|slj\rangle &= \hbar^2 l(l+1)|slj\rangle,\\ \hat{j}^2|slj\rangle &= \hbar^2 j(j+1)|slj\rangle.\end{aligned} \tag{2.7}$$

Thus the expectation value $<\hat{l}\hat{s}>$ becomes

$$<\hat{l}\hat{s}>=\frac{\hbar^2\left[j(j+1)-l(l+1)-s(s+1)\right]}{2}. \tag{2.8}$$

Inserting this relation into Eq. (2.5) leads to

$$V(r)=\frac{-V_0}{1+e^{(r-R)/a}}+V_{ls}(r)\frac{j(j+1)-l(l+1)-s(s+1)}{2}. \tag{2.9}$$

It is known from experiment that $V_{ls}(r)$ takes negative values. The resulting Schrödinger equation cannot be solved analytically, but due to the spherical symmetry of the potential the resulting wave functions can be factorized into a radial wave function $R_{nl}(r)$ and the spherical harmonics $Y_{lm}(\theta,\phi)$, where n, l and m are the radial quantum number, the orbital quantum number and the magnetic quantum number, which can take values from 1 to ∞, 0 to ∞ and $-l$ to l respectively. The spin quantum number s is involved to also take spin-orbital coupling into account. The spin s takes the values $-1/2$ or $+1/2$. It turns out that the energy is independent of the magnetic quantum number m. An energy eigenstate is therefore fixed by the three quantum numbers n, l and s. Usually the states are assigned by n, l and $j=l+s$ instead, where for l the letters s, p, d, f, g, h,... (corresponding to $l=0,1,2,\ldots$) are used. Due to the independence on m, each state is $(2j+1)$ times degenerate. The numerically calculated level schemes for the Woods–Saxon potential with and without spin-orbital coupling are shown in Fig. 2.1.

2.1.2 The Nilsson Model

The above made assumption of a spherical nuclear potential cannot be validated for all nuclei. In general, deformation occurs, leading to changes of the energy eigenvalues, which have to be considered in a more realistic approach. This is especially important for nuclei far from closed shells, as is the case for ^{229}Th. Therefore deformations of the Woods–Saxon potential have to be allowed, which again can only be done numerically. In order to get any insights into the theory of deformed nuclei, the deformed harmonic oscillator potential is considered in the following. This is also the way in which deformation was first introduced into the model by Nilsson [2].

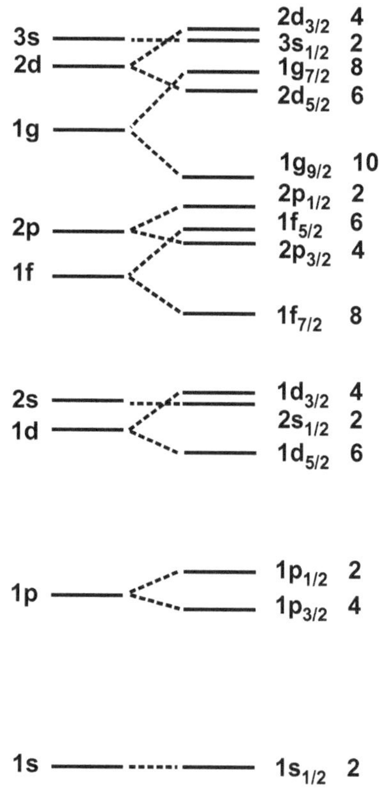

Fig. 2.1 Energy levels of the nuclear single-particle shell model with (right) and without (left) spin-orbital coupling. The corresponding quantum numbers are assigned to each state. The degeneracy of the states are given on the right according to $2j + 1$

The Hamiltonian of the anisotropic harmonic oscillator is of the form [1]

$$\hat{H}_0 = -\frac{\hbar^2}{2m}\Delta + \frac{m}{2}\left(\omega_x^2 x^2 + \omega_y^2 y^2 + \omega_z^2 z^2\right). \tag{2.10}$$

The energy eigenvalues of this Hamiltonian are

$$E_0 = \hbar\omega_x\left(n_x + \frac{1}{2}\right) + \hbar\omega_y\left(n_y + \frac{1}{2}\right) + \hbar\omega_z\left(n_z + \frac{1}{2}\right), \tag{2.11}$$

where n_x, n_y and n_z denote the quantum numbers characterizing the corresponding eigenstates. For the following, axial symmetry with $\omega_x = \omega_y$ is assumed. Further, a constant potential volume $\omega_x\omega_y\omega_z = \text{const.} = \omega_0^3$ is introduced and a deformation parameter δ is introduced as

$$\delta = 2 \cdot \frac{b - a}{b + a}, \tag{2.12}$$

where b denotes the length of the ellipsoid along the symmetry axis and a the length orthogonal to the symmetry axis. Thus negative values of δ correspond to oblate deformations and positive values to prolate deformations. With the help of this deformation parameter, ω_x, ω_y as well as ω_z are expressed as

$$\begin{aligned} \omega_x^2 = \omega_y^2 &= \omega^2(\delta)\left(1 + \frac{2}{3}\delta\right) \\ \omega_z^2 &= \omega^2(\delta)\left(1 - \frac{4}{3}\delta\right). \end{aligned} \tag{2.13}$$

In this case one has

$$\omega_x\omega_y\omega_z = \omega^3(\delta)\left(1 - \frac{4}{3}\delta^2 - \frac{16}{27}\delta^3\right)^{1/2} = \omega_0^3, \tag{2.14}$$

and correspondingly

$$\omega(\delta) = \omega_0\left(1 - \frac{4}{3}\delta^2 - \frac{16}{27}\delta^3\right)^{-1/6}. \tag{2.15}$$

In the original work [2], Nilsson introduces a deformation dependent oscillator length $b(\delta) = (\hbar/m\omega_0(\delta))^{1/2}$ and dimensionless coordinates $\vec{r}\,' = \vec{r}/b$ to transform the Hamiltonian of Eq. (2.10) to

$$\hat{H}_0 = \hbar\omega(\delta)\left(-\frac{1}{2}\Delta' + \frac{r'^2}{2} - \frac{4}{3}\sqrt{\frac{\pi}{5}}\delta r'^2 Y_{20}(\theta', \phi')\right). \tag{2.16}$$

It is convenient to transform to cylindrical coordinates, in which case the eigenstates are characterized by the quantum numbers n_z, n_ρ and m_l, where m_l denotes the projection of the orbital angular momentum onto the symmetry axis and is often denoted as Λ. Correspondingly one obtains

$$\begin{aligned} E_0 &= \hbar\omega_z\left(n_z + \frac{1}{2}\right) + \hbar\omega_{xy}\left(2n_\rho + m_l + 1\right) \\ &= \hbar\omega_0\left[\left(N + \frac{3}{2}\right) + \delta\left(\frac{N}{3} - n_z\right)\right]. \end{aligned} \tag{2.17}$$

Here $N = n_x + n_y + n_z$ was introduced as a new quantum number and $\omega_{xy} = \omega_x = \omega_y$. The axial symmetry of the Hamiltonian causes Λ to be a good quantum number. The same is valid for the z-component of the spin Σ and the z-component of the total angular momentum $\Omega = \Lambda + \Sigma$. For this reason, the eigenstates of the given Hamiltonian are described by the quantum numbers

$$\Omega^{\pi}\left[Nn_z\Lambda\right], \tag{2.18}$$

where $\pi = (-1)^N$ is the parity of the state.

The anisotropic harmonic oscillator Hamiltonian given in Eq. (2.10) is not a realistic nuclear potential. The reason is that no spin-orbital coupling is included and the potential is not realistic for high l values. In order to correct for that, Nilsson introduced two further terms. One spin-orbital coupling term, proportional to $\hat{l}\hat{s}$, and one term proportional to $\hat{l}^2$, leading to a more realistic form of the potential. The full Nilsson-model Hamiltonian reads [1, 2]

$$\hat{H}_{\text{Nilsson}} = \hat{H}_0 + C < \hat{l}\hat{s} > +D\hat{l}^2. \tag{2.19}$$

Here C determines the strength of the spin-orbit coupling and D introduces a level shift. The corresponding Schrödinger equation can only be solved numerically. Solutions for the energy are shown in Fig. 2.2 as a function of the deformation parameter [3]. It can be inferred that each eigenstate of the undeformed Hamiltonian splits in

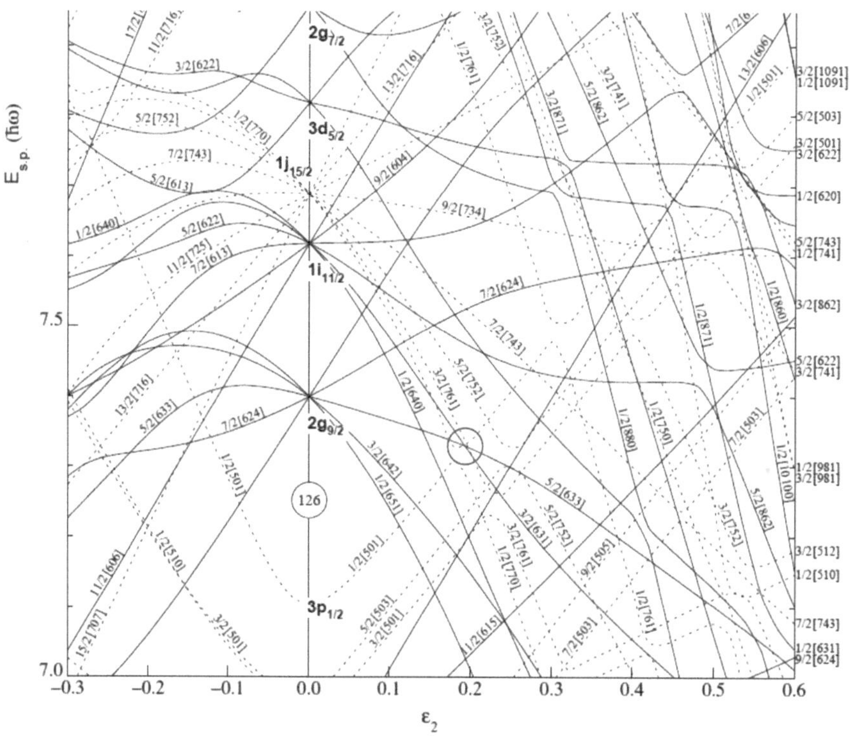

Fig. 2.2 Nilsson diagram for neutron single-particle states, showing states as populated by nuclei with neutron number larger than 126 [3]. The single-particle energy is shown as a function of the deformation parameter ϵ_2. The crossing of the $5/2^{+}633$ ground state of ^{229}Th and the $3/2^{+}631$ isomeric state is marked by a red circle

$(2j+1)/2$ substates during deformation. After introduction of the correction terms $C < \hat{l}\hat{s} >$ and $D\hat{l}^2$, Ω and π are the only remaining good quantum numbers of the states. However, for large deformations, the $\hat{l}\hat{s}$ and $\hat{l}^2$ term are less important and the quantum numbers in Eq. (2.18) are approximately good quantum numbers. For this reason they are denoted as asymptotic quantum numbers and used to describe the Nilsson states.

In case of ^{229}Th, the nuclear ground state is described by the Nilsson quantum numbers $5/2^+633$ and the isomeric state by $3/2^+631$. The deformation parameter is often defined as ϵ_2, the Nilsson quadrupole deformation parameter. ϵ_2 relates to δ as

$$\epsilon_2 = \delta + \frac{1}{6}\delta^2 + \frac{5}{18}\delta^3 + \cdots \tag{2.20}$$

For ^{229}Th, the nuclear deformation can be estimated from the electric quadrupole moment to $\epsilon_2 \approx 0.19$ [4]. This is in good agreement with the crossing point of the $5/2^+633$ ground state and the $3/2^+631$ isomeric state in the Nilsson diagram, as shown as a red circle in Fig. 2.2.

2.1.3 Rotational Bands

In order to understand the nuclear γ-ray spectrum of ^{229}Th, also rotational motion of the nucleus has to be considered. As the nucleus is deformed, rotations around an axis perpendicular to the symmetry axis of the nucleus can occur. In case of cylindrical symmetry, the corresponding Hamilton operator takes the form [1]

$$\hat{H}_{\text{rot}} = \frac{\hat{I}^2 - \hat{I}_z^2}{2\theta_x} + \frac{\hat{I}_z^2}{\theta_z}. \tag{2.21}$$

Here $\hat{I}$ denotes the total angular momentum operator of the rotation, and $\hat{I}_z$ is the corresponding projection onto the z-axis of the body-fixed system. θ_x and θ_z denote the moments of inertia in x and z direction, respectively. Introducing the new quantum number $K = I_z$ as the z-component of the total angular momentum, the rotational energy of the system can be obtained by solving the Schrödinger equation with the given Hamiltonian as

$$E_{\text{rot}} = \frac{\hbar^2}{2\theta_x}\left(I(I+1) - K^2\right) + \frac{\hbar^2}{2\theta_z}K^2. \tag{2.22}$$

Based on symmetry arguments, it is possible to show that in case of axial and reflection symmetry $K = 0$ and $I = 0, 2, 4, \ldots$ holds and the rotational energy is given by [1]

$$E_{\text{rot}} = \frac{\hbar^2}{2\theta_x}\left(I(I+1)\right). \tag{2.23}$$

In accordance with Eq. (2.23), there is a series of excited rotational states, called rotational band, corresponding to each energy level of the Nilsson model. The ground state of each rotational band is denoted as the band head. Besides rotational states also vibrational states can play a role. These will not be discussed, as they are not of importance in the considered context.

2.2 Nuclear γ Decay and Internal Conversion

Two decay channels of $^{229\text{m}}$Th are addressed in this work. These are the γ decay via emission of a photon in the vacuum ultra-violet (VUV) region around 160 nm and the internal conversion (IC) decay under emission of a low-energy electron. The search for an internal conversion decay channel has led to the successful observation of the isomeric decay (see Sect. 5.2.2). In the following, the theoretical background is discussed for both decay channels.

2.2.1 *Nuclear γ Decay*

The theory of nuclear coupling to electromagnetic radiation provides the basis for nuclear γ decay as well as excitation. In the following, the γ-emission rate for a given nuclear transition will be derived. For this purpose, the electromagnetic field is developed in multipoles, closely following the procedure described in Ref. [5]. The starting point are the Maxwell equations, which take the following form (in SI units):

$$\begin{aligned}
\vec{\nabla}\cdot\vec{E}(\vec{r},t) &= \frac{\rho(\vec{r},t)}{\epsilon_0}\\
\vec{\nabla}\cdot\vec{H}(\vec{r},t) &= -\vec{\nabla}\cdot\vec{M}(\vec{r},t)\\
\vec{\nabla}\times\vec{E}(\vec{r},t) &= -\mu_0\frac{\partial(\vec{H}(\vec{r},t)+\vec{M}(\vec{r},t))}{\partial t}\\
\vec{\nabla}\times\vec{H}(\vec{r},t) &= \epsilon_0\frac{\partial\vec{E}(\vec{r},t)}{\partial t}+\vec{J}(\vec{r},t).
\end{aligned} \tag{2.24}$$

These are the usual vacuum Maxwell equations after introduction of a magnetization $\vec{M}(\vec{r},t)$, in order to take the intrinsic angular momenta of the particles into account. $\vec{E}(\vec{r},t)$ denotes the electric field, $\vec{H}(\vec{r},t)$ is the magnetic field and $\vec{J}(\vec{r},t)$ as well as $\rho(\vec{r},t)$ denote the current and charge density, respectively. ϵ_0 and μ_0 are the vacuum permittivity and permeability constants.

Assuming a single source frequency ω, the time dependency can be separated as [6]

$$\begin{aligned}\rho(\vec{r},t) &= \rho(\vec{r})e^{-i\omega t} + \rho^*(\vec{r})e^{i\omega t}\\ \vec{J}(\vec{r},t) &= \vec{J}(\vec{r})e^{-i\omega t} + \vec{J}^*(\vec{r})e^{i\omega t}\\ \vec{M}(\vec{r},t) &= \vec{M}(\vec{r})e^{-i\omega t} + \vec{M}^*(\vec{r})e^{i\omega t}.\end{aligned} \tag{2.25}$$

The same relation then also holds for the electric and magnetic fields. By inserting these relations into Eq. (2.24) and applying the continuity equation

$$\frac{\partial \rho(\vec{r},t)}{\partial t} + \vec{\nabla}\vec{J}(\vec{r},t) = 0, \tag{2.26}$$

the Maxwell equations take the following form

$$\begin{aligned}\vec{\nabla}\cdot\vec{E}(\vec{r}) &= -\frac{i}{k}\sqrt{\frac{\mu_0}{\epsilon_0}}\vec{\nabla}\cdot\vec{J}(\vec{r})\\ \vec{\nabla}\cdot\vec{H}(\vec{r}) &= -\vec{\nabla}\cdot\vec{M}(\vec{r})\\ \vec{\nabla}\times\vec{E}(\vec{r}) &= ik\sqrt{\frac{\mu_0}{\epsilon_0}}\left(\vec{H}(\vec{r}) + \vec{M}(\vec{r})\right)\\ \vec{\nabla}\times\vec{H}(\vec{r}) &= -ik\sqrt{\frac{\epsilon_0}{\mu_0}}\vec{E}(\vec{r}) + \vec{J}(\vec{r}).\end{aligned} \tag{2.27}$$

Here $\omega = c\cdot k$ with k as the wavenumber and $c = 1/\sqrt{\epsilon_0\mu_0}$ was used. Applying the curl operation to the last two equations and making use of the identity

$$\vec{\nabla}\times(\vec{\nabla}\times\vec{V}) = \vec{\nabla}(\vec{\nabla}\cdot\vec{V}) - \vec{\nabla}^2\vec{V}, \tag{2.28}$$

the inhomogeneous Helmholtz equations are derived

$$\begin{aligned}(\vec{\nabla}^2 + k^2)\vec{E}(\vec{r}) &= -ik\sqrt{\frac{\mu_0}{\epsilon_0}}\left(\vec{J}(\vec{r}) + \vec{\nabla}\times\vec{M}(\vec{r})\right) - \frac{i}{k}\sqrt{\frac{\mu_0}{\epsilon_0}}\vec{\nabla}(\vec{\nabla}\cdot\vec{J}(\vec{r}))\\ (\vec{\nabla}^2 + k^2)\vec{H}(\vec{r}) &= -k^2\vec{M}(\vec{r}) - \vec{\nabla}\times\vec{J}(\vec{r}) - \vec{\nabla}(\vec{\nabla}\cdot\vec{M}(\vec{r})).\end{aligned} \tag{2.29}$$

These equations take the form of an inhomogeneous wave equation

$$(\vec{\nabla}^2 + k^2)\vec{\Psi}(\vec{r}) = -\vec{V}(\vec{r}), \tag{2.30}$$

which is solved with the help of the Green's formalism. The solution for the vector component α reads

$$\Psi_\alpha(\vec{r}) = \sum_\beta \int d\vec{r}\,' G_{\alpha\beta}(\vec{r},\vec{r}\,')V_\beta(\vec{r}\,'), \tag{2.31}$$

where the Green's function $G_{\alpha\beta}$ is expressed in terms of vector spherical harmonics as [5]

$$G_{\alpha\beta}(\vec{r},\vec{r}') = \sum_{l=0}^{\infty}\sum_{m=-l}^{l} \frac{ik}{\sqrt{l(l+1)}} h_l^{(1)}(kr) j_l(kr') X_{lm\alpha} X^*_{lm\beta}(\theta',\phi'). \tag{2.32}$$

Here $h_l^{(1)}$ denote the spherical Hankel functions and j_l the spherical Bessel functions. $\vec{X}_{lm}(\theta,\phi) = \vec{L}Y_{lm}(\theta,\phi)/\sqrt{l(l+1)}$ denote the vector spherical harmonics, as developed from the scalar spherical harmonics $Y_{lm}(\theta,\phi)$ with help of the angular momentum operator $\vec{L} = -i\vec{r}\times\vec{\nabla}$. In this way $\vec{X}^*_{lm} = Y^*_{lm}\vec{L}/\sqrt{l(l+1)}$ becomes a differential operator that acts on the inhomogeneity $\vec{V}$. The solution to the inhomogeneous wave equation (2.30) thus reads

$$\vec{\Psi}(\vec{r}) = \sum_{l=0}^{\infty}\sum_{m=-l}^{l} h_l^{(1)}(kr)\vec{X}_{lm}(\theta,\phi)\frac{ik}{\sqrt{l(l+1)}}\int d\vec{r}'\, j_l(kr')Y^*_{lm}(\theta',\phi')\vec{L}'\cdot\vec{V}(\vec{r}'). \tag{2.33}$$

This general solution can be used to solve the inhomogeneous Helmholtz equations (2.29), which will be done in the following only for the first equation with the special inhomogeneity

$$\vec{V}(\vec{r}) = ik\sqrt{\frac{\mu_0}{\epsilon_0}}\left(\vec{J}(\vec{r}) + \vec{\nabla}\times\vec{M}(\vec{r})\right) + \frac{i}{k}\sqrt{\frac{\mu_0}{\epsilon_0}}\vec{\nabla}(\vec{\nabla}\cdot\vec{J}(\vec{r})). \tag{2.34}$$

The solution for the electric field is determined to

$$\vec{E}(\vec{r}) = \sum_{l=0}^{\infty}\sum_{m=-l}^{l} a_{lm}^{(M)} h_l^{(1)}(kr)\vec{X}_{lm}(\theta,\phi), \tag{2.35}$$

with the field amplitude of order l, m

$$\begin{aligned} a_{lm}^{(M)} = \frac{ik}{\sqrt{l(l+1)}} \int d^3r'\, j_l(kr')Y^*_{lm}(\theta',\phi')\vec{L}' \\ \cdot\left[ik\sqrt{\frac{\mu_0}{\epsilon_0}}\left(\vec{J}(\vec{r}') + \vec{\nabla}\times\vec{M}(\vec{r}')\right) + \frac{i}{k}\sqrt{\frac{\mu_0}{\epsilon_0}}\vec{\nabla}(\vec{\nabla}\cdot\vec{J}(\vec{r}'))\right]. \end{aligned} \tag{2.36}$$

Making use of the identities

$$\begin{aligned} &\vec{L}\cdot\vec{V}(\vec{r}) = i\vec{\nabla}\cdot(\vec{r}\times\vec{V}(\vec{r})), \\ &\vec{L}\cdot(\vec{\nabla}\times\vec{V}(\vec{r})) = i\vec{\nabla}^2(\vec{r}\cdot\vec{V}(\vec{r})) - \frac{i\partial(r^2\vec{\nabla}\cdot\vec{V}(\vec{r}))}{r\partial r}, \text{ and} \\ &\vec{L}\cdot\vec{\nabla}s(\vec{r}) = 0, \end{aligned} \tag{2.37}$$

this equation is simplified to

$$a_{lm}^{(M)} = \frac{-ik^2}{\sqrt{l(l+1)}}\sqrt{\frac{\mu_0}{\epsilon_0}} \int d\vec{r}'\, j_l(kr')Y_{lm}^*(\theta',\phi') \cdot \left[\vec{\nabla}\cdot(\vec{r}'\times\vec{J}(\vec{r}')) + \vec{\nabla}^2(\vec{r}'\cdot\vec{M}(\vec{r}')) - \frac{\partial(r'^2\vec{\nabla}\cdot\vec{M}(\vec{r}'))}{r'\partial r'}\right]. \tag{2.38}$$

When applying integration by parts, the $\vec{\nabla}^2$ is replaced by $-k^2$ and by Green's theorem one can cast the radial derivative onto the spherical Bessel function. This results in

$$a_{lm}^{(M)} = \frac{-ik^2}{\sqrt{l(l+1)}}\sqrt{\frac{\mu_0}{\epsilon_0}} \int d\vec{r}'\Big[j_l(kr')Y_{lm}^*(\theta',\phi')\left(\vec{\nabla}\cdot(\vec{r}'\times\vec{J}(\vec{r}')) - k^2(\vec{r}'\cdot\vec{M}(\vec{r}'))\right) + Y_{lm}^*(\theta',\phi')\vec{\nabla}\cdot\vec{M}(\vec{r}')\frac{\partial(r'j_l(kr'))}{\partial r'}\Big]. \tag{2.39}$$

As the source dimensions are small compared to the wavelength ($kr \leq 1$), the argument of the Bessel function is small. In this limit, j_l can be approximated as

$$j_l(kr) \approx \frac{(kr)^l}{(2l+1)!!}. \tag{2.40}$$

Inserting this approximation in Eq. (2.39) and keeping only the lowest orders in kr (leading to the cancellation of the $k^2(\vec{r}'\cdot\vec{M}(\vec{r}'))$ term), one obtains

$$a_{lm}^{(M)} = \frac{-ik^{l+2}}{(2l+1)!!}\sqrt{\frac{l+1}{l}}\sqrt{\frac{\mu_0}{\epsilon_0}}\left(M_{lm} + M'_{lm}\right), \tag{2.41}$$

with the magnetic multipole moments defined as

$$\begin{aligned} M_{lm} &= \frac{1}{l+1}\int d\vec{r}' r'^l Y_{lm}^*(\theta',\phi')\vec{\nabla}\cdot(\vec{r}'\times\vec{J}(\vec{r}')), \text{ and} \\ M'_{lm} &= \int d\vec{r}' r'^l Y_{lm}^*(\theta',\phi')\vec{\nabla}\cdot\vec{M}(\vec{r}'). \end{aligned} \tag{2.42}$$

M_{lm} corresponds to the effective magnetization, while M'_{lm} accounts for the intrinsic magnetization $\vec{M}(\vec{r})$.

A similar procedure can be carried out in order to determine the electric multipole coefficients $a_{lm}^{(E)}$ (Ref. [5]). The result is

$$a_{lm}^{(E)} = \frac{-ick^{l+2}}{(2l+1)!!}\sqrt{\frac{l+1}{l}}[Q_{lm} + Q'_{lm}], \tag{2.43}$$

with the electric multipole moments

$$Q_{lm} = \int d\vec{r}' r'^l Y^*_{lm}(\theta', \phi')\rho(\vec{r}')$$
$$Q'_{lm} = \frac{-ik}{c(l+1)} \int d\vec{r}' r'^l Y^*_{lm}(\theta', \phi')\vec{\nabla} \cdot (\vec{r}' \times \vec{M}(\vec{r}')). \tag{2.44}$$

The Poynting vector $\vec{S}$ is used to infer the angular dependent energy emitted per second and unit area. It is defined as

$$\vec{S}(\vec{r}, t) = \vec{E}(\vec{r}, t) \times \vec{H}(\vec{r}, t). \tag{2.45}$$

In the far field one has $|\vec{H}| = |\vec{B}|/\mu_0 = \sqrt{\epsilon_0/\mu_0}|\vec{E}|$ and the Poynting vector transforms to

$$|\vec{S}(\vec{r}, t)| = \sqrt{\frac{\epsilon_0}{\mu_0}}|\vec{E}(\vec{r}, t)|^2. \tag{2.46}$$

Applying the definition of the electric field (following Ref. [6]) as

$$\vec{E}(\vec{r}, t) = \vec{E}(\vec{r})e^{-i\omega t} + \vec{E}^*(\vec{r})e^{i\omega t}, \tag{2.47}$$

the time average takes the value

$$< |\vec{E}(\vec{r}, t)|^2 >= 2 \cdot |\vec{E}(\vec{r})|^2, \tag{2.48}$$

and thus

$$< |\vec{S}(\vec{r}, t)| >= 2\sqrt{\frac{\epsilon_0}{\mu_0}}|\vec{E}(\vec{r})|^2. \tag{2.49}$$

From Eq. (2.35) we obtain for a single magnetic multipole of order lm

$$\vec{E}^{(M)}_{lm}(\vec{r}) = a^{(M)}_{lm} h^{(1)}_l(kr)\vec{X}_{lm}(\theta, \phi), \tag{2.50}$$

with $h^{(1)}_l$ the spherical Hankel functions and $\vec{X}_{lm}(\theta, \phi)$ the vector spherical harmonics. For far distances from the source ($kr \leq l$), the spherical Hankel functions can be approximated as

$$h^{(1)}_l \approx \frac{e^{i(kr - l\pi/2)}}{kr}. \tag{2.51}$$

Thus the total amount of energy radiated per second is calculated to be

$$\begin{aligned} P^{(M)} &= \int_{\Omega} < |\vec{S}(\vec{r},t)| > d\Omega \\ &= 2\sqrt{\frac{\epsilon_0}{\mu_0}} \int_{\Omega} \frac{|a_{lm}^{(M)}|^2}{k^2} |\vec{X}_{lm}(\theta,\phi)|^2 d\Omega \\ &= 2\sqrt{\frac{\epsilon_0}{\mu_0}} \frac{|a_{lm}^{(M)}|^2}{k^2}, \end{aligned} \tag{2.52}$$

where in the last step the normalization condition of the vector spherical harmonics was used. Inserting Eq. (2.41) for $a_{lm}^{(M)}$ and dividing by $\hbar\omega$, the rate of emission of a photon of frequency ω per unit time is derived.

$$\begin{aligned} A^{(M)} &= \frac{P^{(M)}}{\hbar\omega} = \frac{P^{(M)}}{\hbar c k} \\ &= \frac{2\epsilon_0}{\hbar k^3} \cdot |a_{lm}^{(M)}|^2 \\ &= \frac{2\mu_0}{\hbar} \frac{k^{2l+1}}{[(2l+1)!!]^2} \frac{l+1}{l} B_{eg}(Ml), \end{aligned} \tag{2.53}$$

where $B_{eg}(Ml)$ denotes the nuclear transition matrix element of multipole order l, defined for the transition from the excited state e to the ground state g as

$$B_{eg}(Ml) = \left(M_{lm} + M'_{lm}\right)^2. \tag{2.54}$$

This matrix element can only be determined experimentally, however, Weisskopf derived a rough estimate for $B_{eg}(Ml)$, based on the single-particle nuclear shell model, leading to [6]

$$B_{eg}(Ml) \approx \frac{10}{\pi} \left(\frac{3}{3+l}\right)^2 R^{2l-2} \mu_{\mathrm{N}}^2. \tag{2.55}$$

Here $R = 1.2 \cdot A_N^{1/3}$ fm is the nuclear radius (with A_N the mass number) and $\mu_N = 5.051 \cdot 10^{-27}$ J/T denotes the nuclear magneton. Equation (2.53) is often given in gaussian units, where it has to be multiplied by $4\pi/\mu_0$. The results of Eq. (2.55) can easily deviate by two orders of magnitude from the experimental results. In order to account for this, a correction factor $B_{\mathrm{W.u.}}$ is introduced, corresponding to the transition rate in Weisskopf units. There is no complete agreement for $B_{\mathrm{W.u.}}$ in case of $^{229\mathrm{m}}$Th in literature and the proposed values vary by a factor of 10 between $0.33 \cdot 10^{-2}$ and $4.55 \cdot 10^{-2}$ [7].

Similar to Eq. (2.53), the photon emission probabilities can also be estimated for electric multipole moments. In this case one obtains

$$
\begin{aligned}
A^{(E)} &= \frac{2\mu_0}{\hbar k^3} \cdot |a_{lm}^{(E)}|^2 \\
&= \frac{2}{\hbar \epsilon_0} \frac{k^{2l+1}}{[(2l+1)!!]^2} \frac{l+1}{l} B_{eg}(El),
\end{aligned}
\tag{2.56}
$$

where $B_{eg}(El) = (Q_{lm} + Q'_{lm})^2$ and can be approximated as [6]

$$
B_{eg}(El) \approx \frac{1}{4\pi} \left(\frac{3}{3+l} \right)^2 R^{2l} e^2. \tag{2.57}
$$

During the radiative transition from an initial nuclear state i to a final state f, the overall angular momentum is conserved. This leads to the selection rules for multipole radiation l, m during nuclear transitions as

$$
\begin{aligned}
|j_i - j_f| \leq l &\leq j_i + j_f \\
m_i - m_f &= m.
\end{aligned}
\tag{2.58}
$$

A further selection rule originates from parity conservation. Denoting the parity of the initial and final nuclear states as π_i and π_f, respectively, one obtains

$$
\begin{aligned}
\pi_i &= \pi_f \qquad \text{for} \quad \pi = +1, \\
\pi_i &= -\pi_f \qquad \text{for} \quad \pi = -1,
\end{aligned}
\tag{2.59}
$$

where π denotes the parity of the multipole radiation. From the definition of electric and magnetic multipole radiation (see e.g. Ref. [6]) the parity π is obtained as

$$
\begin{aligned}
\pi &= (-1)^l \qquad \text{for electric multipoles}, \\
\pi &= -(-1)^l \qquad \text{for magnetic multipoles}.
\end{aligned}
\tag{2.60}
$$

In the specific case of $^{229\text{m}}$Th, the nuclear transition occurs between the Nilsson states $3/2^+631$ (excited state) to $5/2^+633$ (ground state). Correspondingly, one obtains $|3/2 - 5/2| \leq l \leq 3/2 + 5/2$ and l can take values between 1 and 4. As decays are dominated by the lowest multipole order, only the multipole order $l = 1$ will be considered. From parity conservation it is seen that $\pi = +1$, which can, under the assumption $l = 1$, only be fulfilled for magnetic multipole radiation. Thus the transition of $^{229\text{m}}$Th to the ground state will be mostly of type $M1$, with a small admixture of $E2$. The photon transition rate is calculated based on Eq. (2.53) to be $A \approx 4.9 \cdot 10^{-5}\ \text{s}^{-1}$, conservatively assuming $B_{\text{W.u.}} = 0.33 \cdot 10^{-2}$. As the purely radiative lifetime τ_γ relates to the photon transition rate via

$$
\tau_\gamma = \frac{1}{A}, \tag{2.61}
$$

this corresponds to a lifetime of $\tau_\gamma \approx 2.0 \cdot 10^4$ s, which is 5.6 h.

2.2.2 Internal Conversion

Besides γ decay of an excited nuclear state, internal conversion (IC) can occur, which is for most nuclear transitions the only relevant competing decay channel. During the internal conversion process, the nuclear excitation energy is transferred to an atomic shell electron, which is subsequently ejected from the atomic shell. Due to energy conservation, the electron's kinetic energy after the IC process (E_e) corresponds to the energy of the nuclear transition E_γ minus the binding energy of the electron E_b: $E_e = E_\gamma - E_b$. Thus internal conversion can only occur if the electron's binding energy is below the nuclear excitation energy. This is of particular importance for the case of ^{229m}Th with an extremely low nuclear excitation energy of about 7.8 eV. In this case it is expected that already for singly charged thorium ions the internal conversion process is energetically forbidden, as the ionization energy of Th^{1+} is 11.9 eV and therefore above the isomeric energy. Only for neutral thorium atoms, with an ionization energy of 6.3 eV, IC is expected to be the dominant decay channel.

It has to be pointed out that, during the IC process, the nucleus couples directly to the atomic shell by means of the exchange of a virtual photon. Internal conversion is thus a second order process, representing a competing decay channel to the first order process of the direct photonic decay (see also Fig. 3.5). In this way, the existence of an internal conversion decay channel leads to a lifetime reduction of the excited nuclear state. Opposed to that, it is not correct to imagine internal conversion as the emission of a (real) photon by nuclear deexcitation followed by a photo effect in the atomic shell. In the latter case the lifetime of the nuclear excited state would not be affected [6].

A complete theoretical derivation of the internal conversion process is involved and beyond the scope of this thesis. The interested reader is referred to the original work of Tralli and Goertzel [8]. Some problems concerning the gauge invariance were later resolved by G. Kramer [9]. For a description of the IC process we follow closely the lines of Rose (Refs. [10, 11]).

As nuclear γ decay and internal conversion are competing decay channels of an excited nuclear state, an internal conversion coefficient α_{ic} is defined as the decay rate via internal conversion (A_{ic}) divided by the γ-decay rate (A_γ). It can be shown [8, 9], that the IC coefficient is calculated (in SI units[1]) as

$$\alpha_{ic} = \frac{A_{ic}}{A_\gamma} = \pi^2 k \lambda_c \alpha \cdot S \left| \int_0^\infty \Psi_f^* \left(\vec{\alpha} \cdot \vec{B}_{lm}(kr) + \phi_{lm}(kr) \right) \Psi_i \, r^2 dr d\Omega \right|^2, \quad (2.62)$$

where $k = E_\gamma/\hbar c$ is the wavenumber corresponding to the nuclear transition energy, $\lambda_c = 2\pi\hbar/m_e c \approx 2.426 \cdot 10^{-12}$ m denotes the Compton wavelength and $\alpha = e^2/4\pi\epsilon_0\hbar c \approx 1/137$ is the finestructure constant. S is an abbreviation for the sum over all unobserved degrees of freedom of the radiation, an average over initial and

[1] Natural units are often used in literature, setting $\hbar = c = m_e = 1$. In this case $\lambda_c = 2\pi$, $\alpha = e^2$ and k and e are dimensionless.

a summation over final magnetic substates of the electron and Ψ_i and Ψ_f represent the electronic wave functions of the initial and final state that will be introduced below. $\vec{\alpha}$ corresponds to the Dirac matrix vector and $\vec{B}_{lm}$ and ϕ_{lm} are the vector and scalar potentials of an outgoing electromagnetic wave, respectively. They take different forms for electric and magnetic radiation of multipole order l. For magnetic radiation one has

$$\begin{aligned} \vec{B}_{lm}^{(M)} &= \sqrt{\frac{2}{\pi}} h_l^{(1)}(kr) \vec{X}_{lm}(\theta, \phi), \\ \phi_{lm}^{(M)} &= 0. \end{aligned} \tag{2.63}$$

Here $h_l^{(1)}$ denote the spherical Hankel function, and $\vec{X}_{lm}$ the vector spherical harmonics. Similarly, for electric multipole radiation, one has

$$\begin{aligned} \vec{B}_{lm}^{(E)} &= \sqrt{\frac{2}{\pi} \frac{1}{l(l+1)}} h_{l-1}^{(1)}(kr) \left(r\vec{\nabla} + \frac{lr}{r} \right) Y_{lm}(\theta, \phi), \\ \phi_{lm}^{(E)} &= i \sqrt{\frac{2}{\pi} \frac{l}{l+1}} h_l^{(1)}(kr) Y_{lm}(\theta, \phi), \end{aligned} \tag{2.64}$$

where Y_{lm} are the scalar spherical harmonics.

In Eq. (2.62), ψ_i and ψ_f represent relativistic electron wave functions of the central field, defined by the quantum numbers L, $J = L \pm 1/2$ and the magnetic quantum number μ. These are solutions of the Dirac equation for the Coulomb potential and can be separated into radial and angular components [10]

$$\psi = \begin{pmatrix} -i\, f_{LJ}(r) \Omega_{J\mu}^{(\pm)}(\theta, \phi) \\ g_{LJ}(r) \Omega_{J\mu}^{(\mp)}(\theta, \phi) \end{pmatrix} \quad \text{for} \quad J = L \pm 1/2. \tag{2.65}$$

Here $\Omega_{J\mu}^{(\pm)}(\theta, \phi)$ correspond to the spin spherical harmonics, representing two-component spinors defined as

$$\Omega_{J\mu}^{(\pm)}(\theta, \phi) = \begin{pmatrix} C_{J\pm 1/2\ \mu-1/2\ 1/2\ 1/2}^{J\mu} Y_{J\pm 1/2\ \mu-1/2}(\theta, \phi) \\ C_{J\pm 1/2\ \mu+1/2\ 1/2\ -1/2}^{J\mu} Y_{J\pm 1/2\ \mu+1/2}(\theta, \phi) \end{pmatrix}, \tag{2.66}$$

where C denotes the corresponding Clebsch–Gordan coefficient and Y_{lm} are the scalar spherical harmonics. The radial spinor wave functions $f(r)$ and $g(r)$ for the Coulomb potential can be calculated explicitly and are given in Ref. [12]. For the integral in Eq. (2.62), one obtains in case of magnetic multipole moments

$$\begin{aligned} K^{(M)} &= \int_0^\infty \Psi_f^* \left(\vec{\alpha} \cdot \vec{B}_{lm}(kr) + \phi_{lm}(kr)\right) \Psi_i r^2 dr d\Omega \\ &= \frac{1}{\sqrt{l(l+1)}} \sqrt{\frac{2}{\pi}} \int_0^\infty \Psi_f^* h_l^{(1)}(kr) \vec{\alpha} \vec{L} Y_{lm}(\theta, \phi) \Psi_i r^2 dr d\Omega, \end{aligned} \tag{2.67}$$

where the definition of the vector spherical harmonics as $\vec{X}_{lm}(\theta, \phi) = 1/\sqrt{l(l+1)}$ $\vec{L}Y_{lm}(\theta, \phi)$ was used. Inserting $\vec{\alpha}$ as

$$\vec{\alpha} = \begin{pmatrix} 0 & \vec{\sigma} \\ \vec{\sigma} & 0 \end{pmatrix} \tag{2.68}$$

and using Eq. (2.65) for Ψ_i and Ψ_f one obtains

$$K^{(M)} = \frac{i}{\sqrt{l(l+1)}} \sqrt{\frac{2}{\pi}} \left[R_1 \int_\Omega \Omega_{J\mu}^{*(\pm)} \vec{\sigma} \vec{L} Y_{lm} \Omega_{J'\mu'}^{(\mp)} d\Omega - R_2 \int_\Omega \Omega_{J\mu}^{*(\mp)} \vec{\sigma} \vec{L} Y_{lm} \Omega_{J'\mu'}^{(\pm)} d\Omega \right]. \tag{2.69}$$

Here the radial integrals R_1 and R_2 were defined as

$$\begin{aligned} R_1 &= \int_0^\infty f_{LJ}(r) h_l^{(1)}(kr) g_{L'J'}(r) r^2 dr \\ R_2 &= \int_0^\infty g_{LJ}(r) h_l^{(1)}(kr) f_{L'J'}(r) r^2 dr. \end{aligned} \tag{2.70}$$

These integrals can only be calculated numerically. Using the identity [10]

$$\vec{\sigma} \cdot \left(\vec{L} Y_{lm}\right) \Omega_{J'\mu'} = \vec{\sigma} \cdot \vec{L} \left(Y_{lm} \Omega_{J'\mu'}\right) - Y_{lm} \vec{\sigma} \cdot \vec{L} \Omega_{J'\mu'}, \tag{2.71}$$

the integrals in Eq. (2.69) can be simplified in the following way

$$\begin{aligned} \int_\Omega \Omega_{J\mu}^{*(\pm)} \vec{\sigma} \vec{L} Y_{lm} \Omega_{J'\mu'}^{(\mp)} d\Omega &= \int_\Omega \vec{\sigma} \vec{L} \Omega_{J\mu}^{*(\pm)} Y_{lm} \Omega_{J'\mu'}^{(\mp)} d\Omega - \int_\Omega \Omega_{J\mu}^{*(\pm)} Y_{lm} \vec{\sigma} \vec{L} \Omega_{J'\mu'}^{(\mp)} d\Omega \\ &= \left(\kappa + \kappa'\right) \int_\Omega \Omega_{J\mu}^{*(\pm)} Y_{lm} \Omega_{J'\mu'}^{(\mp)} d\Omega. \end{aligned} \tag{2.72}$$

In the last step the equations

$$\begin{aligned} \left(\vec{\sigma}\vec{L} + 1\right) \Omega_{J\mu}^{*(\pm)} &= (L - J)(2J + 1) \Omega_{J\mu}^{*(\pm)} \quad \text{for} \quad J = L \pm 1/2 \\ \left(\vec{\sigma}\vec{L} + 1\right) \Omega_{J'\mu'}^{(\mp)} &= -(L' - J')(2J' + 1) \Omega_{J'\mu'}^{(\mp)} \quad \text{for} \quad J' = L' \pm 1/2 \end{aligned} \tag{2.73}$$

were used and κ (and κ') were introduced as

$$\begin{aligned} \kappa &= (L-J)(2J+1) \qquad \text{for} \quad J = L \pm 1/2, \\ \kappa' &= (L'-J')(2J'+1) \qquad \text{for} \quad J' = L' \pm 1/2. \end{aligned} \tag{2.74}$$

Similarly, for the second term in Eq. (2.69) one obtains

$$\begin{aligned} \int_\Omega \Omega^{*(\mp)}_{J\mu} \vec{\sigma}\vec{L} Y_{lm} \Omega^{(\pm)}_{J'\mu'} \, d\Omega &= -\left(\kappa + \kappa'\right) \int_\Omega \Omega^{*(\mp)}_{J\mu} Y_{lm} \Omega^{(\pm)}_{J'\mu'} \, d\Omega \\ &= -\left(\kappa + \kappa'\right) \int_\Omega \Omega^{*(\pm)}_{J\mu} Y_{lm} \Omega^{(\mp)}_{J'\mu'} \, d\Omega. \end{aligned} \tag{2.75}$$

Here, for the last step the square of a Pauli matrix $\sigma^2 = 1$ was introduced and the hermitian property of σ in combination with $\sigma\Omega^{(\pm)}_{J\mu} = -\Omega^{(\mp)}_{J\mu}$ was used [10]. Inserting the results of Eqs. (2.72) and (2.75) into Eq. (2.69) and using that Y_{lm} is a scalar function, the result is

$$K^{(M)} = \frac{i}{\sqrt{l(l+1)}} \sqrt{\frac{2}{\pi}} \left(R_1 + R_2\right) \left(\kappa + \kappa'\right) \int_\Omega Y_{lm} \Omega^{*(\pm)}_{J\mu} \Omega^{(\mp)}_{J'\mu'} \, d\Omega. \tag{2.76}$$

From the definition of $\Omega_{J\mu}$ it is seen that

$$\Omega^{*(\pm)}_{J\mu} \Omega^{(\mp)}_{J'\mu'} = \sum_{\tau=\pm 1/2} C^{J\mu}_{J\pm 1/2\ \mu-\tau\ 1/2\ \tau} C^{J'\mu'}_{J'\mp 1/2\ \mu'-\tau\ 1/2\ \tau} Y^*_{J\pm 1/2\ \mu-\tau} Y_{J'\mp 1/2\ \mu'-\tau}. \tag{2.77}$$

For the following, we use that $L' = J' \mp 1/2$ and define $\bar{L} = J \pm 1/2$. The last equation contains the product of two spherical harmonics, which can be expressed in terms of a single spherical harmonic as [10]

$$Y^*_{\bar{L}\mu-\tau} Y_{L'\mu'-\tau} = \sum_\nu (-1)^{\mu-\tau} \sqrt{\frac{(2\bar{L}+1)(2L'+1)}{4\pi(2\nu+1)}} C^{\nu 0}_{\bar{L}0\ L'0} C^{\nu\ \mu'-\mu}_{\bar{L}\ -\mu+\tau\ L'\ \mu'-\tau} Y_{\nu\ \mu'-\mu}. \tag{2.78}$$

Here the relation $Y^*_{L\,m} = (-1)^m Y_{L\,-m}$ was applied. Inserting this into Eq. (2.77) leads to

$$\Omega^{*(\pm)}_{J\mu} \Omega^{(\mp)}_{J'\mu'} = \sum_\nu \sqrt{\frac{(2\bar{L}+1)(2L'+1)}{4\pi(2\nu+1)}} C^{\nu 0}_{\bar{L}0\ L'0} Y_{\nu\ \mu'-\mu} S_\nu, \tag{2.79}$$

where S_ν was defined as

$$S_\nu = \sum_\tau (-1)^{\mu-\tau} C^{J\mu}_{\bar{L}\ \mu-\tau\ 1/2\ \tau} C^{J'\mu'}_{L'\ \mu'-\tau\ 1/2\ \tau} C^{\nu\ \mu'-\mu}_{\bar{L}\ -\mu+\tau\ L'\ \mu'-\tau}. \tag{2.80}$$

With some calculation, which will not be carried out at this point, it is possible to show that the sum over three Clebsch–Gordan coefficients can be explicitly expressed in terms of Racah coefficients W [10]

$$S_\nu = (-1)^{\bar{L}+L'+\nu+\mu+1/2} \sqrt{(2J+1)(2J'+1)} C_{J\ -\mu\ J'\ \mu'}^{\nu\ \mu'-\mu} W(J\bar{L}J'L';1/2\nu). \tag{2.81}$$

Consequently, for the integral in Eq. (2.76) one obtains

$$\begin{aligned}
\int_\Omega Y_{lm}\Omega_{J\mu}^{*(\pm)}\Omega_{J'\mu'}^{(\mp)}\, d\Omega &= \int_\Omega Y_{lm} \sum_\nu \sqrt{\frac{(2\bar{L}+1)(2L'+1)}{4\pi(2\nu+1)}} C_{\bar{L}0\ L'0}^{\nu 0} Y_{\nu\ \mu'-\mu} S_\nu\, d\Omega \\
&= \sum_\nu \sqrt{\frac{(2\bar{L}+1)(2L'+1)}{4\pi(2\nu+1)}} C_{\bar{L}0\ L'0}^{\nu 0} S_\nu\, (-1)^m \int_\Omega Y_{l\ -m}^* Y_{\nu\ \mu'-\mu} d\Omega \\
&= (-1)^m \sum_\nu \sqrt{\frac{(2\bar{L}+1)(2L'+1)}{4\pi(2\nu+1)}} C_{\bar{L}0\ L'0}^{\nu 0} S_\nu\, \delta_{l,\nu}\delta_{m,\mu-\mu'} \\
&= (-1)^m \sqrt{\frac{(2\bar{L}+1)(2L'+1)}{4\pi(2l+1)}} C_{\bar{L}0\ L'0}^{l\ 0} S_l\, \delta_{\mu,\mu'+m}.
\end{aligned} \tag{2.82}$$

Inserting Eq. (2.76) together with this integral into Eq. (2.62) the result is

$$\alpha_{ic}^{(M)} = \frac{k\lambda_c\alpha}{2} \cdot S \left| \frac{(R_1+R_2)(\kappa+\kappa')}{\sqrt{l(l+1)}} \sqrt{\frac{(2\bar{L}+1)(2L'+1)}{(2l+1)}} C_{\bar{L}0\ L'0}^{l\ 0} S_l\, \delta_{\mu,\mu'+m} \right|^2. \tag{2.83}$$

The sum S, which represents the sum over all final states and the average over all initial states, can be explicitly written as $S = 1/(2l+1)\sum_{\mu\mu' mJ}$ [10], leading to

$$\begin{aligned}
\alpha_{ic}^{(M)} &= \frac{k\lambda_c\alpha}{2} \frac{1}{2l+1} \sum_{\mu\mu' mJ} \left| \frac{(R_1+R_2)(\kappa+\kappa')}{\sqrt{l(l+1)}} \sqrt{\frac{(2\bar{L}+1)(2L'+1)}{(2l+1)}} C_{\bar{L}0\ L'0}^{l\ 0} S_l\, \delta_{\mu,\mu'+m} \right|^2 \\
&= \frac{k\lambda_c\alpha}{2} \frac{2L'+1}{l(l+1)(2l+1)^2} \sum_J |R_1+R_2|^2 (\kappa+\kappa')^2 (2\bar{L}+1)|C_{\bar{L}0\ L'0}^{l\ 0}|^2 \sum_{\mu\mu'} S_l^2 \\
&= \frac{k\lambda_c\alpha}{2} \frac{(2L'+1)(2J'+1)}{l(l+1)(2l+1)^2} \sum_J |R_1+R_2|^2 (\kappa+\kappa')^2 (2\bar{L}+1)(2J+1) \\
&\qquad |C_{\bar{L}0\ L'0}^{l\ 0}|^2 W^2(J\bar{L}J'L';1/2l) \sum_{\mu\mu'} |C_{J\ -\mu\ J'\ \mu'}^{l\ \mu'-\mu}|^2.
\end{aligned} \tag{2.84}$$

Applying now the relation

$$\sum_{\mu\mu'} |C_{J\ -\mu\ J'\ \mu'}^{l\ \mu'-\mu}|^2 = 2l+1 \tag{2.85}$$

leads to the final result for the internal conversion coefficient as

$$\alpha_{ic}^{(M)} = \frac{k\lambda_c\alpha}{2}\frac{(2L'+1)(2J'+1)}{l(l+1)(2l+1)}\sum_J |R_1+R_2|^2\left(\kappa+\kappa'\right)^2(2\bar{L}+1)(2J+1)$$
$$|C^{l\,0}_{\bar{L}0\;L'0}|^2 W^2(J\bar{L}J'L';1/2l). \tag{2.86}$$

Let us now consider the special case of $^{229\text{m}}$Th: Corresponding calculations were for the first time performed by Strizhov and Tkalya [13] and more recently also by Karpeshin and Trzhaskovskaya [14]. Due to the expected energy value of 7.8 eV, internal conversion is expected to occur only in neutral ^{229}Th and only by coupling to the two outermost electrons, populating the $6\text{d}_{3/2}$ and the $7\text{s}_{1/2}$ electronic levels (usually IC electrons are generated in the innermost shells). Calculation reveals, however, that the internal conversion process is dominated by the $7\text{s}_{1/2}$ shell [13]. Thus one can assume $L' = 0$ and $J' = 1/2$. As the nuclear transition is of type $M1$, we have $l = 1$ and Eq. (2.86) simplifies to

$$\alpha_{ic} = \frac{k\lambda_c\alpha}{6}\sum_J |R_1+R_2|^2\left(\kappa+\kappa'\right)^2(2\bar{L}+1)(2J+1)$$
$$|C^{1\,0}_{\bar{L}0\;1/2\;0}|^2 W^2(J\bar{L}1/2\,0;1/2\,1). \tag{2.87}$$

This equation provides the basis for numerical calculations resulting in $\alpha_{ic} \approx 10^9$ [13, 14]. Some minor dependency on the applied atomic shell model was found [14]. Assuming a purely photonic lifetime $\tau_\gamma = 2.0 \cdot 10^4$ s (see Sect. 2.2.1), the expected lifetime of $^{229\text{m}}$Th under internal conversion in the neutral thorium atom is $\tau_{\text{ic}} \approx 20\ \mu$s. This value is in good agreement with the recently measured lifetime $\tau_{\text{ic}} \approx 10\ \mu$s (corresponding to $t_{1/2} \approx 7\ \mu$s) [15].

2.3 Nuclear Laser Excitation

One of the main motivations for studying $^{229\text{m}}$Th is that it is the only known nuclear state that allows for direct nuclear laser excitation using already existing laser technology. In the following, the theory of nuclear laser excitation is discussed. In the first section the optical Bloch equations are derived. In the second section an expression for the number of excited nuclei in saturation is obtained, assuming the low-saturation limit. This equation is applied to the special case of $^{229\text{m}}$Th in the third section. In the last section, the case of $^{235\text{m}}$U, which exhibits the second lowest excitation energy of 76 eV, is discussed for comparison.

2.3.1 The Optical Bloch Equations

Quite generally the optical Bloch equations describe the coupling of coherent light to a two-level system [16]. For an atomic two-level system usually electric dipole transitions are discussed. For nuclear two-level systems, however, higher multipolarities play a more pronounced role and have to be taken into consideration. The optical Bloch equations for a nuclear two-level system consisting of degenerate ground and excited nuclear states will be derived in the following. Starting point for this derivation is the quantum optical master equation in Lindblad form given in the Schrödinger picture as [17, 18]

$$\frac{\partial \hat{\rho}}{\partial t} = -\frac{i}{\hbar}\left[\hat{H}, \hat{\rho}\right] + \mathcal{L}[\hat{\rho}], \tag{2.88}$$

where the Lindblad super-operator $\mathcal{L}[\hat{\rho}]$ is defined as

$$\mathcal{L}[\hat{\rho}] = \sum_{m_g m_e} \Gamma_{m_g m_e} \left(\hat{A}_{m_g m_e} \hat{\rho} \hat{A}^{\dagger}_{m_g m_e} - \frac{1}{2} \hat{\rho} \hat{A}^{\dagger}_{m_g m_e} \hat{A}_{m_g m_e} - \frac{1}{2} \hat{A}^{\dagger}_{m_g m_e} \hat{A}_{m_g m_e} \hat{\rho} \right). \tag{2.89}$$

The first term on the right-hand side of Eq. (2.88) corresponds to the von-Neumann equation, which describes the time propagation of the density operator $\hat{\rho}$ without energy dissipation. The Lindblad super-operator, models the decay of the excited states due to coupling to the environment. The summation is performed over all possible deexcitations from the nuclear excited to the nuclear ground states, taking into consideration the magnetic sublevels. For simplicity we have neglected the decoherences between different sublevels of the ground and excited state. The total angular momentum quantum numbers of the ground and excited states are denoted by j_g and j_e, respectively, with the corresponding magnetic quantum numbers m_g and m_e satisfying $-j_g \leq m_g \leq j_g$ and $-j_e \leq m_e \leq j_e$. Furthermore, $\Gamma_{m_g m_e}$ is the partial decay rate from the excited nuclear sublevel described by the magnetic quantum number m_e to the ground nuclear sublevel described by the magnetic quantum number m_g. The Lindblad operators $\hat{A}_{m_g m_e}$ are the annihilation operators of the excited states, explicitly given as

$$\hat{A}_{m_g m_e} = e^{-iE_g t/\hbar} |j_g, m_g\rangle\langle j_e, m_e| e^{+iE_e t/\hbar}. \tag{2.90}$$

It is assumed that the total Hamiltonian of the system $\hat{H}$ can be expressed as $\hat{H} = \hat{H}_N + \hat{H}_I$, with $\hat{H}_N$ the nuclear Hamiltonian and $\hat{H}_I$ the resonant light-interaction Hamiltonian, which describes the interaction of laser light of angular frequency ω_L with a nuclear transition of the same angular frequency $\omega = \omega_L$.

The Hilbert space vectors of the ground and excited states $|j_g, m_g\rangle$ and $|j_e, m_e\rangle$ are eigenstates to the nuclear Hamiltonian without interaction term $\hat{H}_N|j_i, m_i\rangle = E_i|j_i, m_i\rangle$ with $i \in \{g, e\}$, where E_g and E_e denote the energy of the ground and excited nuclear state. We therefore have $(E_e - E_g)/\hbar = \omega$. The time dependency of the states can be gained back by multiplying with $e^{-iE_g t/\hbar}$ or $e^{-iE_e t/\hbar}$, respectively.

In the Schrödinger picture and for the considered case, the density operator $\hat{\rho}$ takes the explicit form

$$\begin{aligned}\hat{\rho}(t) =& \sum_{m_e} \rho_{ee}(m_e) e^{-iE_e t/\hbar} |j_e, m_e\rangle\langle j_e, m_e| e^{+iE_e t/\hbar} \\ &+ \sum_{m_g m_e} \rho_{ge}(m_g, m_e) e^{-iE_g t/\hbar} |j_g, m_g\rangle\langle j_e, m_e| e^{+iE_e t/\hbar} \\ &+ \sum_{m_g m_e} \rho_{eg}(m_e, m_g) e^{-iE_e t/\hbar} |j_e, m_e\rangle\langle j_g, m_g| e^{+iE_g t/\hbar} \\ &+ \sum_{m_g} \rho_{gg}(m_g) e^{-iE_g t/\hbar} |j_g, m_g\rangle\langle j_g, m_g| e^{+iE_g t/\hbar}.\end{aligned} \tag{2.91}$$

Here $\rho_{ee}(m_e)$ and $\rho_{gg}(m_g)$ denote the population probabilities of the ground and excited nuclear sublevels, respectively and $\rho_{ge}(m_g, m_e) = \rho^*_{eg}(m_g, m_e)$ are the so called coherences. From this it is inferred that the population probability of the excited sublevel characterized by the magnetic quantum number $\tilde{m}_e$ is

$$\rho_{ee}(m_e) = \langle j_e, m_e| e^{+iE_e t/\hbar}\, \hat{\rho}\, e^{-iE_e t/\hbar} |j_e, m_e\rangle. \tag{2.92}$$

For an explicit calculation we will first consider the von-Neumann-term of Eq. (2.88). This transforms to

$$-\frac{i}{\hbar}\left[\hat{H}, \hat{\rho}\right] = -\frac{i}{\hbar}\left[\hat{H}_N + \hat{H}_I, \hat{\rho}\right] \tag{2.93}$$

Calculating the matrix element $\rho_{ee}(m_e)$ of the density operator by making use of Eq. (2.92) and inserting the unity operator $\hat{\mathbb{1}}$ between each operator product of Eq. (2.93) as

$$\begin{aligned}\hat{\mathbb{1}} =& \sum_{m_g} e^{-iE_g t/\hbar} |j_g, m_g\rangle\langle j_g, m_g| e^{+iE_g t/\hbar} \\ &+ \sum_{m_e} e^{-iE_e t/\hbar} |j_e, m_e\rangle\langle j_e, m_e| e^{+iE_e t/\hbar},\end{aligned} \tag{2.94}$$

the von-Neumann-term gives

$$\begin{aligned}&-\frac{i}{\hbar}\langle j_e, m_e| e^{+iE_e t/\hbar}\left[\hat{H}, \hat{\rho}\right] e^{-iE_e t/\hbar} |j_e, m_e\rangle \\ =& -\frac{i}{\hbar}\sum_{m_g}\Big[\rho_{ge}(m_g, m_e) e^{i\omega t}\langle j_e, m_e|\hat{H}_I|j_g, m_g\rangle \\ &- \rho_{eg}(m_g, m_e) e^{-i\omega t}\langle j_g, m_g|\hat{H}_I|j_e, m_e\rangle\Big].\end{aligned} \tag{2.95}$$

Here $\langle j_e, m_e|\hat{H}_I|j_e, m_e\rangle = 0$ was used, as $\hat{H}_I$ describes only transitions between the nuclear ground and excited sublevels.

In a next step, the extra term in Eq. (2.88) as imposed by the Lindblad superoperator is considered. Using Eq. (2.92) for the matrix element $\rho_{ee}(m_e)$ and inserting Eq. (2.90) for $\hat{A}_{m_g m_e}$, a direct calculation reveals that this term is of the form

$$\begin{aligned} &-\frac{i}{\hbar}\langle j_e, m_e|e^{+iE_e t/\hbar}\,\mathcal{L}[\hat{\rho}]\,e^{-iE_e t/\hbar}|j_e, m_e\rangle \\ &= -\,\rho_{ee}(m_e)\sum_{m_g}\Gamma_{m_g m_e}. \end{aligned} \tag{2.96}$$

Performing the same calculation also for the remaining matrix elements $\rho_{gg}(m_g)$ and $\rho_{ge}(m_g, m_e) = \rho^*_{eg}(m_g, m_e)$ leads to the optical Bloch equations for degenerate ground and excited states including damping [19][2](see also Refs. [20, 21])

$$\begin{aligned} \dot{\rho}_{ee}(m_e) &= -\frac{i}{\hbar}\sum_{m_g}\Big[\rho_{ge}(m_g, m_e)e^{i\omega t}\langle j_e, m_e|\hat{H}_I|j_g, m_g\rangle - \\ &\qquad \rho_{eg}(m_g, m_e)e^{-i\omega t}\langle j_g, m_g|\hat{H}_I|j_e, m_e\rangle\Big] - \rho_{ee}(m_e)\sum_{m_g}\Gamma_{m_g m_e}, \\ \dot{\rho}_{ge}(m_g, m_e) &= \frac{i}{\hbar}\Big[\rho_{gg}(m_g) - \rho_{ee}(m_e)\Big]e^{-i\omega t}\langle j_g, m_g|\hat{H}_I|j_e, m_e\rangle - \rho_{ge}(m_g, m_e)\left(i\Delta + \sum_{\tilde{m}_g}\tilde{\Gamma}_{\tilde{m}_g m_e}\right), \\ \dot{\rho}_{gg}(m_g) &= \frac{i}{\hbar}\sum_{m_e}\Big[\rho_{ge}(m_g, m_e)e^{i\omega t}\langle j_e, m_e|\hat{H}_I|j_g, m_g\rangle - \\ &\qquad \rho_{eg}(m_g, m_e)e^{-i\omega t}\langle j_g, m_g|\hat{H}_I|j_e, m_e\rangle\Big] + \sum_{m_e}\rho_{ee}(m_e)\Gamma_{m_g m_e}. \end{aligned} \tag{2.97}$$

In these equations the possibility of detuning of the laser with respect to the transition frequency is taken into account by introducing $\Delta = \omega - \omega_L$, where ω_L denotes the angular frequency of the laser light. The decay rate of the coherences was introduced as $\tilde{\Gamma}_{m_g m_e} = \Gamma_{m_g m_e}/2$.

For a specific multipolarity l, the partial decay rates $\Gamma_{m_g m_e}$ can be explicitly expressed as [1, 19]

$$\Gamma_{m_g m_e} = \sum_{\sigma}\frac{2j_e + 1}{2l + 1}|C^{l\sigma}_{j_g m_g j_e - m_e}|^2\Gamma_{\text{tot}}, \tag{2.98}$$

where $C^{l\sigma}_{j_g m_g j_e - m_e}$ is the corresponding Clebsch–Gordan coefficient, σ denotes the photon angular momentum projection and Γ_{tot} is the total decay rate of the excited

[2]Note, that the sum over $\tilde{m}_g$ is missing in the second equation in Ref. [19].

state, including both radiative decay and internal conversion channels. Using this definition, summation over the final states m_g together with the average over the initial states m_e for the partial decay rates renders the total decay rate Γ_{tot}. From Eq. (2.98) and using the summation properties of the Clebsch–Gordan coefficients, one obtains

$$\begin{aligned}\sum_{m_g} \Gamma_{m_g m_e} &= \sum_{m_g,\sigma} \frac{2j_e+1}{2l+1} |C^{l\sigma}_{j_g m_g j_e -m_e}|^2 \Gamma_{\text{tot}} \\ &= \Gamma_{\text{tot}}.\end{aligned} \tag{2.99}$$

Consequently, also the sum over the decay rates of the coherences can be simplified to

$$\begin{aligned}\sum_{m_g} \tilde{\Gamma}_{m_g m_e} &= \frac{1}{2}\sum_{m_g} \Gamma_{m_g m_e} = \frac{1}{2}\Gamma_{\text{tot}} \\ &= \tilde{\Gamma}_{\text{tot}}.\end{aligned} \tag{2.100}$$

Here the total decay rate of the coherences $\tilde{\Gamma}_{\text{tot}}$ was introduced. A short coherence time of the laser pulse introduces an additional decay rate of the coherence terms in Eqs. (2.97) [22, 23], such that $\tilde{\Gamma}_{\text{tot}} = (\Gamma_{\text{tot}} + \Gamma_L)/2$ where Γ_L denotes the bandwidth of the laser light [21].

In its general form, the optical Bloch equations given in Eq. (2.97) can only be solved numerically. Under certain conditions, however, an analytical solution is possible. The analytical solution for a realistic scenario of nuclear laser excitation will be discussed in the following section.

2.3.2 The Low-Saturation Limit

In this section the optical Bloch equations given in Eq. (2.97) will be analytically solved under the assumption that the nuclear transition is driven with moderate laser intensities and a laser bandwidth broader than the natural linewidth of the nuclear transition. In this case the low-saturation limit ($\sum_{m_g} \rho_{gg}(m_g) \approx 1$ and $\rho_{gg}(m_g) \gg \rho_{ee}(m_e)$) is valid and the initial population of the ground-state magnetic sublevels can be approximated as

$$\rho_{gg}(m_g) \approx \frac{1}{2j_g+1}. \tag{2.101}$$

Further, for a bandwidth of the laser light significantly broader than the linewidth of the nuclear transition ($\Gamma_L \gg \Gamma_{\text{tot}}$) one obtains $\tilde{\Gamma}_{\text{tot}} \approx \Gamma_L/2$ and the coherences ρ_{ge} will relax fast to an equilibrium with $\dot{\rho}_{ge}(m_g, m_e) = 0$. Under these assumptions the equation for $\dot{\rho}_{ge}$ of Eq. (2.97) simplifies to

$$0 = \frac{i}{\hbar} \frac{e^{-i\omega t}}{2j_g + 1} \langle j_g, m_g | \hat{H}_I | j_e, m_e \rangle - \rho_{ge}(m_g, m_e) \left(i\Delta + \tilde{\Gamma}_{\text{tot}} \right), \tag{2.102}$$

which is known as the adiabatic elimination method [24], and one obtains

$$\rho_{ge}(m_g, m_e) = \frac{i}{\hbar} \frac{e^{-i\omega t}}{2j_g + 1} \frac{\langle j_g, m_g | \hat{H}_I | j_e, m_e \rangle}{i\Delta + \tilde{\Gamma}_{\text{tot}}}. \tag{2.103}$$

Inserting this expression into the equation for $\dot{\rho}_{ee}$ leads to

$$\begin{aligned} \dot{\rho}_{ee}(m_e) &= \frac{1}{\hbar^2(2j_g + 1)} \sum_{m_g} \left[\frac{|\langle j_g, m_g | \hat{H}_I | j_e, m_e \rangle|^2}{i\Delta + \tilde{\Gamma}_{\text{tot}}} + \frac{|\langle j_g, m_g | \hat{H}_I | j_e, m_e \rangle|^2}{-i\Delta + \tilde{\Gamma}_{\text{tot}}} \right] - \rho_{ee}(m_e)\Gamma_{\text{tot}} \\ &= \frac{2}{\hbar^2(2j_g + 1)} \frac{\tilde{\Gamma}_{\text{tot}}}{\Delta^2 + \tilde{\Gamma}_{\text{tot}}^2} \sum_{m_g} |\langle j_g, m_g | \hat{H}_I | j_e, m_e \rangle|^2 - \rho_{ee}(m_e)\Gamma_{\text{tot}}. \end{aligned} \tag{2.104}$$

In the rotating wave approximation H_I can be regarded as time independent. Therefore this differential equation is solved for $\rho_{ee}(t = 0) = 0$ by

$$\rho_{ee}(m_e) = \frac{2\tilde{\Gamma}_{\text{tot}} \left[1 - e^{-\Gamma_{\text{tot}} \cdot t} \right]}{\hbar^2 \Gamma_{\text{tot}} (2j_g + 1) \left(\Delta^2 + \tilde{\Gamma}_{\text{tot}}^2 \right)} \sum_{m_g} |\langle j_g, m_g | \hat{H}_I | j_e, m_e \rangle|^2. \tag{2.105}$$

Consequently, from Eq. (2.105) the excited state population density is obtained by summation over m_e and reads

$$\begin{aligned} \rho_{ee} &= \sum_{m_e} \rho_{ee}(m_e) \\ &= \frac{2\tilde{\Gamma}_{\text{tot}} \left[1 - e^{-\Gamma_{\text{tot}} \cdot t} \right]}{\hbar^2 \Gamma_{\text{tot}} (2j_g + 1) \left(\Delta^2 + \tilde{\Gamma}_{\text{tot}}^2 \right)} \sum_{m_g m_e} |\langle j_g, m_g | \hat{H}_I | j_e, m_e \rangle|^2. \end{aligned} \tag{2.106}$$

This excited state population corresponds to a small (arbitrary) frequency fraction $\Delta\omega \leq \Gamma_{\text{tot}}$ of the total frequency width of the laser light Γ_L. In order to calculate the full population density, ρ_{ee} has to be divided by $\Delta\omega$ and integrated over the full laser linewidth Γ_L. Only the overlap with the transition linewidth will contribute to this integral. For this reason, the integration limits can be extended to infinity without changing the value of the integral. Thus we have

$$
\begin{aligned}
\rho_{ee}^{\text{tot}} &= \int \frac{\rho_{ee}}{\Delta\omega} d\omega_L \\
&= \int d\omega_L \frac{2\tilde{\Gamma}_{\text{tot}}\left[1-e^{-\Gamma_{\text{tot}}\cdot t}\right]}{\hbar^2\Gamma_{\text{tot}}\Delta\omega(2j_g+1)\left(\Delta^2+\tilde{\Gamma}_{\text{tot}}^2\right)} \sum_{m_g m_e} |\langle j_g, m_g|\hat{H}_I|j_e, m_e\rangle|^2 \\
&= \frac{2\left[1-e^{-\Gamma_{\text{tot}}\cdot t}\right]}{\hbar^2\Gamma_{\text{tot}}\Delta\omega(2j_g+1)} \sum_{m_g m_e} |\langle j_g, m_g|\hat{H}_I|j_e, m_e\rangle|^2 \int d\omega_L \frac{\tilde{\Gamma}_{\text{tot}}}{\Delta^2+\tilde{\Gamma}_{\text{tot}}^2} \\
&= \frac{2\pi\left[1-e^{-\Gamma_{\text{tot}}\cdot t}\right]}{\hbar^2\Gamma_{\text{tot}}\Delta\omega(2j_g+1)} \sum_{m_g m_e} |\langle j_g, m_g|\hat{H}_I|j_e, m_e\rangle|^2,
\end{aligned}
\tag{2.107}
$$

where in the last step the integral

$$
\frac{1}{\pi}\int d\omega_L \frac{\tilde{\Gamma}_{\text{tot}}}{\Delta^2+\tilde{\Gamma}_{\text{tot}}^2} = 1 \tag{2.108}
$$

was applied.

The interaction matrix element $\langle j_g, m_g|\hat{H}_I|j_e m_e\rangle$ can be expressed in terms of the reduced interaction matrix element $\langle j_g\|\hat{H}_I\|j_e\rangle$ by means of the Wigner–Eckart theorem [25]

$$
\langle j_g, m_g|\hat{H}_I|j_e, m_e\rangle = C_{j_e m_e\, l\sigma}^{j_g m_g}\langle j_g\|\hat{H}_I\|j_e\rangle. \tag{2.109}
$$

From the summation properties of the Clebsch–Gordan coefficients one has [26]

$$
\sum_{m_g, m_e} |C_{j_e m_e\, l\sigma}^{j_g m_g}|^2 = \frac{2j_g+1}{2l+1} \tag{2.110}
$$

and therefore

$$
\begin{aligned}
\sum_{m_g, m_e} |\langle j_g, m_g|\hat{H}_I|j_e, m_e\rangle|^2 &= \sum_{m_g, m_e} |C_{j_e m_e\, l\sigma}^{j_g m_g}|^2 |\langle j_g\|\hat{H}_I\|j_e\rangle|^2 \\
&= \frac{2j_g+1}{2l+1}|\langle j_g\|\hat{H}_I\|j_e\rangle|^2.
\end{aligned}
\tag{2.111}
$$

Defining the Rabi frequency Ω as

$$
\Omega = \frac{2|\langle j_e\|\hat{H}_I\|j_g\rangle|}{\sqrt{2l+1}\hbar} = \sqrt{\frac{2j_g+1}{2j_e+1}}\,\frac{2|\langle j_g\|\hat{H}_I\|j_e\rangle|}{\sqrt{2l+1}\hbar} \tag{2.112}
$$

and inserting Eq. (2.111) into Eq. (2.107), the resulting expression for ρ_{ee}^{tot} reads

$$
\rho_{ee}^{\text{tot}} = \frac{\pi\Omega^2}{2\Gamma_{\text{tot}}\Delta\omega}\frac{2j_e+1}{2j_g+1}. \tag{2.113}
$$

A detailed description for the derivation of the explicit form of the interaction matrix element $\langle j_g, m_g|\hat{H}_I|j_e m_e\rangle$ is given in Refs. [1, 27] and will not be carried out at this point. The result for magnetic multipole radiation is [19][3]

$$|\langle j_g, m_g|\hat{H}_I|j_e, m_e\rangle| = \frac{E_0\sqrt{2\pi}}{c}\sqrt{\frac{(2l+1)(l+1)}{l}}\frac{k^{l-1}}{(2l+1)!!}\sqrt{\frac{2j_e+1}{2j_g+1}}|C^{j_g m_g}_{j_e m_e\ l\sigma}|\sqrt{B_{eg}(Ml)}, \tag{2.115}$$

similarly, for electric multipole radiation, we have

$$|\langle j_g, m_g|\hat{H}_I|j_e, m_e\rangle| = E_0\sqrt{2\pi}\sqrt{\frac{(2l+1)(l+1)}{l}}\frac{k^{l-1}}{(2l+1)!!}\sqrt{\frac{2j_e+1}{2j_g+1}}|C^{j_g m_g}_{j_e m_e\ l\sigma}|\sqrt{B_{eg}(El)}. \tag{2.116}$$

Here E_0 is the amplitude of the electric driving field and approximations for the transition matrix elements $B_{eg}(Ml)$ and $B_{eg}(El)$ were given in Eqs. (2.55) and (2.57), respectively.

In the following, only the magnetic multipole term will be discussed. By comparison with Eq. (2.109), it is clear that the reduced interaction Hamiltonian is obtained from Eq. (2.115) by dropping the Clebsch–Gordan coefficient. According to Eq. (2.112), the Rabi frequency is thus inferred to be

$$\Omega = \frac{\sqrt{8\pi}E_0}{\hbar c}\sqrt{\frac{(l+1)}{l}}\frac{k^{l-1}}{(2l+1)!!}\sqrt{B_{eg}(Ml)}. \tag{2.117}$$

The transition matrix element $B_{eg}(Ml)$ can be related to the transition rate $A^{(M)}$ by means of Eq. (2.53) as

$$B_{eg}(Ml) = \frac{\hbar}{2\mu_0}\frac{l[(2l+1)!!]^2}{l+1}\left(\frac{1}{k}\right)^{2l+1}A^{(M)}. \tag{2.118}$$

---

[3]Note, that compared to Ref. [19] a further factor of $\sqrt{2l+1/2j_g+1}$ occurs. This difference originates from a different Clebsch–Gordan coefficient. Using the notation of Ref. [19], redefining $-\sigma$ to σ and interchanging the ground and excited state, one obtains

$$|C(j_g j_e l; m_g - m_e\sigma)| = |C^{l\sigma}_{j_g m_g j_e - m_e}| = \sqrt{2l+1}\left|\begin{pmatrix} j_g & j_e & l \\ m_g & -m_e & -\sigma\end{pmatrix}\right| = \sqrt{2l+1}\left|\begin{pmatrix} j_g & j_e & l \\ -m_g & m_e & \sigma\end{pmatrix}\right|$$
$$= \sqrt{2l+1}\left|\begin{pmatrix} j_e & l & j_g \\ m_e & \sigma & -m_g\end{pmatrix}\right| = \frac{\sqrt{2l+1}}{\sqrt{2j_g+1}}|C^{j_g m_g}_{j_e m_e l\sigma}|. \tag{2.114}$$

Here some relations for the Wigner-3j symbols were used [26]. Only if the Clebsch–Gordan coefficient is in the final form, the Wigner–Eckart theorem (2.109) can be applied to define the reduced interaction Hamiltonian.

For the derivation of Eq. (2.115) the electric field was assumed to be of the form [27]

$$E(x,t) = E_0\left(e^{i(k_0x-\omega_0t)} + e^{-i(k_0x-\omega_0t)}\right) = 2E_0\cos(k_0x-\omega_0t). \tag{2.119}$$

The time-averaged energy per unit volume of the electromagnetic wave is consequently given as

$$\begin{aligned}\rho &= \frac{\epsilon_0\overline{|E(x,t)|^2}}{2} + \frac{\overline{|B(x,t)|^2}}{2\mu_0} \\ &= \epsilon_0\overline{|E(x,t)|^2} = 2\epsilon_0E_0^2,\end{aligned} \tag{2.120}$$

where $B(x,t) = E(x,t)/c$ and $\overline{\cos^2(k_0x-\omega_0t)} = 1/2$ was used. In order to obtain E_0 from the spectral energy density per frequency interval ρ^ω, one has to divide ρ by the arbitrary but small frequency fraction $\Delta\omega \leq \Gamma_{\text{tot}}$.

$$\rho^\omega = \frac{\rho}{\Delta\omega}. \tag{2.121}$$

Correspondingly E_0 can be expressed as

$$E_0 = \sqrt{\frac{\rho^\omega\Delta\omega}{2\epsilon_0}}. \tag{2.122}$$

Using $k = \omega/c$, where ω denotes the frequency of the nuclear transition and inserting Eqs. (2.122) and (2.118) in Eq. (2.117), one obtains for the Rabi frequency

$$\Omega = \sqrt{\frac{2\pi c^3\rho^\omega\Delta\omega A^{(M)}}{\hbar\omega^3}}. \tag{2.123}$$

This expression is inserted in Eq. (2.113) in order to infer the total number of excited nuclei in saturation for a given spectral energy density ρ^ω of the laser light

$$\begin{aligned}N_{\text{sat}} &= \rho_{ee}^{\text{tot}}\cdot N_0 \\ &= \frac{\pi\Omega^2N_0}{2\Gamma_{\text{tot}}\Delta\omega}\frac{2j_e+1}{2j_g+1} \\ &= \frac{\rho^\omega\pi^2c^3A^{(M)}N_0}{\Gamma_{\text{tot}}\hbar\omega^3}\frac{2j_e+1}{2j_g+1}.\end{aligned} \tag{2.124}$$

Here N_0 denotes the total number of irradiated nuclei. Also the time dependency can be regained. From Eq. (2.105) one obtains

$$N(t) = N_{\text{sat}}\left(1-e^{-\Gamma_{\text{tot}}t}\right). \tag{2.125}$$

The same result also holds for electric multipole transitions, however, with $A^{(M)}$ interchanged by $A^{(E)}$. We therefore write for the radiative transition rate

$$A^{(M)} = A^{(E)} = A = \frac{1}{\tau_\gamma}, \tag{2.126}$$

where τ_γ denotes the radiative lifetime of the excited nuclear state. Opposed to that, Γ_{tot} corresponds to the total natural decay rate of the transition, which also includes potential non-radiative decay branches. In case of a potential internal conversion decay channel of the excited state, this means

$$\Gamma_{\text{tot}} = (1 + \alpha_{\text{ic}})A, \tag{2.127}$$

where α_{ic} denotes the internal conversion coefficient.

It should be pointed out that the same result Eq. (2.125) can also be obtained via the Einstein rate equations, which are valid for the considered case of the low-saturation limit and a transition linewidth smaller than the width of the laser light. The Einstein rate equations under resonant irradiation read [28]

$$\begin{aligned} \dot{N}_g &= (1 + \alpha_{\text{ic}})AN_e + \rho^\omega B^\omega N_e - \rho^\omega B^\omega \frac{g_e}{g_g} N_g \\ \dot{N}_e &= -(1 + \alpha_{\text{ic}})AN_e - \rho^\omega B^\omega N_e + \rho^\omega B^\omega \frac{g_e}{g_g} N_g, \end{aligned} \tag{2.128}$$

where N_g and N_e are the population numbers of the ground and excited state, respectively, A and B^ω denote the Einstein coefficients corresponding to spontaneous emission (A) as well as stimulated emission and absorption (B^ω) [28, 29]. $g_g = 2j_g + 1$ and $g_e = 2j_e + 1$ are the degeneracies of the ground and excited state and ρ^ω denotes the spectral energy density of the electromagnetic field which drives the transition. Solving this equation leads to the number of excited nuclei given as

$$N_e(t) = \frac{\rho^\omega B^\omega \frac{g_e}{g_g} N_0}{(1 + \alpha_{\text{ic}})\, A + (1 + \frac{g_e}{g_g})\rho^\omega B^\omega} \left(1 - \mathrm{e}^{-\left[(1+\alpha_{\text{ic}})A + \left(1 + \frac{g_e}{g_g}\right)\rho^\omega B^\omega\right]t}\right), \tag{2.129}$$

The Einstein B^ω coefficient is related to A via [28]

$$B^\omega = \frac{\pi^2 c^3}{\hbar\omega^3} \cdot A. \tag{2.130}$$

In case that $B^\omega \rho^\omega \ll A$ is fulfilled, Eq. (2.129) transforms to

$$N_e(t) = \frac{\rho^\omega B^\omega \frac{g_e}{g_g} N_0}{(1+\alpha_{\text{ic}})\,A}\left(1 - \mathrm{e}^{-(1+\alpha_{\text{ic}})At}\right)$$
$$= \frac{\rho^\omega \pi^2 c^3 A N_0}{\Gamma_{\text{tot}} \hbar \omega^3}\,\frac{2j_e+1}{2j_g+1}\left(1 - \mathrm{e}^{-\Gamma_{\text{tot}} t}\right), \tag{2.131}$$

where in the last step Eq. (2.130) was inserted for B^ω and $\Gamma_{\text{tot}} = (1+\alpha_{\text{ic}})\,A$ was used. This result is identical with Eq. (2.125), derived via the optical Bloch equations, and will be applied to ^{229m}Th in the following.

2.3.3 Laser Excitation of ^{229m}Th

In the previous section it was shown that, during laser irradiation, the number of excited nuclei in saturation N_{sat} can be calculated via

$$N_{\text{sat}} = \frac{\rho^\omega \pi^2 c^3 A N_0}{\Gamma_{\text{tot}} \hbar \omega^3}\,\frac{2j_e+1}{2j_g+1} \tag{2.132}$$

where ρ^ω denotes the spectral energy density of the laser light, $A = 1/\tau_\gamma$ is the radiative transition rate, $\Gamma_{\text{tot}} = (1+\alpha_{\text{ic}})A$ the total transition rate with α_{ic} as the internal conversion coefficient and ω the angular frequency of the nuclear transition. It is concluded that N_{sat} is independent of the lifetime of the excited state. However, the time after which saturation is reached is lifetime dependent. From Eq. (2.131) the time dependence is known to be of the form

$$N_e(t) = N_{\text{sat}}\left(1 - e^{-\Gamma_{\text{tot}} t}\right). \tag{2.133}$$

Thus the characteristic saturation time τ_{sat} is

$$\tau_{\text{sat}} = \frac{1}{\Gamma_{\text{tot}}} = \frac{1}{(1+\alpha_{\text{ic}})A} = \frac{\tau_\gamma}{1+\alpha_{\text{ic}}}. \tag{2.134}$$

Based on Eq. (2.133) the nuclear excitation rate can be approximated as

$$R \approx \dot{N}_e(t)|_{t=0} = \Gamma_{\text{tot}} N_{\text{sat}} = \frac{\rho^\omega \pi^2 c^3 A N_0}{\hbar \omega^3}\,\frac{2j_e+1}{2j_g+1}. \tag{2.135}$$

The spectral energy density ρ^ω can be related to the laser intensity I of the laser light via

$$\rho^\omega = \frac{I}{c\Gamma_L}, \tag{2.136}$$

with Γ_L the bandwidth of the laser light. In the nuclear clock concept it is envisaged to excite a single ion ($N_0 = 1$) in a Paul trap. The laser excitation should be carried out with a single mode of a frequency comb with a bandwidth of approximately $\Gamma_L = 2\pi$ Hz and the nuclear excitation rate should be of about the same value $R \approx \Gamma_L$. Thus the condition for the required laser intensity for the development of a single-ion nuclear clock reads

$$I \approx \frac{\hbar\omega^3\Gamma_L^2}{\pi^2c^2A}\frac{2j_g+1}{2j_e+1}. \tag{2.137}$$

Inserting the values of $A = 10^{-4}$ Hz [30], $\omega = 1.19 \cdot 10^{16}$ Hz, $j_g = 5/2$ and $j_e = 3/2$ for ^{229m}Th results in $I = 0.012$ W/cm^2. Assuming that the single mode of the frequency comb could be focused onto an area of 10 μm^2, the required mode power for the development of a nuclear clock based on ^{229}Th would be $\rho \approx 12$ nW. This is a realistic power for a single mode of a frequency comb.

A similar result can also be obtained via the Rabi frequency Ω. From the theory of Rabi oscillations it is known, that these can occur under the condition that $\tilde{\Gamma}_{tot}/2 < \Omega$ [31]. Here $\tilde{\Gamma}_{tot} = (\Gamma_{tot} + \Gamma_L)/2$ denotes the total decay rate of the coherences, with Γ_{tot} the linewidth of the transition and Γ_L the bandwidth of the laser light. From Eq. (2.123) in combination with Eq. (2.121) and $\rho = I/c$, the Rabi frequency is obtained as

$$\Omega = \sqrt{\frac{2\pi c^2 I A}{\hbar\omega^3}}. \tag{2.138}$$

Accordingly, the condition for the observation of Rabi oscillations can be written as

$$\sqrt{\frac{2\pi c^2 I A}{\hbar\omega^3}} > \frac{\max(\Gamma_{tot}, \Gamma_L)}{4}. \tag{2.139}$$

In the case of ^{229}Th ions in a Paul trap the internal conversion process is energetically suppressed and thus $\alpha_{ic} = 0$ holds (see Sect. 2.2.2). Therefore we have $\Gamma_L \gg \Gamma_{tot}$ and the above condition is securely fulfilled for $\Omega = \Gamma_L$. This leads to a required laser intensity in order to drive Rabi oscillations of

$$I \approx \frac{\hbar\omega^3\Gamma_L^2}{2\pi c^2 A}. \tag{2.140}$$

For ^{229}Th we obtain the same result as above of $I = 0.012$ W/cm^2. Changes to these considerations could originate from the fact that, even if internal conversion is suppressed, electronic bridge effects might still be present (see Sect. 3.5.1) and have to be considered by a further electronic bridge coefficient α_{eb} that cannot be set to zero. Values of α_{eb} between 10 (Th^{3+}) and 10^5 (Th^{1+} in case of resonance) have to be expected, depending on the charge state and transition energy under consideration [32–34]. These effects will, however, only play a role if the nuclear transition linewidth would be broadened beyond the bandwidth of the laser light used for nuclear excitation. Further, the electric field strength at the point of the nucleus may not equal the field strength of the incoming laser wave, as (anti-)shielding effects from the atomic shell will play a role (e.g. in case of Th^{4+} a magnetic dipole anti-shielding factor of 217.3 was found [35]).

From the above discussions it is clear that nuclear laser excitation of ^{229m}Th appears possible, especially as there are strong ongoing developments of VUV-laser technology. The generation of coherent VUV light providing an average output power of 0.5 mW at 149 nm via intra-cavity 7th harmonic generation in noble gas of the 1040 nm light produced by an Yb-fiber laser system was recently reported [36]. Considering that $\sim 10^5$ comb modes are generated in this way, the estimated average mode power is 5 nW, sufficiently large to drive the ^{229}Th nuclear transition when focused to an area of about 10 μm^2.

The discussed laser excitation scheme requires a precise knowledge of the isomer's transition energy and does not allow for a search of the resonance frequency, given the large uncertainty in transition energy of about 1 eV. This has led to the general assumption that the direct nuclear laser excitation of ^{229m}Th requires a better knowledge of the isomer's excitation energy. In the following, a new laser excitation scheme for ^{229m}Th is presented that allows to circumvent this requirement, in this way providing the basis for a first nuclear laser excitation of ^{229m}Th. The presented method is making use of the short-lived internal conversion decay channel, thereby allowing for nearly background-free detection of electrons emitted in the isomeric decay [37].

A tunable and pulsed VUV laser system with an energy of $E_L \approx 10\ \mu$J per pulse, $R_L = 10$ Hz repetition rate, 5 ns pulse duration at a beam area of $A_L = 1$ mm^2 and a linewidth of $\Delta\nu_L = 10$ GHz as is already available [38] is assumed for irradiation of a thin (2.5 nm) layer of ^{229}Th atoms on a surface of 1 mm^2. Under this condition, the isomeric decay will occur by internal conversion with a half-life of 7 μs, resulting in an IC decay factor of $\alpha_{\text{ic}} \approx 10^9$ [15]. As the spectral energy density of a single laser pulse is significantly larger than the time-averaged laser energy density and also the number of irradiated thorium atoms is large ($N_0 \approx 7.6 \cdot 10^{13}$), a significant amount of nuclei can be expected to be excited during a single laser pulse. For the spectral energy density of the laser pulse one obtains

$$\rho^\omega = \frac{E_L}{cT_L A_L \cdot \Delta\omega_L} \approx 1.1 \cdot 10^{-10}\ \text{Js/m}^3. \tag{2.141}$$

Here $T_L = 5$ ns denotes the pulse duration, the remaining laser parameters were taken as stated above. As the laser linewidth of 10 GHz is still significantly larger than the IC-broadened nuclear transition linewidth of about 15.9 kHz, Eq. (2.132) for the number of excited nuclei in saturation is valid and one obtains

$$N_{\text{sat}} = \frac{2}{3}\frac{\rho^\omega \pi^2 c^3 N_0}{(1+\alpha_{\text{ic}})\hbar\omega^3} \approx 4.2 \cdot 10^3. \tag{2.142}$$

As these nuclear excitations will decay with a characteristic half-life of about 7 μs after the laser pulse via internal conversion, it is possible to trigger the detector in accordance with the laser pulses in order to allow for a nearly background-free detection of the emitted electrons. These could be detected with an MCP detector, providing about 50% detection efficiency if the electrons are post-accelerated to about 300 eV kinetic energy. The only source of background that has to be considered are electrons, which are generated by the laser light on the surface. Electrons produced by the photo effect, however, appear on the timescale of nano-seconds and can therefore be distinguished from the comparatively long-lived isomeric decay. Due to the relatively low laser intensity, no thermal electron emission can be expected to occur.

In the described way, it should be possible to detect the isomeric decay by means of its characteristic lifetime within one single laser pulse and thus on a micro-second timescale. Even when assuming 100 laser pulses per scan step, the total time required for scanning the large energy range of 1 eV would only be $2.4 \cdot 10^5$ s, corresponding to 2.7 days. This detection scheme would pave the way for a first direct nuclear laser excitation of $^{229\mathrm{m}}$Th without the requirement of an improved isomeric energy value. The described detection technique can be considered as laser-driven conversion-electron Mössbauer spectroscopy (CEMS) [39] in the optical region. It could also provide the basis for a CEMS-based solid-state nuclear clock, however, with an IC-enlarged natural transition linewidth of expectedly about 15.9 kHz (corresponding to 7 μs half-life) [15].

2.3.4 A Comparison to ^{235m}U

The second lowest nuclear excitation, which is known by today, is the isomeric state of ^{235}U with an excitation energy of about 76 eV [40]. This low excitation energy has led to the conclusion that also $^{235\mathrm{m}}$U could be a candidate for direct nuclear laser excitation, as soon as laser technology has evolved to deliver intense laser light at the required wavelength by high-harmonic generation [41, 42]. In this case, the long half-life of $^{235\mathrm{m}}$U of about 26 min would even make the isomer a good candidate for a nuclear clock.

Such considerations, however, do not take the multipolarity of the ^{235}U isomer-to-ground-state transition into account. Being a transition from the spin and parity values $1/2^+$ (excited state) to $7/2^-$ (ground state), the multipolarity is $E3$. According to the Weisskopf estimate Eqs. (2.56) and (2.57), the radiative transition rate $A^{(E)}$ is

$$\begin{aligned} A^{(E)} &\approx \frac{1}{2\pi\hbar\epsilon_0}\frac{k^{2l+1}}{[(2l+1)!!]^2}\frac{l+1}{l}\left(\frac{3}{3+l}\right)^2 R^{2l}e^2 \\ &= \frac{\omega^7 R^6 e^2}{66150\pi\hbar\epsilon_0 c^7} = 2.60 \cdot 10^{-23}\ \mathrm{s}^{-1}. \end{aligned} \tag{2.143}$$

This corresponds to a lifetime of $\tau_\gamma = 3.8 \cdot 10^{22}$ s or $1.2 \cdot 10^{15}$ years. The reason for the actually measured lifetime of $2.25 \cdot 10^3$ s is a huge internal conversion coefficient of estimatedly $\alpha_{\mathrm{ic}} \approx 1.7 \cdot 10^{19}$. According to Eq. (2.140) the laser intensity required to drive the transition with a single mode of a frequency comb $\Gamma_L = 2\pi$ Hz can be estimated to be

$$I \approx \frac{\hbar\omega^3\Gamma_L^2}{2\pi c^2 A} = 4.3 \cdot 10^{19}\ \mathrm{W/cm^2}. \tag{2.144}$$

Here the angular frequency $\omega = 1.15 \cdot 10^{17}$ Hz for $^{229\mathrm{m}}$U was used. This laser intensity is unrealistically large. The reason is, that only the radiative decay rate A enters the calculation and not the IC-broadened nuclear decay rate Γ_{tot}, which corresponds to the comparatively short half-life of 26 min.

The nuclear transition of third lowest energy is contained in ^{110}Ag and possesses an excitation energy of 1.1 keV ($t_{1/2} = 660$ ns) [43], which could be excited with the help of free-electron lasers, similar as the 14.4 keV Mössbauer transition of ^{57}Fe [44]. This technology, however, can by today not provide linewidths sufficiently small to allow for the development of nuclear frequency standards. For this reason, $^{229\mathrm{m}}$Th is currently the only known nucleus that would allow for the development of a nuclear clock applying already existing technology. This situation may change in the future, taking the fast development of laser technology into account [45–48].

References

1. Ring P, Schuck P (1980) The nuclear many-body problem. Springer, New York
2. Nilsson SG (1955) Binding states of individual nucleons in strongly deformed nuclei. Dan Mat Fys Medd 16:29
3. Firestone RB, Shirley VS (1996) Table of isotopes. Wiley, New York
4. Swanberg E (2012) Searching for the decay of $^{229\mathrm{m}}$Th, Ph.D. thesis, University of California, Berkeley
5. Jackson JD (1975) Classical electrodynamics. Wiley, New York
6. Blatt JM, Weisskopf VF (1952) Theoretical nuclear physics. Wiley, New York
7. Tkalya EV et al (2015) Radiative lifetime and energy of the low-energy isomeric level in ^{229}Th. Phys Rev C 92:054324
8. Tralli N, Goertzel G (1951) The theory of internal conversion. Phys Rev 83:399–404
9. Kramer G (1956) Zur Theorie der Konversionskoeffizienten. Zeitschrift für Physik 146:187–204 (in German)
10. Rose ME (1955) Multipole fields. Wiley, New York
11. Rose ME (1955) Theory of internal conversion in Beta- and gamma-ray spectroscopy. In: Siegbahn K (ed) North-Holland Publishing Company Amsterdam
12. Rose ME (1937) Relativistic wave functions in the continuous spectrum for the Coulomb field. Phys Rev 51:484–485
13. Strizhov VF, Tkalya EV (1991) Decay channel of low-lying isomer state of the ^{229}Th nucleus. Possibilities of experimental investigation. Sov Phys JETP 72:387
14. Karpeshin FF, Trzhaskovskaya MB (2007) Impact of the electron environment on the lifetime of the ^{229}Th^{m} low-lying isomer. Phys Rev C 76:054313
15. Seiferle B et al (2017) Lifetime measurement of the ^{229}Th nuclear isomer. Phys Rev Lett 118:042501
16. Loudon R (2003) The quantum theory of light. Oxford University Press, New York
17. Lindblad G (1976) On the generators of quantum dynamical semigroups. Commun Math Phys 48:119
18. Pearle P (2012) Simple derivation of the Lindblad equation. Eur J Phys 33:805–822

19. Pálffy A, et al (2008) Electric-dipole-forbidden nuclear transitions driven by super-intense laser fields. Phys Rev C 77:044602
20. Kazakov GA et al (2012) Performance of a ^{229}Thorium solid-state nuclear clock. New J Phys 14:083019
21. Kazakov GA et al (2013) Atomic clock with nuclear transition: current status in TU Wien. http://arxiv.org/abs/1110.0741
22. Bürvenich TJ, Evers J, Keitel CH (2006) Nuclear quantum optics with x-ray laser pulses. Phys Rev Lett 96:142501
23. Steck DA, Quantum and Atom Optics. Available online at http://steck.us/teaching (revision 0.11.6, 24 Feb 2017)
24. Shore BW (2008) Coherent manipulation of atoms using laser light. Acta Physica Slovaca 58:243–486
25. Varshalovich DA, Moskalev AN, Khersonskii VK (1988) Quantum theory of angular momentum. World Scientific Publishing, Singapore
26. Edmonds AR (1996) Angular Momentum in Quantum Mechanics. Princeton University Press, Princeton
27. Liao WT (2013) Coherent control of nuclei and X-rays. Ph.D. thesis, University of Heidelberg
28. Hilborn RC (1982) Einstein coefficients, cross sections, f values, dipole moments and all that. Am J Phys 50:982–986
29. Einstein A (1917) Zur quantentheorie der Strahlung. Physikalische Zeitschrift 18:121–128 (in German)
30. Minkov N, Pálffy A (2017) Reduced transition probabilities for the γ-decay of the 7.8 eV isomer in ^{229}Th. Phys Rev Lett 118:212501
31. Noh HR, Jhe W (2010) Analytic solutions of the optical Bloch equations. Opt Comm 283:2353–2355
32. Porsev SG, Flambaum VV (2010) Effect of atomic electrons on the 7.6 eV nuclear transition in $^{229m}Th^{3+}$. Phys Rev A 81:032504
33. Porsev SG, Flambaum VV (2010) Electronic bridge process in $^{229}Th^{+}$. Phys Rev A 81:042516
34. Porsev SG et al (2010) Excitation of the isomeric ^{229m}Th nuclear state via an electronic bridge process in $^{229}Th^{+}$. Phys Rev Lett 105:182501
35. Feiock FD, Johnson WR (1969) Atomic susceptibilities and shielding factors. Phys Rev 187:39–50
36. Ozawa A et al (2015) High average power coherent VUV generation at 10 MHz repition frequency by intracavity high harmonic generation. Opt Express 23:15107
37. von der Wense L et al (2017) A laser excitation scheme for ^{229m}Th, Phys Rev Lett 119:132503
38. Hanna SJ et al (2009) A new broadly tunable (7.4–10.2 eV) laser based VUV light source and its first application to aerosol mass spectrometry. Int J Mass Spectrom 279:134–146
39. Salvat F, Parellada J (1984) Theory of conversion electron Mössbauer spectroscopy (CEMS). Nucl Instrum Meth B 1:70–84
40. Panov AD (1997) Quantitative conversion spectroscopy of the ultrasoft isomeric transition of uranium-235 and the electronic structure of uranium oxides. JETP 85:313–324
41. Alexandrov VS et al (2011) Nuclear transitions and new standards of length and time. Acta Physica Polonica B 42:853–858
42. Flambaum VV (2016) Enhancing the effect of Lorentz invariance and Einstein's equivalence principle violation in nuclei and atoms. Phys Rev Lett 117:072501
43. NNDC Interactive Chart of Nuclides. http://www.nndc.bnl.gov/chart [2016, March 22nd], Brookhaven National Laboratory, Brookhaven
44. Chumakov A (2016) Information was presented at the XFELO Science Workshop. Results are still unpublished, USA, SLAC National Accelerator Laboratory
45. Gohle C et al (2005) A frequency comb in the extreme ultraviolet. Nature 436:234–237
46. Cingöz A et al (2012) Direct frequency comb spectroscopy in the extreme ultraviolet. Nature 482:68–71
47. Pupeza I et al (2013) Compact high-repetition-rate source of coherent 100 eV radiation. Nat Photonics 7:608–612
48. Cavaletto SM et al (2014) Broadband high-resolution X-ray frequency combs. Nat Photonics 8:520–523

Chapter 3
The History of ^{229m}Th

This chapter provides a complete review of theoretical and experimental investigations on ^{229m}Th. It was a central aspect to also give a rather complete list of existing literature on the topic, which could serve as a reference for future investigations. It is subdivided into seven sections discussing in detail: the first prediction (Sect. 3.1) and further evidence (Sect. 3.2) of the isomer's existence, experiments that have been conducted to constrain the isomeric transition energy (Sect. 3.3), potential applications for ^{229m}Th (Sect. 3.4), theoretical investigations on the isomer's excitation and decay (Sect. 3.5), the challenge of a direct detection of the isomer's decay (Sect. 3.6) as well as further experimental investigations (Sect. 3.7).

3.1 First Prediction of Existence

The existence of an excited nuclear state of low energy in ^{229}Th was first proposed by Kroger and Reich in 1976 [1] (see also [2]). In their work they performed a precise study of the low-energy γ-ray spectrum of ^{229}Th, as produced in the α decay of ^{233}U. A set of excited nuclear levels, identified by their γ decay and located close to the ground-state rotational band of ^{229}Th, had already been established to possess a rotational-band-like character in earlier studies [3]. However, at that time the spin and parity assignments of the corresponding states were still under discussion. As detailed in the following, Kroger and Reich were able to show that the 29.19 keV state of the rotational band has spin and parity values $j^{\pi} = 5/2^{+}$. The corresponding band structures as well as their inter- and intra-band transitions are shown in Fig. 3.1.

The fact that the parity of this rotational band had to be positive was deduced from the multipolarities of γ rays, which are emitted during the decay of higher-lying states into the rotational band. Especially the 245.33 keV γ ray, which was deduced from its line intensity to possess $M1$ multipolarity, was assigned to the decay of the 317.15 keV state ($5/2^{+}$) to the 71.82 keV state of the rotational band. Also the

L. Von der Wense, *On the Direct Detection of ^{229m}Th*,
Springer Theses, https://doi.org/10.1007/978-3-319-70461-6_3

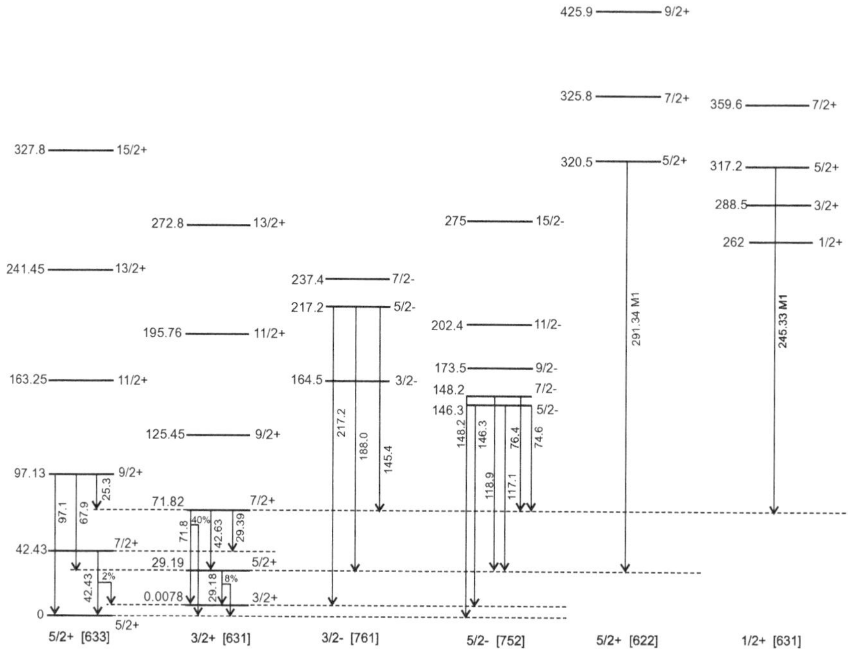

Fig. 3.1 Low energy rotational-band structure of ^{229}Th. The spin and parity values as well as Nilsson quantum numbers for each band head are assigned below the bands. Each state is given together with its energy value in keV as well as its spin and parity assignments. Transitions between the energy levels are only given if relevant for the identification of the ^{229}Th isomeric state and indicated as arrows between the corresponding energy levels. The transition energies are also given. The ^{229}Th ground-state rotational band is shown on the left side together with the $5/2^+$ [633] ground state. The isomeric state is the $3/2^+$ [631] band head of the second rotational band. For this state, the energy gap to the ground state is not shown to scale to improve clarity

291.34 keV γ ray was found to be of $M1$ multipolarity and could be assigned to the transition between the 320.53 keV state $(5/2^+)$ and the 29.19 keV rotational-band state.

The spin of the 29.19 keV state was determined making use of the 67.98 keV γ ray, which was assigned to the inter-band transition between the 97.13 keV $(9/2^+)$ state of the ground-state rotational band and the 29.19 keV state of the second rotational band. A spin of 7/2 or larger for this state would be inconsistent with the energy-level spacing, further, a spin value of 3/2 or smaller can be ruled out by the lifetime of the 97.13 keV $(9/2^+)$ state, which would be longer when assuming a multipolarity of $E4$ or $M3$ for the corresponding transition. Thus, the spin and parity of $5/2^+$ is the only consistent choice for the 29.19 keV state.

While the spin and parity assignment of this rotational band was well established in this way, the location of the band head was still in question. The straightforward assumption would be that the 29.19 keV $5/2^+$ level is the band head of the rotational band. There is, however, strong indication that this interpretation is incorrect and instead the band head is of spin and parity $3/2^+$ and is located close to the ground state. This assumption is based on three independent observations:

1. Based on the single-particle Nilsson state model, there is no rotational band of spin and parity assignment $5/2^+$ expected that close to the ground-state. Only the $3/2^+$ assignment can easily be interpreted in terms of the Nilsson model, in which case the Nilsson quantum numbers [631] are assigned to the rotational band (Fig. 3.1).
2. The assumption that there is a $3/2^+$ state close to the ground state allows to explain several features of the ^{229}Th γ-ray spectrum, which are otherwise unexplained. This concerns the decay of negative-parity levels to the positive-parity levels located close to the ground state. In the following only one example is given, which is provided by the E1 decay of the 146.36 keV state. This state reveals three E1 transitions: to the 29.19 keV $5/2^+$ level, to the 71.82 keV $7/2^+$ level and to the level which was formerly thought to be the ground state. However, no transition to the $7/2^+$ state of the ground-state band is observed. This is surprising, as a decay to the band head, as well as to a $7/2^+$ level of a different rotational band is present. By introducing the new state, close to the ground state and of spin and parity $3/2^+$, this behavior can easily be explained when assuming that the 146.36 keV decay does not populate the ground state, but the first excited $3/2^+$ state instead. Similarly, features in the decay of the 148.16 keV and the 217.13 keV state can be explained.
3. A further support for this interpretation is provided by the ^{230}Th(d,t)^{229}Th reaction. It is stated by Kroger and Reich [1], that in this reaction the four rotational-band members of lowest energy of the $3/2^+$ [631] band have been observed. The energy spacings were found to be in agreement with the expectations. At the time of publication, however, this information was based on private communication and the underlying data was unpublished. While no publication of the underlying data is known to us by today, a new measurement, published by D.G. Burke et al. in 1990 [4], supports the assumption of the existence of an excited state of low energy in ^{229}Th and is discussed in the following section. Early investigations of the ^{229}Th level scheme based on the α decay of ^{233}U were also performed by Canty [5] and later by Barci [6].

3.2 Further Evidence for ^{229m}Th

Already in 1976, Kroger and Reich referred to a measurement of the ^{230}Th(d,t)^{229}Th reaction [1], which was supposed to give further evidence for an excited state of low energy in ^{229}Th. However, to our knowledge, this data has never been published and the first publication giving additional evidence for the excited state based on the ^{230}Th(d,t)^{229}Th reaction is that of D.G. Burke et al. in 1990 [4]. They triggered the ^{230}Th(d,t)^{229}Th reaction by collision of 17 MeV deuterons with a target consisting of ^{230}Th, which was produced by direct deposition from an isotope separator onto

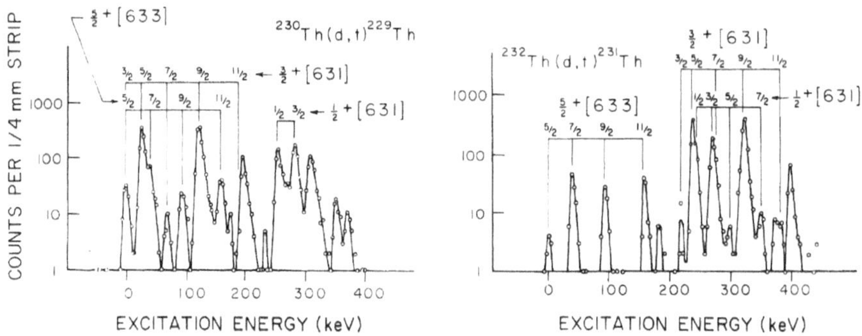

Fig. 3.2 Triton energy-spectrum of the ^{230}Th(d,t)^{229}Th reaction (left) as well as the ^{232}Th(d,t)^{231}Th reaction (right) for a fixed angle of 60°. The 5/2$^+$[633] ground-state band as well as the 3/2$^+$[631] band are clearly visible in both spectra. For ^{229}Th, the band heads of the 5/2$^+$[633] and 3/2$^+$[631] bands cannot be energetically separated and lead to a single peak of twice the expected height. With kind permission of The American Physical Soceity [4]

a carbon foil. Tritons, as produced in the nuclear reaction, were measured energy-resolved at 17 angles between 5° and 80°, in order to draw conclusions on the angular momenta of the states. The results are shortly discussed in the following.

The triton energy-spectrum of the ^{230}Th(d,t)^{229}Th reaction at a fixed angle of 60° is shown in Fig. 3.2 together with the ^{232}Th(d,t)^{231}Th spectrum [4]. There is good evidence that the peaks are correctly assigned, as the excitation energies are consistent with those of previously known levels. Considering the angular dependent shapes of the triton spectra of the 29 and 125 keV levels, angular momenta of $l = 2$ and $l = 4$ are deduced, respectively. This is in agreement with the previous spin and parity assignments of 5/2$^+$ (29 keV) and 9/2$^+$ (125 keV) by Kroger and Reich.

Comparing the peak intensities in the ^{230}Th(d,t)^{229}Th energy spectrum with those occurring in the ^{232}Th(d,t)^{231}Th spectrum provides further evidence for the correct band assignments. In the ^{232}Th(d,t)^{231}Th reaction the band assignments are known with some confidence. From this spectrum it is inferred that the 5/2$^+$ and 9/2$^+$ members of the 3/2$^+$ [631] band reveal relatively large cross sections, while all other peaks are weaker. The same pattern is observed in the ^{230}Th(d,t)^{229}Th spectrum, thus giving evidence that the 5/2$^+$ and 9/2$^+$ members of the 3/2$^+$ [631] band are correctly assigned. A quantitative discussion and a comparison with calculated values further support this interpretation.

While the 5/2$^+$ [633] and the 3/2$^+$ [631] rotational band heads are clearly separated in the ^{232}Th(d,t)^{231}Th reaction, they cannot be distinguished in case of ^{230}Th(d,t)^{229}Th. However, the signal strength, as observed for the ground-state doublet in ^{229}Th, is similar to the sum of the two band head signals in ^{232}Th.

The 29 keV level is inferred to be the second state of the 3/2$^+$[631] rotational band. Burke et al. also discuss possible alternative interpretations of the obtained data. Since the spin and parity of the 29 keV level seems to be 5/2$^+$, the only different possibility would be to assign it as the band head of the 5/2$^+$[622] rotational band. This interpretation, however, is unlikely, as, based on the Nilsson model, this band head is expected at higher energy. Also several features of the low-energy γ spectrum of ^{229}Th, as discussed in the previous section, would not be resolved by this interpretation. The non-observation of the band head is explained by its vicinity to the ground state, however, it could not be concluded which band head has the lowest energy. An improved version of essentially the same measurement was published in 2008 [7]. Further investigations of the ^{229}Th level structure were performed by Coulomb excitation [8] and by β^- decay of ^{229}Ac [9–11].

3.3 Constraining the Transition Energy

As soon as the existence of a low energy state in ^{229}Th was established, the determination of its energy became an important experimental objective. While the energy determination based on indirect measurements has convincingly shown that the isomeric energy must be extremely low compared to usual nuclear excitation energies (most likely below 10 eV), up to today no precise value for the isomer's energy is available. It was emphasized that small energy differences of the ground and first excited nuclear states are a peculiarity of the $N = 139$ isotones [12]. However, even in this context ^{229m}Th exhibits a unique position.

3.3.1 *First Energy Constraints*

In their 1976 paper, Kroger and Reich estimated an upper limit of 100 eV for the transition energy, solely based on the non-observation of the excited states' direct decay [1]. In a further study, published in 1990, Reich and Helmer inferred an energy value of higher precision based on the differences of nuclear levels of larger energies, populating the ground and the low-energy state, respectively [13] (see also [14]). First preparatory measurements had already been published in 1984 [15]. Three different energy combinations were used in this study, leading to an energy constraint of (-1 ± 4) eV. The differences of γ-ray transition energies, which were used in this study, are:

$$\begin{aligned}
\Delta_1 &= E(97.1) - E(71.8) - E(25.3),\\
\Delta_2 &= E(97.1) - E(67.9) - E(29.1),\\
\Delta_3 &= [E(148.1) - E(146.3)] - [E(118.9) - E(117.1)].
\end{aligned} \tag{3.1}$$

The 42.43 keV γ-ray transition was not used in the presented work, as this line could not be resolved from the 42.63 keV line at the time of the measurement.[1] From this study, the authors concluded that the energy separation between the ground state and the first excited state of ^{229}Th is smaller than the precision of the presented measurement and "almost certainly less than 10 eV". Also a first half-life estimate for a tentative 1 eV $M1$ transition between the excited state and the ground state was given as about 7 hours, placing the excitation to be a relatively long-lived isomer. In their 1990 study, Reich and Helmer also further discussed the hypothetical assumption that the energy of the 3/2+ [631] level is actually below the 5/2+ [633] state. This would make the latter state to be an excited isomeric state of ^{229}Th, which is known to decay predominantly via α decay with a half-life of 7932 years. The half-life of the $M1$ transition to the ground state would thus have to be significantly longer than this value in order to allow for this scenario. While not entirely excluded at the time of the study, this turns out to be very unlikely, given the predicted half-life of 7 hours.

The work of Reich and Helmer represents the first study in which a nuclear transition was considered to possess an extremely low energy of only a few eV. While Reich and Helmer themselves did not propose any applications for this nuclear state in their publication, their work was the basis for an increasing interest leading to the proposal of many interesting applications in the subsequent years.

3.3.2 *Improved Energy Determination*

Driven by the increasing interest in the newly discovered low-energy excited nuclear state, Helmer and Reich aimed for an improved energy determination of this excitation. This led to a publication in 1994, in which an energy value of (3.5 ± 1.0) eV was presented[2] [17]. The techniques applied in this study were in principle the same as used in their earlier work, this time, however, with improved statistics and more accurate values of γ-ray energies of higher lying levels. For their analysis they used the same three γ-ray transition energy differences as before (Eq. 3.1) together with one further difference:

$$\Delta_4 = [E(148.1) - E(146.3)] - [E(76.4) - E(74.6)]. \qquad (3.2)$$

A level scheme of all γ transitions used in this study is shown in Fig. 3.3. The energy value of (3.5 ± 1.0) eV, as presented based on this study, was the most accepted one until 2007. This value is below the ionization potential of thorium of 6.3 eV. For this reason, internal conversion, as a potential decay channel, was expected to be energetically forbidden, leading to an expectedly enhanced radiative decay and

[1]The 29.18 keV γ line, which also has an unresolved overlap with the 29.39 keV γ transition, could be used without restrictions, as the latter is of significantly smaller intensity, corresponding to an inter-band transition.

[2]A preliminary value of (4.5 ± 1.0) eV is given in Ref. [16].

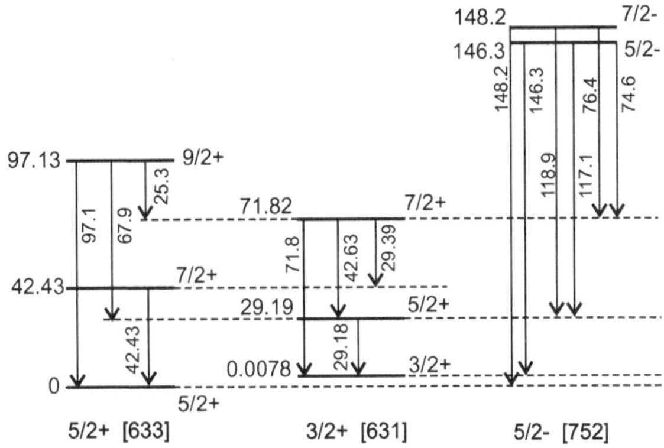

Fig. 3.3 ^{229}Th level scheme containing only levels that were used in the isomer's energy determination of Helmer and Reich in 1994 [17]. The energy gap between the 5/2$^+$ [633] ground state and the 3/2$^+$ [631] first excited state was calculated from this scheme in four independent ways. The resulting energy value was given as (3.5 ± 1.0) eV

an increased half-life of 20 to 120 hours (assuming no coupling to the electronic environment) [17]. These assumptions had to be corrected following further energy investigations. Helmer and Reich assumed already in their 1994 work that no "unique half-life" for ^{229m}Th might exist, as this will depend on the electronic environment of the sample.

In 2005, motivated by an improved understanding of the ^{229}Th level-scheme branching ratios [6, 10] and the non-observation of the direct ^{229m}Th γ ray, Guimarães-Filho and Helene published a re-analysis of the spectroscopy data obtained by Helmer and Reich [18]. The central technique applied in this work was the same as in the 1994 analysis of Helmer and Reich, however, this time an improved matrix formalism was used, including many reference lines to obtain better statistics. Also improved branching ratios were applied for the 29.18 keV and the 71.8 keV lines. These were assumed to decay by 100% branching ratio into the 3/2+ isomeric state in the 1994 work of Helmer and Reich. However, more recent work proposed that the decay of these states might populate the ground state by branching ratios of 25 and 40%, respectively. In this re-analysis, the γ-ray transition energies were also corrected for recoil effects, leading to the different value of (5.5 ± 1.0) eV for the isomeric energy.

3.3.3 A Corrected Energy Value

The value of 5.5 eV, as obtained in 2005 [18], was still below the threshold of the first ionization potential of thorium and an internal-conversion decay of the isomeric state was therefore expected to be suppressed. In 2007, however, a new measurement was published by Beck et al., which suggested the significantly larger energy value of (7.6 ± 0.5) eV [19]. This measurement made use of a different detection technique, using a cryogenically cooled microcalorimeter spectrometer with a resolution of about 30 eV. By applying this significantly improved resolution, it was possible to

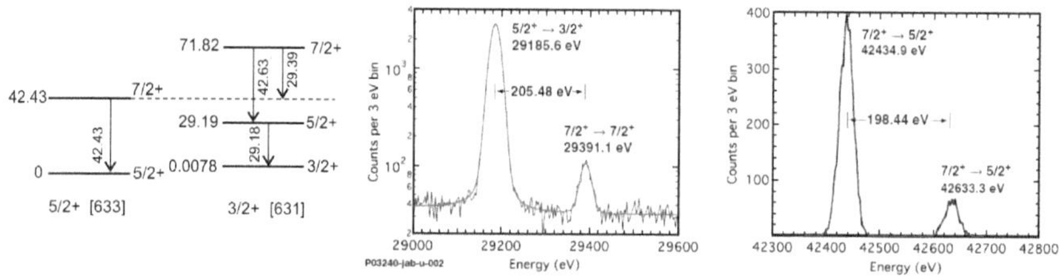

Fig. 3.4 Partial ^{229}Th level scheme (left) as used for the improved energy determination by Beck et al. in 2007 [19]. The measured data taken from Ref. [19] are shown on the right. The γ-ray doublets around 29 keV as well as 42 keV were spectroscopically resolved for the first time. This was made possible by employing the NASA microcalorimeter spectrometer system XRS, allowing for a spectral resolution of 30 eV. With kind permission of The American Physical Soceity [19]

resolve the closely spaced γ-ray lines of 29.18 and 29.36 keV as well as 42.43 and 42.63 keV. This in turn allowed to use a new transition energy difference for the energy determination of the isomeric state (Fig. 3.4):

$$\Delta_5 = [E(29.39) - E(29.18)] - [E(42.63) - E(42.43)]. \tag{3.3}$$

Further, also a correction for the branching ratio of the 29.19 keV to ground-state decay was included, which was estimated to be 1/13 (as opposed to 1/4, assumed in Ref. [18]). The value of (7.6 ± 0.5) eV, which was deduced in this way, poses a significant change in technology required for the direct detection of the isomeric decay. As the transition energy is placed above the ionization potential of neutral thorium of about 6.3 eV, internal conversion is allowed as an isomeric decay channel. Therefore, any significant chance to detect a photonic decay is only given for charged ^{229}Th. In this case, the isomeric half-life was suggested to be about 5 hours by Beck et al. The wavelength of photons emitted in the isomeric decay was placed in the vacuum ultra-violet (VUV) region at (163 ± 11) nm by this measurement. Formerly, the wavelength was expected to be around 350 nm (corresponding to 3.5 eV) in the near ultra-violet. This wavelength shift leads to the requirement of applying vacuum ultra-violet optics when searching for a potential photonic isomeric decay and partly explains the failure of former experiments to detect any isomeric decay signal.

A minor correction to this value was introduced in 2009 by the same group [20]. While a possible non-zero branching ratio for the 29.19 keV to ground-state transition was already included in their first publication, this time also a non-zero branching ratio for the 42.43 keV to ^{229m}Th inter-band transition was introduced. The estimated value of this branching ratio is 2%, leading only to a small correction for the isomeric energy value to (7.8 ± 0.5) eV. This is today's most accepted energy value, however, it has been argued that the actual error of this measurement might be significantly larger than originally proposed by the authors [21]. Recently a detailed re-evaluation of the Beck et al. data was performed, essentially resulting in a confirmation of the isomeric energy value [22].

3.4 Potential Applications

Shortly after the publication of Reich and Helmer in 1990 [13], in which the existence of a nuclear isomeric state with an energy in the range of only a few eV was claimed, the interest in this state was rising. The first publication, which appeared after the work of Reich and Helmer, was a theoretical paper of Strizhov and Tkalya in 1991, which aimed at a description of the different decay channels [23]. In their publication, they predicted an increasing interest of physicists from other disciplines like "*optics, solid-state physics, lasers, plasma, and others*". In the following years the list of potential applications continued to grow. These include the development of a nuclear clock [24, 25], the search for temporal variations in fundamental constants [24, 26] and the development of a nuclear laser [27, 28]. The most important ones will be detailed in the following.

3.4.1 *A Nuclear Clock Based on* $^{229\mathrm{m}}$*Th*

A first hint towards a potential use of the isomeric state for metrology was given by Tkalya et al. as early as 1996 [29]: "*The existence of such a level opens a unique possibility for investigation of some very interesting and important physical problems. The main of them are the following: a process of nuclear isomer excitation by a laser beam; a decay of nuclear isomeric level via electronic bridge; the investigation of the chemical environment and electronic properties of solids by means of measuring the half-life time of isomeric levels and energies of emitted photons; development of a high stability nuclear source of light for metrology; creation of* γ*-laser in the optical range, and so on.*" While this proposal does not include the development of a nuclear clock based on ^{229}Th, it is already a remark pointing at an expected high stability.

Tkalya also provided first calculations of the resonance cross section for direct laser excitation of the isomeric state, which was estimated to be $\sigma \approx 10^{-25}$ cm^2 for usual laser parameters [29, 30]. This cross section was inferred to be too small to allow for direct laser excitation, given the low precision of the isomer's energy value. A further calculation of the cross section for direct laser excitation of the ^{229}Th nucleus was published in 1999 by Karpeshin et al., leading to a similar result [31]. This predicted small cross section for direct laser coupling to the nucleus has led to the generally accepted requirement of an improved knowledge of the isomer's transition energy.

The potential for a nuclear clock based on $^{229\mathrm{m}}$Th was highlighted in 2003 by Peik and Tamm [24]. This proposal was largely driven by the interest to find a reference system for an optical clock with improved accuracy and stability. In their work, Peik and Tamm proposed to perform nuclear laser spectroscopy in the optical region with $^{229}\mathrm{Th}^{3+}$. The nuclear excitation with laser light was proposed to be probed using the hyperfine shift of the ^{229}Th shell, as induced by the change of nuclear spin and magnetic moment during the excitation of the nuclear isomeric state (double-resonance method). It was proposed to use the 3+ charge state, as this charge state corresponds to a convenient electronic configuration of a radon core plus one valence electron, which possesses a closed level scheme for applying the double-resonance method and further allows for direct laser cooling. Also estimates for the expected stability of such a nuclear clock were given, proposing a very high performance as will be detailed in the following.

A resonator of high frequency (but not too high to still allow for optical access) with a high Q-value is the basis for all modern atomic clocks. ^{229m}Th with an energy of about 7.8 eV, corresponding to a frequency of $f \approx 1.88$ PHz, provides such a high-Q resonator, as the radiative isomeric half-life is expected to be in the range of minutes to hours. Assuming a radiative lifetime of about 10^4 s, the linewidth of $\Delta f \approx 10^{-4}$ Hz would lead to a Q-value of about $f/\Delta f \approx 10^{19}$. At this level, however, the clock performance is limited by external perturbations, mostly introduced into the system by electric and magnetic fields. The sensitivity of the clock resonator to these perturbations is the limiting factor for the stability of all modern optical atomic clocks. Typical perturbing shifts, which have to be considered, are listed in Table 3.1.

The most important advantage of a nuclear clock compared to usual atomic clocks is that, due to the small nuclear moments, direct coupling of external fields to the nucleus is small.[3] However, nuclear-shell coupling via hyperfine interactions still has to be taken into account. Assuming an LS-coupling scheme, all electronic quantum numbers remain constant during the nuclear transition. Therefore, no shifts occur which entirely depend on the electronic quantum numbers (shifts due to static electric fields, electromagnetic radiation or collisions). Only a small black-body radiation shift of 10^{-19} at room temperature is expected due to the hyperfine Stark shift [24]. Further, by electronic state selection, the quantum numbers of the electronic shell could be chosen in an appropriate way to obtain "good" quantum numbers for the entire system (shell plus nucleus). Such a choice would avoid frequency shifts due to the linear Zeeman effect, the tensor part of the quadratic Stark effect and atomic quadrupole interactions.

While this first proposal of a nuclear clock can be seen as the pioneering step towards nuclear clock technology, there are still problems remaining. One problem is that the quadratic Zeeman effect is still existent and estimated as 1 kHz at 0.1 mT, which is comparable to that of usual atomic clocks. Further, the required choice of an electronic level leaves us with the metastable $7s^2S_{1/2}$ electronic state in ^{229}Th^{3+} as the only appropriate choice. This state, however, has an expected lifetime of only 1 s (compared to the up to 10^4 s expected nuclear isomeric lifetime), which significantly reduces the expected Q-value of the entire system.

[3] The magnetic dipole moments of the ground and excited state of ^{229}Th were estimated to be $\mu_g = 0.36 \cdot \mu_N$ [32] and $\mu_m \approx -0.08 \cdot \mu_N$ [25], where μ_N is the nuclear magneton $\mu_N = 5.05 \cdot 10^{-27}$ J/T, while typical atomic magnetic moments are in the order of the Bohr magneton $\mu_B = 9.27 \cdot 10^{-24}$ J/T. The ^{229}Th nuclear electric quadrupole moments of ground and excited state are $Q_g = 3.11|e|b$ [32] and $Q_m = 1.74|e|b$ [25], where the conventional unit of electron-barn is used (1 b = 10^{-28} m^2). Compared to that, the quadrupole moments of electronic states are in the range of up to hundreds of ea_0^2, where a_0 is the Bohr radius (1 $a_0^2 = 2.80 \cdot 10^{-21}$ m^2).

Table 3.1 Expected systematic shifts and uncertainties of a $^{229m}Th^{3+}$ single-ion nuclear clock [25]

Type of shift	Shift ($\times 10^{-20}$)	Uncertainty ($\times 10^{-20}$)
Excess micromotion	10	10
Gravitational	0	10
Cooling laser Stark	0	5
Electric quadrupole	3	3
Secular motion	5	1
Linear doppler	0	1
Linear zeeman	0	1
Background collisions	0	1
Blackbody radiation	0.013	0.013
Clock laser stark	0	$\ll$0.01
Trapping field stark	0	$\ll$0.01
Quadratic zeeman	0	0
Total	18	15

In the same 2003 paper [24], Peik and Tamm also proposed the development of a solid-state optical clock based on Mössbauer spectroscopy of ^{229m}Th in the optical region. This proposal was largely based on the earlier idea of Tkalya to directly excite the nucleus by laser light in a dielectric medium [33]. In 2009, also frequency shifts of a solid-state nuclear clock were estimated by Peik et al. to be beyond 10^{-15} for cryogenically cooled crystals [34]. While less accurate than the trapped ion version, such a solid-state device would provide high stability and would have the advantage of being technologically simpler and mechanically robust, leading to many potential applications. Also the number of ^{229}Th atoms in the crystal lattice environment is significantly larger than that of usual optical lattice clocks (up to 10^{19} ^{229}Th atoms per cm^3), reducing the statistical uncertainties. Experimental investigations towards the development of a solid-state nuclear clock are driven by the University of California and the Technical University of Vienna [35, 36].

In 2012, an alternative approach for a nuclear clock based on ^{229m}Th was proposed by Campbell et al. [25]. This proposal aims at a solution of the above mentioned problems of the earlier clock approach [24] of (i) a quadratic Zeeman effect being comparable to the one of ordinary optical atomic clocks and (ii) a reduced Q-value due to the required atomic excitation. In their work, they propose to use a pair of stretched nuclear hyperfine states for the clock transition, while $^{229}Th^{3+}$ remains in its $5f^2F_{5/2}$ electronic ground level. By a detailed analysis, partly based on numerical simulations, they were able to show that such a nuclear clock has the potential to outperform all existing atomic clock technology. The expected systematic error budget of this clock is shown in Table 3.1, as taken from Ref. [25]. A review on the experimental status of the nuclear-clock development was published in 2015 by Peik and Okhapkin [37].

3.4.2 *Search for Temporal Variations of Fundamental Constants*

Already in their 2003 work, Peik and Tamm proposed that nuclear laser spectroscopy with ^{229}Th would open a new possibility to probe for potential temporal variation of fundamental constants [24]. Such temporal variations are predicted by some theories beyond the standard model [38]. The idea is to compare the nuclear frequency standard to atomic-shell based frequency standards. The frequency standards are expected to show a completely different dependence on the fine structure constant as well as the dimensionless strong interaction parameter.

A first quantitative analysis was performed by Flambaum and published in 2006 [26]. The result predicts an extremely high sensitivity for potential temporal changes of the fine structure constant α ($\dot{\alpha}$) as well as the dimensionless strong interaction scale parameter m_q/Λ_{QCD}. It was highlighted in [26] that the sensitivity to these variations may be below 10^{-20} per year and thus at least three orders of magnitude more sensitive than constraints based on atomic-shell transitions (which by today pose the most stringent limits on such variations).

The reason for the predicted enhancement in sensitivity is that, from nuclear structure theory, the energies of the nuclear ground and excited states are proposed to be dominated by two individual high energy (MeV) terms which, by coincidence, cancel each other down to the eV range when subtracted in the very special case of the ^{229}Th nucleus. The first term is the Coulomb-energy term, which is responsible for the sensitivity to variations in the fine structure constant α, the second term results from various contributions of the strong interaction and thus leads to a sensitivity with respect to the strong interaction scale parameter m_q/Λ_{QCD}. The ratio between the typical nuclear energy scale and the energy scale of ^{229m}Th can then be directly transferred to a sensitivity enhancement for changes in fundamental constants. The estimate for the variation of the ^{229m}Th transition frequency as a function of the variation of α and Λ_{QCD} is given as follows [26, 39]:

$$\frac{\delta\omega}{\omega} \approx 10^5 \left(2\frac{\delta\alpha}{\alpha} + 0.5\frac{\delta X_q}{X_q} - 5\frac{\delta X_s}{X_s}\right)\frac{7\text{ eV}}{\omega}, \tag{3.4}$$

with $X_q = m_q/\Lambda_{\text{QCD}}$ and $X_s = m_s/\Lambda_{\text{QCD}}$, where $m_q = (m_u + m_d)/2$ and m_s denote the light quark mass (~5 MeV) and the strange quark mass (~120 MeV), respectively. This expression has already been corrected for the new energy value of ~7.8 eV [39] and contains an expected enhancement in sensitivity by five orders of magnitude for changes in the two fundamental constants (based on the Walecka model).

Opposed to that, Hayes et al. came to the result that no significant enhancement of sensitivity for potential temporal variations in fundamental constants could be achieved from a nuclear frequency standard [40, 41]. Based on the Feynman–Hellmann Theorem, which is a fundamental theorem of quantum-field theory, they derived the simple formula

$$\frac{\delta\omega}{\omega} = \frac{\Delta V_C}{\omega}\frac{\delta\alpha}{\alpha} \tag{3.5}$$

for the α-dependence of the relative nuclear frequency shift [40]. In this equation, ΔV_C denotes the Coulomb energy difference between the ground and the excited nuclear state. Taking the Nilsson model as the basis for nuclear energy calculations, there is no Coulomb interaction included, leading to the prediction of $\Delta V_C = 0$ and thus no expected sensitivity for a potential $\dot{\alpha}$.

The same authors inferred a similar relation for the dependence of the frequency on the nucleon mass M_N. The equation which is, however, based on significant simplifications, reads:

$$\frac{\delta\omega}{\omega} = \frac{\Delta T}{\omega}\frac{\delta M_N}{M_N} + \cdots. \tag{3.6}$$

Here ΔT denotes the difference in the kinetic energies of the ground and excited state. Again, applying the asymptotic Nilsson model, $\Delta T = 0$ is inferred, leading to no expected enhancement in sensitivity for a variation of M_N. A variation with respect to m_q/Λ_{QCD} was not performed for reasons of complexity.

Given these controverse results, He and Ren reexamined the problem based on the Feynman–Hellmann theorem, however, using the more complex model of relativistic mean-field theory to determine a value for the Coulomb energy difference ΔV_C [42] (see also Refs. [43, 44]). Their estimates lead to an enhanced sensitivity for temporal changes of α of at least four orders of magnitude. Also the sensitivities for changes of the nucleon and meson masses are found to be enhanced by five to six orders of magnitude. In their work, they argue that the earlier publications of Hayes et al. [40, 41] neglect the correlation between the strong interaction and the Coulomb interaction in an unallowed way.

In 2009 four papers on this topic appeared, which were published by V. Flambaum and co-workers [45–48]. In Ref. [45], Flambaum and Wiringa reconfirmed an expected enhancement for temporal changes of the dimensionless strong interaction parameter m_q/Λ_{QCD} by about five orders of magnitude. In their work, they used the variational Monte Carlo (VCM) method, as opposed to the Walecka model or relativistic mean-field theory, which was used in earlier work [26, 42]. In a different publication, Flambaum et al. also performed a new estimate for a potential enhanced variation of the fine structure constant α with the result of a potential enhancement by four orders of magnitude [46]. In this work, the enhancement based on a polarization contribution caused by a valence neutron in the nucleus was estimated. Litvinova et al. carried out detailed nuclear structure calculations in order to infer the potential sensitivity of the ^{229m}Th nuclear transition to temporal variations in α [47]. The result is, that such sensitivity calculations heavily depend on the applied nuclear model and the therein achievable precision and therefore no reliable prediction could be made, based on current nuclear models (see also Ref. [49]).

As a consequence, in Ref. [48], it was proposed to measure the nuclear charge radii as well as the electric quadrupole moments of ^{229}Th and ^{229m}Th, respectively, by laser spectroscopy of the atomic shells of both nuclear states. A technique is presented, which would allow to deduce V_C for both states from these parameters and thus to determine the expected enhancement in sensitivity for potential variations in α. As this method is directly based on experimental data, it would be independent of any particular choice of the nuclear structure model.

In 2010, a review article on the search for potential temporal variations of fundamental constants with ^{229}Th was published by Berengut and Flambaum [50]. Since then three further publications on this topic appeared. One also in 2010 by Rellergert et al. [35], who highlighted the potentials of a solid-state nuclear clock with respect to searches for fundamental constant variation. In 2012, Berengut and Flambaum proposed that, besides temporal variations, also spatial variations of fundamental constants could be probed [51]. Most recently it was proposed that also effects of Lorentz invariance and equivalence principle violation could be probed with the help of ^{229m}Th [52].

3.4.3 A ^{229m}Th-Based Nuclear Laser

A nuclear laser based on ^{229m}Th was conceptually proposed by Oganessian and Karamian in a publication from 1995, in which the thorium isomer was discussed in an individual section within the more general framework of nuclear γ-ray lasers [27].

The working principle of a nuclear laser would be the same as for atomic-shell based lasers, however, using nuclear transitions instead. While there are significant problems to overcome when developing a high energy γ-ray laser based on nuclear transitions (see for example Ref. [53]), the isomeric state in ^{229}Th could allow to develop a first proof-of-principle nuclear laser [27]. The main important aspect for developing such a laser is to achieve population inversion between the ground and the isomeric first excited state. Oganessian and Karamian proposed to achieve this inversion by temperatures of 10^4 K in combination with an NEET process, however, at a time when the isomeric energy was still assumed to be at about 3.5 eV and thus the internal-conversion decay channel of the isomer was expected to be suppressed in neutral ^{229}Th. An alternative way to achieve population inversion was proposed by Karpeshin et al. via nuclear excitation by electron transition (NEET) [54].

A quantitative analysis of the possibility of a ^{229m}Th-based nuclear laser was performed by Tkalya in 2011 [28]. In this work, a nuclear laser based on ^{229}Th doped solid-state crystals is investigated. The population inversion is proposed to be achieved in a two-step approach: First, the isomeric state is populated via direct laser excitation. As the nuclear ground and isomeric state provide only a two-level system, no population inversion can be achieved in this way. However, it is still possible to excite a significant amount of nuclei. In a second step, it is proposed to apply a strong magnetic field (of up to 100 T), in order to achieve a Zeeman splitting of the nuclear ground and excited state into corresponding sublevels. By further cooling the crystal into the temperature region of about 0.01 K, the nuclei are expected to predominantly populate the energetically lowest Zeeman sublevels of each nuclear state. In this way, a population inversion is achieved between the lower Zeeman sublevels of the excited state and the upper Zeeman sublevels of the ground state. The transition between these sublevels would allow for light amplification by stimulated emission in the nucleus.

Alternatively, line splitting into nuclear sublevels could also be achieved via electric quadrupole splitting and is also discussed in Ref. [28]. In 2013, an alternative to the cooling method in order to achieve population inversion was proposed by Tkalya and Yatsenko [55]. In this approach, a narrowband laser is used to drive individual transitions in the Zeeman-splitted nuclear multiplet. A nuclear laser based on ^{229m}Th could be a proof-of-principle device for the development of high-energy nuclear γ-ray lasers.

3.4.4 *Further Potential Applications*

The half-life of the ^{229}Th nuclear isomeric state is expected to significantly depend on the electronic surrounding. This has motivated the idea to use the isomer's decay as a probe for material or surface structure as early as 1991 [23]. While the 76 eV isomeric state of ^{235}U has long been in use for such purposes, ^{229m}Th is expected to be advantageous, as the valence $6d_{3/2}$ and $7s_{1/2}$ electronic subshells are expected to play an important role for the isomeric decay. These valence subshells are significantly influenced by the surrounding electronic environment, as opposed to the filled electronic subshells, which play the dominant role in the internal-conversion decay of ^{235m}U [29].

A further advantage of ^{229}Th compared to ^{235}U was thought to be the photonic decay channel, which was expected to be dominant even for neutral ^{229m}Th at the time of the proposal in 1996 [29]. However, this expectation of a dominant photonic decay channel is not fulfilled in view of recent research.

In 1998 it was proposed that the exponential character of the radioactive decay law could be probed using ^{229m}Th [56]. One expected advantage when using ^{229}Th is, that, in a solid-state sample, a large number of isomers could be produced, e.g., by direct laser excitation. It was shown that, assuming the radiative decay to be a significant decay channel, ^{229m}Th activities of up to 10^{14} Bq could be achieved within a solid-state sample of about 1 mg. Further, these high activities would not require any special shielding, as the deexcitation occurs in the form of VUV photons. Even after a time period of $50 \cdot t_{1/2}$, the activity level would be 0.1 Bq, which could easily be probed and therefore used as a test for the decay law. This idea is based on the assumption that a strong optical excitation of the ^{229}Th isomeric state is easily achievable to a significant amount in a solid sample. So far, however, all experiments performed to detect this excitation were unsuccessful and point to a significant non-radiative decay channel.

Due to the expected high resilience of ^{229m}Th against external influences, it was proposed to use the isomeric state for nuclear quantum-logic studies and entanglement with enhanced coherence times [24] and as a qubit for quantum computing

[57]. In 2009 it was even proposed to implant ^{229m}Th into solar cells as a proof-of-principle for direct nuclear energy transformation [58]. This approach, however, cannot be seen as realistic, given that no radiative isomeric decay has been observed until today.

Besides the potential application of an ultra-precise clock based on ^{229m}Th for the detection of time variations of fundamental constants [26], such clocks could be used as gravitational detectors for geodesy and earthquake detection [59, 60], as well as gravitational wave detection [61] and (topological) dark matter search [62].

3.5 ^{229m}Th Excitation and Decay

In parallel to the potential applications, the theoretical interest in the isomeric state of ^{229}Th was growing. This includes the energy dependent theoretical study of the isomeric half-life and decay channels under different conditions, as well as the potential for its excitation. In the following, a short review of these theoretical investigations is presented.

3.5.1 *Basic Theoretical Investigations*

The first theoretical work that appeared after the isomeric energy had been constrained to below 10 eV in 1990 was the one of Strizhov and Tkalya [23]. The three possible decay channels, internal conversion (IC), electronic bridge (EB), which is a bound internal conversion process accompanied by photon emission,[4] and radiative decay (γ) were discussed as a function of the isomeric energy (see Fig. 3.5 for diagrammatic visualization). As the energy value was not securely constrained, all three decay channels were considered as equally possible. It was found that, in case that the isomeric energy is above the ionization potential of thorium (for which 6.03 eV was the best value at that time), internal conversion would be the dominant decay channel, reducing the isomeric lifetime into the μs region. Opposed to that, for an energy of below 1.3 eV, not even electronic bridge processes would be allowed, leading to a purely radiative lifetime of up to 27 days. The energy region inbetween is hard to discuss, as the isomer's lifetime and decay channels would heavily depend on the exact isomeric energy. In case that the energy would be of the same value as any of the shell transitions, a fast decay via an electronic bridge could occur. However, if no such energy agreement existed, EB could be suppressed resulting in a prolonged lifetime. Therefore the lifetime could vary from microseconds up to tens of hours, depending on the exact energy value.

[4] The expression "electronic bridge" was originally introduced by V.A. Krutov [63–65]. The process had already been used to describe laser-assisted nuclear decay of ^{235m}U by Zon and Karpeshin [66].

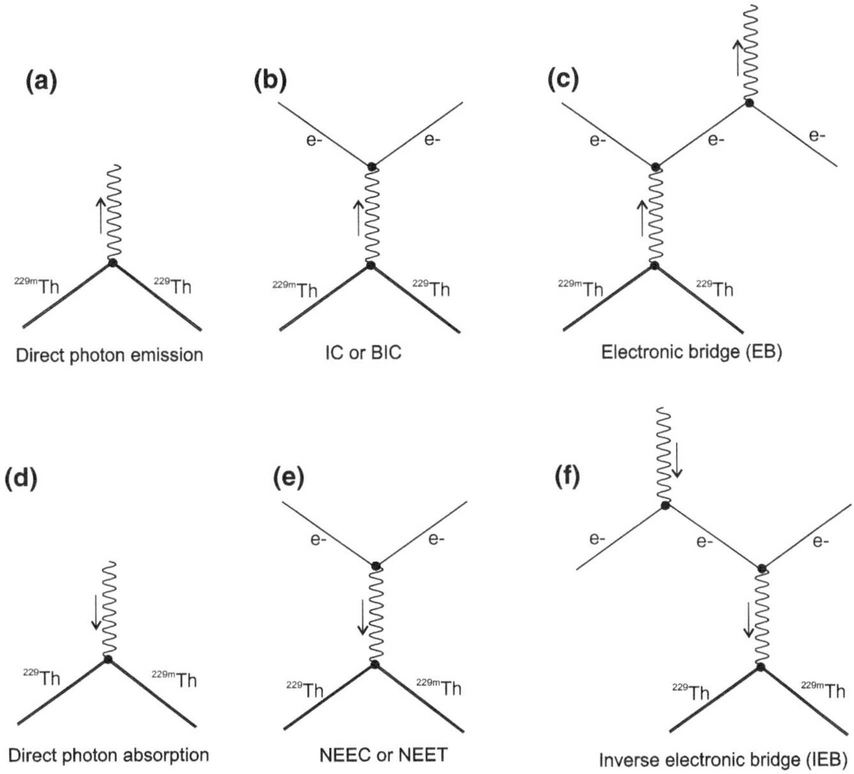

Fig. 3.5 Different fundamental processes of isomeric decay (first row) and excitation (second row). **a** The nuclear deexcitation via direct γ emission is a first-order process. **b** The second-order process corresponds to shell-nucleus coupling. In this case, an electron is excited by the process of nuclear deexcitation, leading either to electron emission (internal conversion) or to the transformation of the electron into an excited bound state (bound internal conversion). **c** In case that a bound internal conversion (BIC) process occurs, the electron will undergo deexcitation back to the ground state, accompanied by photon emission. The combined process is a third-order process, introduced as electronic bridge (EB). **d** The reverse of process **a** is the process of isomer excitation by direct photon absorption, which has to be employed for the nuclear clock concept. **e** The reverse of process **b** is called nuclear excitation by electron transition (NEET) in case that the electron was bound; and nuclear excitation by electron capture (NEEC) in case of an unbound electron. **f** The reverse of process **c** is typically referred to as inverse electronic bridge (IEB). It is investigated as a potentially strong excitation channel for ^{229m}Th. Note, that the use of notation is not uniform in literature and this process is sometimes referred to as (reverse) BIC

Already in this early theoretical work, it was proposed to probe for the internal-conversion decay of the isomeric state as a first step [23], which by today has led to the successful direct detection of the isomeric state. It was also proposed to probe for surface influences as well as laser triggered IC, which have become again topics of current experimental investigations.

Also a way to produce the isomeric nuclei via plasma excitation was proposed. At that time, the population of ^{229m}Th in the ^{233}U α decay was assumed to be too low (a fraction of a percent) to allow for the isomer's detection. The dominant process of nuclear excitation in a plasma was inferred to be nuclear excitation by electron capture (NEEC), which is the reverse of the internal conversion process. Isomer population via direct laser excitation was not discussed.

The next three articles, which were published by Tkalya (all in 1992) [30, 67, 68], contain a detailed discussion of the isomer's excitation via the reverse of the electronic bridge process, typically referred to as inverse EB (IEB). In the IEB process an atomic-shell transition is excited by a light source. The excited state then transfers its energy to the nucleus, thereby exciting the ground to the isomeric state. Opposed to the process of nuclear excitation by electron transition (NEET), in the IEB process the electron hole, produced by shell excitation, can be virtual. Thus the energy of the irradiated light can be chosen in accordance with the nuclear transition. The IEB process was discussed for the non-resonant case, for which the energy of the excited electronic-shell state differs from the energy of the nuclear isomeric state. The concept was found to be more efficient than the direct laser excitation of the nucleus, for which the cross section was also calculated in the second publication [30].

The process of laser-induced isomeric deexcitation (resonance internal conversion) was studied by Karpeshin et al. [69]. The basic idea of the effect is to use laser light in order to introduce the missing energy value into the shell-nucleus system, which is required to fulfill the resonance condition between the nuclear transition and a correspondingly chosen excited electronic state. It is inferred that the deexcitation probability could be enhanced by a factor of 10^3 or more in this way. The same effect was later also discussed by Typel and Leclercq-Willain [70].

A process of nuclear excitation via a reverse electronic bridge, accompanied by photon emission, was studied by Kálmán et al. in 1994 [71, 72]. This process is similar to the inverse electronic bridge process discussed in Ref. [30] and corresponds to the reverse of the laser assisted decay process studied in Ref. [69]. In 1996 Karpeshin et al. proposed to also excite the isomeric state via nuclear excitation by electron transition (NEET) [54], in which case the energy introduced into the system corresponds to the excitation energy of a shell state in order to produce an on-shell electron hole. In the same year, the optical excitation of ^{229m}Th via IEB was reviewed by Tkalya et al. [29, 73]. This work had the focus on the resonant excitation, as opposed to earlier publications, in which the nonresonant case was discussed [30, 67, 68]. A review of the theoretical investigations until 1998 was provided by Matinyan [74]. Still it is stated that "*the situation for the direct use of laser radiation [for nuclear excitation] remains hopeless, even in the case of very low lying levels*". A detailed discussion of different nuclear excitation processes via NEET is given in Ref. [31]. This paper also introduces the reverse process (inverse NEET or TEEN) and contains a comparison between the nuclear excitation by electron transition (NEET) process as proposed in [54] and the inverse electronic bridge (IEB) mechanism [29, 73]. A more recent comparative discussion of the IEB mechanism (denoted as reverse BIC) and the NEET mechanism for optical pumping of ^{229m}Th can be found in Ref. [75].

In the following years, the same decay mechanisms were revisited, applying different theoretical models or new computational packages. In 2001, the process of bound internal conversion (discrete internal conversion) was discussed, applying a different theoretical model [76]. In 2006, Ruchowska et al. inferred a radiative half-life for ^{229m}Th of about 10 hours, based on a quasi-particle-plus-phonon model [11]. In 2005 and 2006 the electronic bridge process was reconsidered by Karpeshin and Trzhaskovskaya with a special focus on experimental investigations [77, 78]. It was proposed to search for low energy photons as emitted from the thorium shell during the isomeric decay via the electronic bridge mechanism. Also the lifetime shortening was investigated. These investigations were revised, also containing a discussion of the internal conversion process in 2007 [79], shortly after the publication of the corrected energy value by Beck et al. [19]. This was the first time that the internal-conversion decay channel was reconsidered since the first theoretical publication on the topic in 1991 [23], as in the meantime this decay channel was expected to be energetically forbidden. An overview over the different isomeric decay channels and corresponding half-lives is given in Ref. [80, 81].

The transition probability for the ^{229m}Th γ-decay was recently re-investigated resulting in a radiative lifetime of about 10^4 s [82]. This value is significantly larger than most of the previously reported values that assume an isomeric energy of 7.8 eV (see [80] and references therein).

3.5.2 Excitation and Decay Under Special Conditions

After the main decay channels of ^{229m}Th had been theoretically investigated, theoretical work focused on the isomer's excitation and decay under special conditions. These investigations were partly driven by experimental studies, and after 2007 partly by the corrected isomeric energy value [19].

The first paper along this line discusses the non-radiative isomeric decay in a metal [83]. It is emphasised that, even if the internal conversion process would be energetically forbidden in the free thorium atom, the isomer could transfer its energy to the conduction electrons of the metal surface if the thorium atom is attached to it. The electron could leave the metal surface, if the surface work function is below the energy of the isomeric state. Such effects are expected to shorten the isomer's lifetime and could be used for the energy determination of the isomeric state [83]. More comprehensive investigations of the influence of the chemical environment on the isomeric decay were published in 2000 [33, 84, 85]. These publications also contain the deexcitation of the isomeric state in dielectric media, in which case the radiative decay is expected to dominate as long as the material's band gap is larger than the isomeric energy. The work provided the basis for the later proposal of a solid-state nuclear frequency standard [24]. A compact overview of the theoretical investigations at that time can be found in Ref. [86]. This paper also contains a detailed list of potential applications.

Following the revision of the ^{229m}Th energy value in 2007, the electronic bridge mechanism, which was previously discussed as the dominant decay channel in the neutral thorium atom, was shifted to thorium ions. This process was discussed in Refs. [87, 88] for ^{229}Th^{3+} and in Ref [89] for ^{229}Th^{+}. The reverse process, the nuclear excitation via the inverse electronic bridge (IEB) in ^{229}Th^{+}, was also discussed and proposed as the basis for an experimental investigation of ^{229m}Th [90].

The direct two-photon excitation of the nuclear isomeric state in thorium ions was proposed in Ref. [91]. Recently, also the two photon isomeric excitation via the electronic shell was revisited, partly as a reaction on the non-observation of the isomer's deexcitation [92, 93]. In 2017 an experimental discrepancy between the theoretically expected and experimentally observed lifetime of the isomeric state in ^{229}Th^{+} was published [94, 95]. Some theoretical work was published, dealing with this discrepancy [96].

Further, it was proposed to investigate the internal conversion decay of the isomer in thorium ions, in which an excited shell state was populated via laser irradiation. This proposal corresponds to laser-triggered internal conversion decay [97]. Also the determination of the isomeric energy via laser-induced electronic bridge was proposed [98]. This proposal transfers an idea of that was developed by Karpeshin et al. for neutral thorium atoms to ^{229}Th^{+} [69].

3.5.3 Other Processes of Isomer Excitation and Decay

A few other processes of excitation and decay of ^{229m}Th have been discussed in literature. In 1996 it was proposed by Varlamov et al. to excite the isomeric state via surface plasmons [99]. In the same year also the possibility of a ^{229m}Th α decay was discussed [100]. It was found that the lifetime of ^{229m}Th with respect to α decay is shortened by a factor of 2 to 4, compared to the lifetime of the ground state. It is proposed in [100] to use the α decay of the isomeric state for its direct identification. Given the ground-state's half-life of 7932 years, this would, however, require a long isomeric lifetime (only obtainable if the IC decay channel is suppressed) and a large number of excited nuclei.

The special case of the isomeric decay in hydrogen-like ^{229}Th (Th^{89+}) was considered by Karpeshin et al. in 1998 [101]. An experiment to probe the special predictions, making use of the ESR storage ring at GSI, is proposed. Hydrogen-like ^{229}Th had already been discussed earlier in the context of nuclear spin mixing [102, 103]. Here also a storage ring experiment was proposed and the case of muonic thorium was discussed. The isomeric state in ^{229}Th^{89+} and ^{229}Th^{87+}, as well as in muonic thorium, has recently been re-investigated by Tkalya in several independent publications [104–106]. Further, it was proposed in 2011 to excite the ^{229}Th isomer via Coulomb excitation [107, 108].

3.5.4 *Coherent Control of Nuclei*

The general concept of coherent nuclear control has been developed independently of the special case of the ^{229}Th isomer [109–111]. In this approach, strong coherent X-ray sources (e.g. free-electron lasers) are used in combination with accelerated target nuclei to boost the light field. As the nuclear matrix elements are small, typically ultra-intense laser fields are required to drive coherent population transfer.

The special case of ^{229}Th in this context was discussed by Liao and Das et al. [112, 113]. It was proposed to identify the ^{229}Th isomeric decay via electromagnetically modified nuclear forward scattering [112, 113].

Recently, an optomechanical microlever was proposed to bridge between an optical laser and x-rays [114]. It was emphasized that such a device could have significant implications for ^{229m}Th and, in particular, could be applied for the isomer's energy determination.

3.6 Search for a Direct Decay

After the prediction of the existence of ^{229m}Th in 1976 [1], the isomer's energy value started to be constrained via high resolution γ-ray spectroscopy. Until 1990 the energy was constrained to below 10 eV, based on the comparison of γ lines of higher energy [13]. Later, the value of (3.5 ± 1.0) eV was determined, placing the transition in the optical region [17]. Since then, this uniquely low nuclear excitation energy and its potential applications have triggered a significant amount of experimental efforts aiming at the direct identification of the isomeric decay.

In 2007, the energy value was corrected first to (7.6 ± 0.5) eV [19] and in a re-analysis in 2009 to (7.8 ± 0.5) eV [20], corresponding to a wavelength of about 159 nm. This new energy value imposes different requirements on the techniques to be applied in the search for a direct isomeric decay. Not only because the wavelength of photons, potentially emitted in the decay, was shifted into the vacuum ultra-violet (VUV) region, but also as the new energy value is above the first ionization potential of thorium, rendering non-radiative internal conversion (IC) the dominant decay channel in the neutral thorium atom.

The experiments performed in the search for the isomer's direct detection can therefore roughly be divided into investigations performed before and after the correction of the energy value to about 7.8 eV in 2007 and 2009 [19, 20].

3.6.1 *First Claim of a Direct Detection*

A detection of light emitted in the decay of ^{229m}Th was for the first time reported in 1997 by Irwin and Kim [115]. In their work, they detected light emitted from a ^{233}U sample with the help of a photomultiplier tube (PMT). A monochromator was used in order to acquire the energy spectra. An emission around 3.5 eV energy was observed, which was interpreted as the light emitted in the direct decay of the ^{229}Th isomeric state. As a reaction on the direct observation stated in [115], theoretical work was published, proposing a way to experimentally determine the isomer's half-life [116] or coming to the conclusion that there is reason to doubt the interpretation of the experimental result [117, 118] (The topic was later also discussed by Kálmán [72]).

The same spectral features were reobserved by Richardson et al. [119], where a liquid ^{233}U source was used and comparisons between ^{233}U and ^{232}U were performed. As opposed to the earlier work of Irwin and Kim, Richardson et al. obtained a better spectroscopic resolution and a substructure of the 3.5 eV line became visible. The assignment of this line structure to the ^{229}Th isomeric decay was stated possible, but not unambiguous, as several other origins of the spectral feature could not be excluded [119].

In 1999 the light emitted from the ^{233}U samples could be unambiguously shown to originate from α-particle induced fluorescence of nitrogen [120, 121]. While Utter et al. were able to show that the lines disappear under vacuum conditions [120], the lines were found to be also present in nitrogen discharge lamps [121]. In the same year, Young et al. were also able to show that a spectroscopic feature around 520 nm, which was previously assumed to originate from the isomeric decay via the electronic bridge channel [115], can be attributed to luminescence of the uranyl ion [122]. Therefore, the problem of directly detecting the isomeric decay in ^{229}Th remained unsolved [123]. A review on these early attempts of ^{229}Th isomer detection can be found in Ref. [124].

3.6.2 Search for ^{229m}Th via α Decay

A completely different approach of searching for the direct decay of ^{229m}Th was investigated between 2003 and 2009. In 2003 Mitsugashira et al. published a study, which investigated the α decay of ^{229m}Th [125] that had been proposed earlier by Dykhne et al. [100]. ^{229m}Th was expected to be produced by the (γ, n) reaction on a ^{230}Th target, the observation of ^{229m}Th α decay with a half-life of (13.9 ± 3) hours was asserted. A similar study was published in 2005 by the same group, that time ^{229m}Th was expected to be produced from the ^{229}Ac β decay, following the ^{232}Th$(\gamma, p2n)$ reaction [126, 127]. α decays were seen, which were interpreted as originating from ^{229m}Th, no value for the half-life was inferred.

These measurements are, however, inconsistent with an earlier indirect study performed by Browne et al. [128]. In this study, freshly produced ^{229}Th was radiochemically separated from a relatively large amount of ^{233}U of 25 g. Searches were then performed for a growing activity of the ^{229}Th ground state as expected to occur due to the delayed population of the ground state from the isomer. The ^{229}Th activity was monitored by detection of the 193 keV γ ray emitted in the ^{229}Th α decay. No activity ingrowth could be detected, resulting in a ^{229m}Th life-time of less than 6 hours or more than 20 days [128].

Kasamatsu et al. (of the same group as Mitsugashira) also performed experiments searching for photons emitted in the isomeric decay. In these measurements, however, no decay signal was observed [129]. Given the inconclusive results, in 2009 a further study was published by this group [130]. This time ^{229m}Th was populated from the α decay of ^{233}U. While no α decay was observed which could clearly be assigned to ^{229m}Th, still an upper half-life limit of 2 hours could be inferred [130].

In 2003 Inamura et al. proposed to follow a different experimental line and to excite the isomeric state via the NEET process, using light as emitted by a hollow-cathode electric discharge lamp [131, 132]. The ^{229}Th isomeric decay was proposed to be detected by means of its α decay and photon emission [133]. First results of these measurements were published in 2009 [134], which were interpreted as an indication of the isomer's population via NEET, followed by the observation of the ^{229m}Th α decay with a half-life between 1 and 3 min. However, some uncertainties corresponding to the signal interpretation remained [134].

3.6.3 *Search for ^{229m}Th in VUV Transparent Material*

After some early attempts to observe a signal from the isomeric decay in a liquid solution of ^{229}Th had failed [135, 136], a new class of experiments was developed, taking into account the corrected energy value of 7.6 (and later 7.8) eV [19, 20]. This class of experiments can be subdivided into two approaches. In the first approach, ^{229m}Th is populated by the 2% decay branch in the α decay of ^{233}U. In the second approach, broad-band light sources are employed for the isomer's excitation.

For the first approach, typically a thin layer of ^{233}U with a large surface area is used for the production of ^{229}Th. A significant amount of the ^{229}Th α-recoil isotopes, produced in the ^{233}U decay, can leave the source material and is implanted into an absorber plate, which typically consists of CaF_2 or MgF_2 in order to provide transparency in the vacuum-ultraviolet region around 159 nm [136]. Afterwards, searches for photon emission due to the isomeric decay are performed. Photons are expected to be emitted in the deexcitation process, as the band gaps of the materials are large, thus suppressing the internal-conversion decay channel. Such experiments were carried out at the Physikalisch-Technische Bundesanstalt (PTB) in Braunschweig, Germany [136], at the Lawrence Livermore National Laboratory (LLNL), USA [137, 138], at the Los Alamos National Laboratory (LANL), USA [139] and, in an improved version, by our own group at the Maier-Leibnitz-Laboratory (MLL) in Garching, Germany [94, 140–144]. When applying this method, some background effects have to be considered, which were discussed, e.g., in Refs. [145, 146]. The LANL group reported a direct detection of the isomeric decay by this method in 2012 [139]. This result is, however, subject to controversial discussions within the community [138, 145, 147] and could so far not be reproduced by any other group.

A slightly different experiment of the first approach has recently been proposed by Hehlen et al. [148] and was then further developed by Stellmer et al. [146]. Here the idea is to grow ^{233}U directly into a VUV transparent crystal. Detailed studies of the occurring Cherenkov background were carried out, allowing the group to show that, even in this case, there is the chance to detect the ^{229}Th isomeric decay. Experiments along this line are still ongoing.

For the second approach, ^{229}Th is typically grown into large band-gap crystals. These crystals, with a ^{229}Th doping concentration of up to $4.1 \cdot 10^{17}$ cm^{-3}, are then irradiated with broad-band VUV light as, for example, provided by synchrotrons or D_2 lamps, in order to excite the isomeric decay. The search for the isomer's deexcitation is then again performed in the photonic decay channel. Experiments along this line were first proposed in published form in 2010 [35]. In the following years, the development of ^{229}Th-doped crystals was driven by two groups, located at the University of California, USA and at the Technical University of Vienna, Austria. These crystals are expected to also provide the basis for solid-state nuclear frequency standards. Recently, significant progress was made in crystal development and theoretical understanding [36, 148–154]. A first experimental result using synchrotron radiation has been reported, which was, however, negative and provided half-life constraints only [155].

Results of a slightly different experiment were recently published [156]. Here ^{229}Th was adsorbed onto a CaF_2 surface and irradiated with undulator radiation. No photons in the expected wavelength region could be observed, which tentatively can be explained by means of the chemical structure of thorium on the CaF_2 surface [157]. It was also proposed to excite the 29 keV state of ^{229}Th by synchrotron radiation. ^{229m}Th is populated to a significant amount from the decay of this state. Following the excitation of the 29 keV state, the subsequently occurring ^{229m}Th decay is therefore expected to be detectable [33, 124, 158]. Results of a similar experiment, however without using ^{229}Th-doped crystals, were already reported in 2005 [159].

Recently it was proposed to investigate the isomer following its excitation in the ^{229}Ac β^- decay and implantation into a crystal lattice [160]. The possibility of annealing the crystal in order to remove lattice defects after the ^{229}Ac implantation but before the ^{229}Ac β decay could make an advantage for the search of a radiative decay channel in this approach.

3.6.4 Search for ^{229m}Th in a Paul Trap

The ^{229}Th isomeric decay can be investigated independently of any surface influences by storing ^{229}Th ions in a linear Paul trap. Experiments along this line were performed at the Georgia Institute of Technology, USA and at PTB in Braunschweig, Germany. A further ^{229m}Th Paul trap experiment is currently in preparation in Moscow, Russia. The basic approach is to populate the isomeric state via the inverse electronic bridge (IEB) process (first proposed by Tkalya [30] and transferred to Th$^+$ by Porsev et al. [90]) and then detect the photons as expectedly emitted during the isomeric decay.

Detailed theoretical investigation of the thorium electronic shell was performed in order to lay the foundations for the experimental investigation of the IEB process [161, 162]. The thorium ions are produced by laser ablation [163] and stored in a linear Paul trap [136]. Laser cooling and trapping of $^{229}Th^{3+}$, as required in the nuclear-clock concept, was successfully performed at Georgia Tech [164–169]. At PTB, $^{229}Th^{+}$ ions are stored without laser cooling. A pulsed laser system has been developed to excite the ^{229}Th ion shell. Subsequently emitted photons are detected by a PMT [170–172]. Two Paul traps are in operation in parallel, one for ^{229}Th and one for ^{232}Th, as an isomeric signal can only be identified by performing comparisons. Alternatively, at Georgia Tech, the population of the isomeric state can be probed via interrupts in the laser fluorescence of one of the cooling transitions [173].

Reported experimental results include the observation of the 717 nm electric quadrupole transition in $^{229}Th^{3+}$ [173], the detection of 43 previously unknown energy levels in $^{232}Th^{+}$ [172] and the observation of an unexpected negative isotope shift in $^{229}Th^{+}$ [174].

In the Moscow experiment a linear Paul trap has been developed to provide the basis for nuclear laser spectroscopy of triply charged thorium ions [175–179]. Laser ablation and inductively coupled plasma techniques were studied for the production of ^{229}Th ions [180]. New efforts on ^{229m}Th are also underway in Australia [60].

3.6.5 Search for ^{229m}Th via Internal Conversion

Since the revision of the isomeric energy value to 7.6 eV in 2007 [19], it was clear that, in the neutral thorium atom, the isomeric state will predominantly decay by internal conversion (IC). IC branching ratios of up to 10^9 were predicted for this case [79]. Early considerations of an IC decay of ^{229m}Th were already made in 1991 by Strizhov and Tkalya [23] and experimental investigations followed in 1995 [181]. These experiments, however, assumed an IC-decay half-life in the range of hours and there was no chance for the successful observation of the isomeric decay. In the following years this decay channel was not further investigated, as it was thought to be energetically forbidden due to the expected isomeric energy of only 3.5 eV [17].

A first theoretical investigation of a potential electronic decay channel in a metal was given by Tkalya in 1999 [33, 83]. Here the isomeric state couples to the conduction electrons of the metal and, if the work function of the metal is below the isomer's energy, electrons could leave the surface. Also the potential to infer an energy value by varying the substrate material was pointed out [83].

The obvious advantage in the search for an IC decay channel is that experimental conditions, under which IC will dominate, can easily be prepared. Therefore one does not have to care about other potential decay branches, which are not covered experimentally, as is the case in the search for radiative decay. After 2007, experiments in the search for an IC decay channel were carried out at the Lawrence Livermore National Laboratory (LLNL), Livermore, USA [138, 182]. These experiments covered the huge isomeric half-life region of 13 orders of magnitude between $4 \cdot 10^{-8}$ and $2 \cdot 10^5$ seconds. The experiments will be shortly discussed in the following.

Three different experimental techniques were applied in order to cover the enormous half-life region [138]. In order to search for half lives from 1 second to two days, a moving catcher method was applied. In this approach, ^{229}Th α-recoil isotopes are implanted into a catcher foil. Following the implantation, the foil is moved in front of an MCP detector in order to search for any electron emission. The half-life that can be probed by this method is limited by the time it takes to move the catcher foil.

In order to also probe for shorter half-lives between 2 ms and 4.5 s, a mechanical shutter method was used. In this experiment, the ^{233}U source, used for α-recoil production, was placed on one side of a mechanical shutter and the catcher foil on the other. The lower half-life limit in this method is determined by the shutter closing time.

The third method is an α-coincidence method, allowing to probe for half-lives between 40 ns and 1 ms. In this approach, a ^{233}U-coated mylar foil was used as a source. A silicon detector was placed on one side and a catcher foil on the other side of the source. The MCP was placed under an angle with a direct line-of-sight to the catcher foil. The MCP acquisition was triggered in accordance with the ^{233}U decays, which could be recorded with the help of the Si detector.

None of the presented experiments has led to the detection of an isomeric IC decay signal. Given today's knowledge about the isomer's half-life of about 7 μs [95], it is clear that only the last method would have conceptionally allowed for the isomer detection. One reason that the experiment failed in observing the isomeric decay might have been the kinetic energy of ^{229}Th-recoil isotopes of about 84 keV. This kinetic energy leads to an implantation depth of the recoil nuclei of several atomic layers into the catcher foil, which may have hindered the IC decay electrons from effectively leaving the catcher foil. While this is just speculation, there are plans to repeat an experiment of similar type at the Technical University of Vienna.

A first successful experiment, leading to the detection of the IC decay channel of neutral ^{229m}Th, was recently presented by our group [94, 183]. While also populating the isomeric state via the 2% decay branch in the α decay of ^{233}U, a low energy, purified ^{229}Th ion beam of α-recoil isotopes was produced. This was achieved by stopping the α-recoil ions with the help of a buffer-gas stopping-cell [184, 185]. The low energy ion beam was directly accumulated on the surface of a micro-channel plate (MCP) detector. Charge capture on the MCP surface leads to neutralization of the ions, thus triggering the IC decay of the isomeric state. The electrons produced in this process were detected by the MCP. Based on this detection method the isomeric half-life under IC decay could be determined on a nickel alloy surface to be 7 ± 1 μs [95]. Further, recently a new direct laser excitation scheme for ^{229m}Th was proposed, which would allow to pin down the isomeric energy value to 10^{-6} eV [186]A new proposal, based on the same detection method, was recently published [187].

3.7 Further Experimental Investigations

There are further ongoing experimental investigations that do not aim for a direct detection of the ^{229m}Th isomeric decay. These include an improved determination of the isomer's energy using state-of-the-art magnetic microcalorimeters, probing the ^{229m}Th hyperfine structure and the investigation of the isomeric state in ion storage rings.

3.7.1 *An Improved Energy Determination*

By today, the most accepted isomeric energy value was determined to 7.8 eV using a cryogenic microcalorimeter [19, 20]. The resolution of this device was in the range of 30 eV full-width at half-maximum (FWHM), therefore not allowing for the individual resolution of the decay lines from the 29 keV second ^{229}Th excitation level to the ground and excited state, respectively. In the meantime, the technology of cryogenic microcalorimeters has undergone considerable development, which improved the energy resolution by a factor of 10, reaching 3 eV FWHM and below. A state-of-the-art microcalorimeter could therefore directly resolve these closely spaced γ-ray lines around 29 keV and thus open the way for the determination of a significantly improved isomeric energy value. Such an experiment was proposed in 2014 by Kazakov et al. [188]. It can be seen as an improved version of the Beck et al. measurement from 2007 [19] and is currently prepared at the Kirchhoff-Institut für Physik in Heidelberg [189].

The approach of an improved energy determination via indirect methods is an alternative to the investigation of the isomeric state by its direct detection. The energy determination via the direct decay detection of ^{229m}Th making use of the microcalorimetric detection technique is envisaged at the Lawrence-Livermore National Laboratory. Successful observation of the ^{235m}U decay has already been reported [190, 191]. An energy determination based on the IC decay channel, using high-resolution electron spectroscopy, was proposed by our group and is currently investigated [183, 192].

Recently also different approaches for exploring the isomer's excitation energy were proposed by the Moscow collaboration. Here the ideas are to perform low-energy ion scattering spectroscopy on ^{229}Th surfaces [193] or to excite the isomeric state via electron-beam surface irradiation [194]. Corresponding preparatory chemical surface studies with ^{229}Th have already been performed [193–197].

3.7.2 *Probing the ^{229m}Th Hyperfine Structure*

It was proposed by Tordoff et al. in 2006 to probe the ^{229m}Th hyperfine structure via collinear laser spectroscopy [198]. Such an experiment would not only give further evidence of the isomer's existence, providing an alternative to its direct detection, but it is also an important step towards a nuclear clock. In the nuclear clock concept, as proposed in 2003, it is foreseen to probe for the isomer's excitation by the double-resonance method, which requires the knowledge of the isomer's hyperfine structure [24]. Also important information about the nuclear structure could be inferred, which is a requirement for a quantitative estimate of the sensitivity-enhancement factor predicted to occur when using ^{229m}Th to probe for potential time variations of fundamental constants [48]. In 2014 it was emphasized that even an improved isomeric energy value could be inferred by probing the ^{229}Th^{3+} hyperfine structure [199].

Strong efforts in the direction of thorium collinear laser spectroscopy are made by a collaboration of the University of Jyväskylä and the University of Mainz. In this approach, ^{229}Th ions are extracted from a ^{233}U source by a buffer-gas stopping-cell in order to form an ion beam [200, 201]. Experimental results include the detection of 20 previously unknown states in the ^{232}Th level scheme, as well as numerous auto-ionizing states [57] and the measurement of the ground-state hyperfine structure in neutral ^{229}Th [202]. An experimental overview can be found in Ref. [203]. A further proposal along this line was made in 2015. Here it was proposed to probe the thorium hyperfine structure in Th^{+} ions [204].

Very recently a successful observation of the ^{229m}Th hyperfine structure was reported by a collaboration of the PTB and our group [205]. This observation allowed for a determination of the nuclear moments, isomeric shifts, nuclear charge radius and a confirmation of the 2% population efficiency in the ^{233}U α decay.

3.7.3 The Search for ^{229m}Th at Storage Rings

Several theoretical proposals to investigate hydrogen-like or muonic ^{229}Th at storage rings can be found in the literature (e.g. Refs. [101–103, 105]). Currently, one experiment along this line is in preparation. In this experiment, nuclear excitation by electron capture (NEEC), which is the reverse of the internal conversion process, is used to investigate the transition energy of ^{229m}Th [206–208]. In this process, ^{229}Th ions are stored in a high-energy storage ring (like, e.g., the ESR at GSI in Germany or the CSR at the IMP in Lanzhou/China). When these ions catch electrons that fulfill the resonance condition, namely that the electron's kinetic energy plus their binding energy after recombination equals the energy of the isomeric state, there is an enhanced probability for exciting ^{229}Th into its isomeric state [209, 210]. By tuning the energy of an electron beam and monitoring the number of recombinations by detecting the ions' charge states, it is possible to find the resonance and thus to determine the isomer's energy. For shell processes, the method is known as dielectronic recombination (DR) [206]. Successful excitation and investigation of nuclear isomeric states was performed for ^{234m}Pa^{88+} and ^{235m}U^{89+} [207].

References

1. Kroger LA, Reich CW (1976) Features of the low energy level scheme of ^{229}Th as observed in the α decay of ^{233}U. Nucl Phys A 259:29
2. Kroger LA (1972) Level-scheme studies in ^{229}Th and ^{231}Th from the alpha-decay of ^{233}U and ^{235}U. Ph.D. thesis, University of Wyoming, USA
3. Baranov SA, Gadzhiev MK, Kulakov VM, Shatinskii VM (1967) New data on the α decay of U^{233}. Yad Fiz 5:518; Sov J Nucl Phys 5:365
4. Burke DG et al (1990) Additional evidence for the proposed excited state at ≤ 5 eV in ^{229}Th. Phys Rev C 42:499
5. Canty MJ, Connor RD, Doha DA, Pople Eb (1977) The decay of ^{233}U. J Phys G Nucl Phys 3:421
6. Barci V et al (2003) Nuclear structure of ^{229}Th from γ-ray spectroscopy study of ^{233}U α-particle decay. Phys Rev C 68:034329
7. Burke DG et al (2008) Nuclear structure of ^{229,231}Th studied with the ^{230,232}Th(d, t) reactions. Nucl Phys A 809:129–170
8. Bemis CE et al (1988) Coulomb excitation of states in ^{229}Th. Phys Scr 38:657–663
9. Chayawattanangkur K, Herrmann G, Trautmann N (1973) Heavy isotopes of actinium: ^{229}Ac, ^{230}Ac, ^{231}Ac, ^{232}Ac. J Inorg Nucl Chem 35:3061
10. Gulda K et al (2002) The nuclear structure of ^{229}Th. Nucl Phys A 703:45
11. Ruchowska E et al (2006) Nuclear structure of ^{229}Th. Phys Rev C 73:044326
12. Kotthaus T et al (2012) Energy splitting between the first two quasi-particle states in ^{230}Pa in comparison with the isotone ^{229}Th. Phys Lett B 718:460–464
13. Reich CW, Helmer R (1990) Energy separation of the doublet of intrinsic states at the ground state of ^{229}Th. Phys Rev Lett 64:271
14. Reich CW (1990) An interesting finding in ^{229}Th. In: Invited paper presented at the 40th all-union conference on nuclear spectroscopy and nuclear structure, Leningrad, USSR, 10–13 April 1990. https://www.osti.gov/scitech/servlets/purl/6159406
15. Reich CW, Helmer RG, Baker JD, Gehrke RJ (1984) Emission probabilities and energies of γ-ray transitions from the decay of ^{233}U. Int J Appl Radiat Isot 35:185–188
16. Reich CW, Helmer RG (1993) Proceedings of the international symposium on nuclear physics of our times, World Scientific, Singapore, p 474
17. Helmer R, Reich CW (1994) An excited state of ^{229}Th at 3.5 eV. Phys Rev C 49:1845
18. Guimaraes-Filho ZO, Helene O (2005) Energy of the $3/2^{+}$ state of ^{229}Th reexamined. Phys Rev C 71:044303
19. Beck BR et al (2007) Energy splitting of the ground-state doublet in the nucleus ^{229}Th. Phys Rev Lett 109:142501
20. Beck BR et al (2009) Improved value for the energy splitting of the ground-state doublet in the nucleus ^{229m}Th. LLNL-PROC-415170
21. Sakharov SL (2010) On the energy of the 3.5 eV level in ^{229}Th. Phys At Nucl 73:1
22. Kazakov GA et al (2017) Re-evaluation of the Beck et al data to constrain the energy of the ^{229}Th isomer. arXiv:1702.00749 [nucl-ex]
23. Strizhov VF, Tkalya EV (1991) Decay channel of low-lying isomer state of the ^{229}Th nucleus. Possibilities of experimental investigation. Sov Phys JETP 72:387
24. Peik E, Tamm C (2003) Nuclear laser spectroscopy of the 3.5 eV transition in ^{229}Th. Eur Phys Lett 61:181
25. Campbell CJ et al (2012) Single-Ion nuclear clock for metrology at the 19th decimal place. Phys Rev Lett 108:120802
26. Flambaum VV (2006) Enhanced effect of temporal variation of the fine structure constant and the strong interaction in ^{229}Th. Phys Rev Lett 97:092502
27. Oganessian YT, Karamian SA (1995) On the question of a gamma-ray laser on nuclear levels. Laser Phys 5:336–342
28. Tkalya EV (2011) Proposal for a nuclear gamma-ray laser of optical range. Phys Rev Lett 106:162501

29. Tkalya EV, Varlamov VO, Lomonosov VV, Nikulin SA (1996) Processes of the nuclear isomer ^{229m}Th(3/2$^+$, 3.5$\pm$1.0 eV) resonant excitation by optical photons. Phys Scr 53:296–299
30. Tkalya EV (1992) Cross section for excitation of the low-lying ($\leq$ 5 eV) ^{229}Th isomer with laser radiation by the inverse electron bridge. Sov J Nucl Phys 55:1611–1617
31. Karpeshin FF et al (1999) 3.5 eV isomer of ^{229m}Th: how it can be produced. Nucl Phys A 654:579–596
32. Safronova MS et al (2013) Magnetic dipole and electric quadrupole moments of the ^{229}Th nucleus. Phys Rev A 88:060501
33. Tkalya EV et al (2000) Decay of the low-energy nuclear isomer ^{229}Th^m(3/2$^+$, 3.5$\pm$1.0 eV) in solids (dielectrics and metals): a new scheme of experimental research. Phys Rev C 61:064308
34. Peik E et al (2009) Prospects for a nuclear optical frequency standard based on thorium-229. In: Proceedings of 7th symposium on frequency standards and metrology, 5–11 October 2008. pp 532–538
35. Rellergert WG et al (2010) Constraining the evolution of the fundamental constants with a solid-state optical frequency reference based on the ^{229}Th nucleus. Phys Rev Lett 104:200802
36. Kazakov GA et al (2012) Performance of a ^{229} Thorium solid-state nuclear clock. New J Phys 14:083019
37. Peik E, Okhapkin M (2015) Nuclear clocks based on resonant excitation of γ-transitions. C R Phys 16:516–523
38. Uzan JP (2003) The fundamental constants and their variation: observational and theoretical status. Rev Mod Phys 75:403
39. Flambaum VV (2007) Variation of fundamental constants: theory and observations. Int J Mod Phys A 22:4937–4950
40. Hayes AC, Friar JL (2007) Sensitivity of nuclear transition frequencies to temporal variation of the fine structure constant or the strong interaction. Phys Lett B 650:229–232
41. Hayes AC et al (2008) Splitting sensitivity of the ground and 7.6 eV isomeric states of ^{229}Th. Phys Rev C 78:024311
42. He X, Ren Z (2008) Temporal variation of the fine structure constant and the strong interaction parameter in the ^{229}Th transition. Nucl Phys A 806:117–123
43. He XT, Ren ZZ (2007) Enhanced sensitivity to variation of fundamental constants in the transition of Th-229 and Bk-249. J Phys G Nucl Part Phys 37:1611–1619
44. He XT, Ren ZZ (2008) Enhanced sensitivity to variation of fundamental constants in the narrow nuclear transitions (II). J Phys G Nucl Part Phys 35:035106
45. Flambaum VV, Wiringa RB (2009) Enhanced effect of quark mass variation in Th229 and limits from Oklo data. Phys Rev C 79:034302
46. Flambaum VV et al (2009) Coulomb energy contribution to the excitation energy in ^{229}Th and enhanced effect of α variation. Europhys Lett 85:50005
47. Litvinova E et al (2009) Nuclear structure of lowest ^{229}Th states and time-dependent fundamental constants. Phys Rev C 79:064303
48. Berengut JC et al (2009) Proposed experimental method to determine α sensitivity of splitting between ground and 7.6 eV isomeric states in ^{229}Th. Phys Rev Lett 102:210801
49. Feldmeier H et al (2015) Variation of fundamental constants and ^{229}Th. In: WSPC Proceedings
50. Berengut JC, Flambaum VV (2010) Testing time-variation of fundamental constants using a ^{229}Th nuclear clock. Nucl Phys News 20:19–22
51. Berengut JC, Flambaum VC (2012) Manifestation of a spatial variation of fundamental constants in atomic and nuclear clocks, Oklo, meteorites, and cosmological phenomena. EPL 97:20006
52. Flambaum VV (2016) Enhancing the effect of Lorentz invariance and Einstein's equivalence principle violation in nuclei and atoms. Phys Rev Lett 117:072501
53. Rivlin LA (2007) Nuclear gamma-ray laser: the evolution of the idea. Quantum Electron 37:723–744
54. Karpeshin FF et al (1996) Optical pumping ^{229m}Th through NEET as a new effective way of producing nuclear isomers. Phys Lett B 372:1–7

55. Tkalya EV, Yatsenko LP (2013) Creation of inverse population in the ^{229}Th ground-state doublet by means of a narrowband laser. Laser Phys Lett 10:105808
56. Dykhne AM, Tkalya EV (1998) ^{229m}Th(3/2$^+$, 3.5 eV) and a check of the exponentiality of the decay law. JETP Lett 67:549–552
57. Raeder S et al (2011) Resonance ionization spectroscopy of thorium isotopes - towards a laser spectroscopic identification of the low-lying 7.6 eV isomer of ^{229}Th. J Phys B 44:165005
58. Liakos JK (2009) Thorium-229m1/m2-powered heterojunction GRPVCs: an interface between nuclear and semiconductor physics. Semicond Sci Technol 24:065001
59. Safronova M (2016) Elusive transition spotted in thorium. Nature 533:44–45
60. Piotrowski M, Gensemer S (2016) A network of nuclear atomic clocks for precision gravity measurement. In: Joint 13th Asia pacific physics conference and 22nd Australian institute of physics congress (APPC-AIP) abstracts, Brisbane Convention and Exhibition Centre
61. Kolkowitz S et al (2016) Gravitational wave detection with optical lattice atomic clocks. Phys Rev D 94:124043
62. Derevianko A, Pospelov M (2014) Hunting for topological dark matter with atomic clocks. Nat Phys 10:933–936
63. Krutov VA, Fomenko VN (1968) Influence of electronic shell on gamma radiation of atomic nuclei. Ann Phys 21:291
64. Crasemann B (1973) Some aspects of atomic effects in nuclear transitions. Nucl Instrum Methods 112:33–39
65. Krutov VA (1990) Internal conversion in the field of an "electronic bridge". JETP Lett 52:584–588
66. Zon BA, Karpeshin FF (1990) Acceleration of the decay of $^{235\mathrm{m}}$U by laser-induced resonant internal conversion. Sov Phys JETP 70:224–227
67. Tkalya EV (1992) Excitation of low-lying isomer level of the nucleus ^{229}Th by optical photons. JETP Lett (USSR) 55:211
68. Tkalya EV (1992) Probability of nonradiative excitation of nuclei in transitions of an electron in an atomic shell. Sov Phys JETP 75:200–209
69. Karpeshin FF et al (1992) Study of ^{229}Th through laser-induced resonance internal conversion. Phys Lett B 282:267–270
70. Typel S, Leclercq-Willain C (1996) Nuclear excitation by laser-assisted electronic transitions. Phys Rev A 53:2547–2561
71. Kálmán P, Keszthelyi T (1994) Laser-driven inverse electronic bridge process: an optical tool for determining low-energy separation of nearby nuclear states. Phys Rev C 49:324–328
72. Kálmán P, Bükki T (2001) Deexcitation of ^{229}Th^m: direct γ decay and electronic-bridge process. Phys Rev C 63:027601
73. Tkalya EV et al (1996) Various mechanisms of the resonant excitation of the isomeric level ^{229m}Th ((3/2)$^+$, 3.5 eV) by photons. Phys Atomic Nuclei 59:779
74. Matinyan S (1998) Lasers as a bridge between atomic and nuclear physics. Phys Rep 298:199
75. Karpeshin FF, Trzhaskovskaya MB (2017) Bound internal conversion versus nuclear excitation by electron transition: revision of the theory of optical pumping of the $^{229\mathrm{m}}$Th isomer. Phys Rev C 95:034310
76. Band IM (2001) Discrete conversion of gamma rays in ^{229}Th and in highly ionized ^{125}Te^{Q} ions. J Exp Theor Phys 93:948–956
77. Karpeshin FF, Trzhaskovskaya MB (2005) Resonance conversion as the predominant decay mode of ^{229m}Th. Hyperfine Interact 162:125–132
78. Karpeshin FF, Trzhaskovskaya MB (2006) Resonance conversion as a dominant decay mode for the 3.5-eV isomer in ^{229m}Th. Phys At Nucl 69:571
79. Karpeshin FF, Trzhaskovskaya MB (2007) Impact of the electron environment on the lifetime of the ^{229}Th^m low-lying isomer. Phys Rev C 76:054313
80. Tkalya EV et al (2015) Radiative lifetime and energy of the low-energy isomeric level in ^{229}Th. Phys Rev C 92:054324
81. Tkalya EV et al (2017) Erratum: Radiative lifetime and energy of the low-energy isomeric level in ^{229}Th. Phys Rev C 95:039902(E); Phys Rev C 92:054324 (2015)

82. Minkov N, Pálffy A (2017) Reduced transition probabilities for the γ-decay of the 7.8 eV isomer in ^{229}Th. Phys Rev Lett 118:212501
83. Tkalya EV (1999) Nonradiative decay of the low-lying nuclear isomer ^{229m}Th (3.5eV) in a metal. JETP Lett 70:371–374
84. Tkalya EV (2000) Spontaneous emission probability for M1 transition in a dielectric medium: ^{229m}Th(3/2$^+$, 3.5$\pm$1.0 eV) decay. JETP Lett 71:311
85. Koltsov VV (2000) Effect of dielectric properties of a medium on the probability of the isomeric transition with energy 3.5 eV in the ^{229m}Th nucleus. Bull Rus Acad Sci Phys 64:447
86. Tkalya EV (2003) Properties of the optical transition in the ^{229}Th nucleus. Phys Usp 46:315–324
87. Porsev SG, Flambaum VV (2010) Effect of atomic electrons on the 7.6 eV nuclear transition in ^{229m}Th^{3+}. Phys Rev A 81:032504
88. Müller RA et al (2017) Theoretical anaylsis of the electron bridge process in ^{229}Th^{3+}. Nucl Instrum Methods B 408:84–88
89. Porsev SG, Flambaum VV (2010) Electronic bridge process in ^{229}Th$^+$. Phys Rev A 81:042516
90. Porsev SG et al (2010) Excitation of the isomeric ^{229m}Th nuclear state via an electronic bridge process in ^{229}Th$^+$. Phys Rev Lett 105:182501
91. Romanenko VI et al (2012) Direct two-photon excitation of isomeric transition in thorium-229 nucleus. Ukr J Phys 57:1119–1131
92. Karpeshin FF, Trzhaskovskaya MB (2015) Excitation of the ^{229m}Th nuclear isomer via resonance conversion in ionized atoms. Phys Atom Nucl 78:715–719
93. Karpeshin FF, Trzhaskovskaya MB (2016) The ^{229}Th isomer line as a reference for a high-precision frequency standard. Meas Tech 85:29–35
94. von der Wense L (2016) On the direct detection of ^{229m}Th. Ph.D. thesis, Ludwig-Maximilians-Universität München, Germany
95. Seiferle B et al (2017) Lifetime measurement of the ^{229}Th nuclear isomer. Phys Rev Lett 118:042501
96. Karpeshin FF, Trzhaskovskaya MB, Vitushkin LF (2017) Impact of the ionization of the atomic shell on the lifetime of the ^{229m}Th isomer. arXiv:1701.05340
97. Bilous PV et al (2017) Internal conversion from excited electronic states of ^{229}Th ions. Phys Rev A 95:032503
98. Bilous PV, Peik E, Pálffy A (2017) Laser induced electronic bridge for characterization of the ^{229m}Th $\rightarrow$ ^{229g}Th nuclear transition with a tunable optical laser. arXiv:1709.04364 [atom-ph]
99. Varlamov VO et al (1996) Excitation of a ^{229m}Th ((3/2)$^+$, 3.5 eV) isomer by surface plasmons. Phys Doklady 41:47
100. Dykhne AM et al (1996) Alpha decay of the first excited state of the Th-229 nucleus. JETP Lett 64:345–349
101. Karpeshin FF et al (1998) Rates of transitions between the hyperfine-splitting components of the ground-state and the 3.5 eV isomer in ^{229}Th^{89+}. Phys Rev C 57:3085
102. Wycech S, Zylicz J (1993) Predictions for nuclear spin mixing in magnetic fields. Acta Phys Pol B 24:637
103. Pachucki K et al (2001) Nuclear spin mixing oscillations in ^{229}Th^{89+}. Phys Rev C 64:064301
104. Tkalya EV (2016) Anomalous magnetic hyperfine structure of the ^{229}Th ground-state doublet in muonic atoms. Phys Rev A 94:012510
105. Tkalya EV, Nikolaev AV (2016) Magnetic hyperfine structure of the ground-state doublet in highly charged ions ^{229}Th$^{89+,87+}$ and the Bohr–Weisskopf effect. Phys Rev C 94:014323
106. Tkalya EV (2017) Low-energy E0 transition between the components of the ground-state doublet in the muonic atom ^{229}Th. Phys Rev A 95:042512
107. Alexandrov VS et al (2011) Nuclear transitions and new standards of length and time. Acta Phys Pol B 42:853–858
108. Isakov VI (2017) On the properties of the ^{229}Th isotope. arXiv:1706.02417 [nucl-th]
109. Bürvenich TJ et al (2006) Nuclear quantum optics with X-ray laser pulses. Phys Rev Lett 96:142501

110. Pálffy A et al (2008) Electric-dipole-forbidden nuclear transitions driven by super-intense laser fields. Phys Rev C 77:044602
111. Liao WT (2013) Coherent control of nuclei and X-rays. Ph.D. thesis, University of Heidelberg
112. Liao WT et al (2012) Coherence-enhanced optical determination of the ^{229}Th isomeric transition. Phys Rev Lett 109:262502
113. Das S et al (2013) Quantum interference effects in an ensemble of ^{229}Th nuclei interacting with coherent light. Phys Rev C 88:024601
114. Liao WT, Pálffy A (2017) Optomechanically induced transparency of x-rays via optical control. Sci Rep 7:321
115. Irwin GM, Kim KH (1997) Observation of electromagnetic radiation from deexcitation of the ^{229}Th isomer. Phys Rev Lett 79:990–993
116. Dykhne AM, Tkalya EV (1998) Matrix element of the anomalously low-energy (3.5±0.5 eV) transition in ^{229}Th and the isomer lifetime. JETP Lett 67:251
117. Karpeshin FF et al (1999) On the question of electron bridge for the 3.5 eV isomer of ^{229}Th. Phys Rev Lett 83:1072
118. Karpeshin FF et al (1999) Role of the electron bridge in de-excitation of the 3.5 eV ^{229}Th isomer. Bull Rus Acad Sci Phys 63:30
119. Richardson DS et al (1998) Ultraviolet photon emission observed in the search for the decay of the ^{229}Th isomer. Phys Rev Lett 80:3206–3208
120. Utter SB et al (1999) Reexamination of the optical gamma ray decay in ^{229}Th. Phys Rev Lett 82:505–508
121. Shaw RW et al (1999) Spontaneous ultraviolet emission from ^{233} Uranium/^{229} Thorium samples. Phys Rev Lett 82:1109–1111
122. Young JP et al (1999) Radioactive origin of emissions observed from uranium compounds and their silica cells. Inorg Chem 38:5192–5194
123. Irwin GM, Kim KH (1999) Irwin and Kim reply. Phys Rev Lett 83:1073
124. Kim KH, Irwin GM (2011) Excitation of the low-lying isomeric level in ^{229}Th. Fizika B (Zagreb) 20:319–330
125. Mitsugashira T et al (2003) Alpha-decay from the 3.5 eV isomer of ^{229}Th. J Radioanal Nucl Chem 255:63–66
126. Kikunaga H et al (2005) Search for α-decay of ^{229m}Th produced from ^{229}Ac β-decay following ^{232}Th (γ, $p2n$) reaction. Radiochim Acta 93:507–510
127. Kikunaga H (2007) Search for an isomer state of ^{229}Th with extremely low energy using alpha-spectrometry. Ph.D. thesis, Kanazawa University, Japan
128. Browne E et al (2001) Search for decay of the 3.5-eV level in ^{229}Th. Phys Rev C 64:014311
129. Kasamatsu Y et al (2005) Search for the decay of ^{229m}Th by photon detection. Radiochim Acta 93:511–514
130. Kikunaga H et al (2009) Half-life estimation of the first excited state of ^{229}Th by using α-particle spectrometry. Phys Rev C 80:034315
131. Inamura TT, Karpeshin FF, Trzhaskovskaya MB (2003) Bridging atomic and nuclear states in ^{229}Th. Czech J Phys Suppl B 53:349
132. Inamura TT (2005) Laser methods in the study of nuclei, atoms and molecules. Phys Scr 71:C1–C4
133. Inamura TT, Mitsugashira T (2005) Pumping ^{229m}Th by hollow-cathode discharge. Hyperfine Interact 162:115–123
134. Inamura TT, Haba H (2009) Search for a "3.5-eV isomer" in ^{229}Th in a hollow-cathode electric discharge. Phys Rev C 79:034313
135. Moore ID et al (2004) Search for a low-lying 3.5-eV isomeric state in ^{229}Th. ANL Phys Dev Rep PHY-10990-ME-2004
136. Zimmermann K (2010) Experiments towards optical nuclear spectroscopy with thorium-229. Ph.D. thesis, University of Hannover, Germany
137. Burke JT et al (2010) ^{229}Th the bridge between nuclear and atomic interactions. Lawrence Livermore National Laboratory Technical report LLNL-TR-463538

138. Swanberg E (2012) Searching for the decay of ^{229m}Th. Ph.D. thesis, University of California, Berkeley
139. Zhao X et al (2012) Observation of the deexcitation of the ^{229m}Th nuclear isomer. Phys Rev Lett 109:160801
140. Thirolf PG et al (2007) Optical access to the lowest nuclear transition in ^{229m}Th. Annual Report of the Maier–Leibnitz Laboratory, Garching, p 18
141. von der Wense L et al (2013) Towards a direct transition energy measurement of the lowest nuclear excitation in ^{229}Th. JINST 8:P03005
142. Thirolf PG et al (2013) Towards an all-optical access to the lowest nuclear excitation in ^{229m}Th. Acta Phys Pol B 44:391–394
143. Seiferle B (2015) Setup of a VUV detection system for the direct identification of the fluorescence radiation of ^{229m}Th. Master thesis, Ludwig-Maximilians-Universität München, Germany
144. Seiferle B et al (2016) A VUV detection system for the direct photonic identification of the first excited isomeric state of ^{229}Th. Eur Phys J D 70:58
145. Bilous PV, Yatsenko LP (2015) Analysis of parasitic signals in the method of recoil nuclei applied to direct observation of the ^{229}Th isomeric state. Ukr J Phys 60:376–381
146. Stellmer S et al (2016) Feasibility study of measuring the ^{229}Th nuclear isomer transition with ^{233}U-doped crystals. Phys Rev C 94:014302
147. Peik E, Zimmermann K (2013) Comment on "Observation of the deexcitation of the ^{229m}Th nuclear isomer". Phys Rev Lett 111:018901
148. Hehlen MP et al (2013) Optical spectroscopy of an atomic nucleus: progress toward direct observation of the ^{229}Th isomer transition. J Lumin 133:91–95
149. Jackson RA et al (2009) Computer modelling of thorium doping in $LiCaAlF_6$ and $LiSrAlF_6$: application to the development of solid state optical frequency devices. J Phys Condens Matter 21:325403
150. Rellergert WG et al (2010) Progress towards fabrication of ^{229}Th-doped high energy band-gap crystals for use as a solid-state optical frequency reference. IOP Conf Ser Mater Sci Eng 15:012005
151. Kazakov GA et al (2013) Atomic clock with nuclear transition: current status in TU Wien. arXiv:1110.0741
152. Dessovic P et al (2014) ^{229} Thorium-doped calcium fluoride for nuclear laser spectroscopy. J Phys Condens Matter 26:105402
153. Ellis JK et al (2014) Investigation of thorium salts as candidate materials for direct observation of the ^{229m}Th nuclear transition. Inorg Chem 53:6769–6774
154. Stellmer S et al (2015) Radioluminescence and photoluminescence of Th:CaF_2 crystals. Sci Rep 5:15580
155. Jeet J et al (2015) Results of a direct search using synchrotron radiation for the low-energy ^{229}Th nuclear isomeric transition. Phys Rev Lett 114:253001
156. Yamaguchi A et al (2015) Experimental search for the low-energy nuclear transition in ^{229}Th with undulator radiation. New J Phys 17:053053
157. Borisyuk PV et al (2015) Band structure and decay channels of thorium-229 low lying isomeric state for ensemble of thorium atoms adsorbed on calcium fluoride. Phys Status Solidi C 12:1333–1337
158. Stellmer S et al (2016) Towards a measurement of the nuclear clock transition in ^{229}Th. J Phys Conf Ser 723:012059
159. Gangrsky YuP et al (2005) Search for light radiation in decay of ^{229}Th isomer with anomalously low excitation energy. Bull Rus Acad Sci Phys 69:1857
160. van Duppen P et al (2017) Characterization of the low-energy ^{229m}Th isomer. Letter of intent to the ISOLDE and Neutron Time-of-flight Committee. http://cds.cern.ch/record/2266840
161. Dzuba VA, Flambaum VV (2010) Exponential increase of energy level density in atoms: Th and Th II. Phys Rev Lett 104:213002
162. Safronova MS, Safronova UI, Clark W (2014) Relativistic all-order calculations of Th, Th^+, and Th^{2+} atomic properties. Phys Rev A 90:032512

163. Zimmermann K, Okhapkin MV, Herrera-Sancho OA, Peik E (2012) Laser ablation loading of a radiofrequency ion trap. Appl Phys B 107:883–889
164. Steele AV (2008) Barium ion cavity QED and triply ionized thorium ion trapping. Ph.D. Thesis, Georgia Institute of Technology, Georgia
165. Campbell CJ et al (2009) Multiply charged thorium crystals for nuclear laser spectroscopy. Phys Rev Lett 102:233004
166. Campbell CJ et al (2011) Wigner Crystals of ^{229}Th for optical excitation of the nuclear isomer. Phys Rev Lett 106:223001
167. Campbell CJ (2011) Trapping, laser cooling, and spectroscopy of thorium IV. Ph.D. thesis, Georgia Institute of Technology, USA
168. Churchill LR, DePalatis MV, Chapman MS (2011) Charge exchange and chemical reactions with trapped Th^{3+}. Phys Rev A 83:012710
169. Radnaev AG (2012) Towards quantum telecommunication and a thorium nuclear clock. Ph.D. thesis, Georgia Institute of Technology, USA
170. Herrera-Sancho OA et al (2012) Two-photon laser excitation of trapped $^{232}Th^{+}$ ions via the 402-nm resonance line. Phys Rev A 85:033402
171. Herrera-Sancho OA (2012) Laser excitation of 8-eV electronic states in Th^{+}: a first pillar of the electronic bridge toward excitation of the Th-229 nucleus. Ph.D. thesis, University of Hannover, Germany
172. Herrera-Sancho OA et al (2013) Energy levels of Th^{+} between 7.3 and 8.3 eV. Phys Rev A 88:012512
173. Radnaev AG, Campbell CJ, Kuzmich A (2012) Observation of the 717-nm electric quadrupole transition in triply charged thorium. Phys Rev A 86:060501
174. Okhapkin MV et al (2015) Observation of an unexpected negative isotope shift in $^{229}Th^{+}$ and its theoretical explanation. Phys Rev A 92:020503
175. Troyan VI et al (2014) Quadrupole Paul ion trap in the multi-charged thorium ions optical spectroscopy system for development of nuclear frequency standard. Meas Tech 57:777
176. Troyan VI et al (2015) The development of nuclear frequency standard with the use of ion crystal manipulation system. Phys Procedia 72:245–248
177. Troyan VI et al (2015) Multisectional linear ion trap and novel loading method for optical spectroscopy of electron and nuclear transitions. Eur J Mass Spectrom 21:1–12
178. Borisyuk PV et al (2017) Trapping, retention and laser cooling of Th^{3+} ions in a multisectional linear quadrupole trap. Quantum Electron 47:406–411
179. Borisyuk PV et al (2017) Mass selective laser cooling of $^{229}Th^{3+}$ in a multisectional linear Paul trap loaded with a mixture of thorium isotopes. Eur J Mass Spectrom 23:136–139
180. Troyan VI et al (2013) Generation of thorium ions by laser ablation and inductively coupled plasma techniques for optical nuclear spectroscopy. Laser Phys Lett 10:105301
181. Vorykhalov OV, Koltsov VV (1995) Search for an isomeric transition of energy below 5 eV in ^{229}Th nucleus. Bull Rus Acad Sci Phys 59:20–24
182. Wakeling MA (2014) Charge states of ^{229m}Th: path to finding the half-life. Thesis performed at LLNL, Livermore
183. von der Wense L et al (2016) Direct detection of the ^{229}Th nuclear clock transition. Nature 533:47–51
184. von der Wense L et al (2015) Determination of the extraction efficiency for ^{233}U source α-recoil ions from the MLL buffer-gas stopping cell. Eur Phys J A 51:29
185. von der Wense L et al (2016) The extraction of $^{229}Th^{3+}$ from a buffer-gas stopping cell. Nucl Instrum Methods B 376:260–264
186. von der Wense L et al (2017) A laser excitation scheme for ^{229m}Th. Phys Rev Lett 119:132503
187. Gusev YuI et al (2016) Studying the decay of thorium-229 isomer by means of conversion spectroscopy. Bull Rus Acad Sci Phys 80:875–879
188. Kazakov GA et al (2014) Prospects for measuring the ^{229}Th isomer energy using a metallic magnetic microcalorimeter. Nucl Instrum Methods A 735:229–239
189. Schneider P (2016) Spektroskopische Messungen an Thorium-229 mit einem Detektor-Array aus metallischen magnetischen Kalorimetern. Master thesis, Ruprecht-Karls-University, Heidelberg, German

190. Friedrich S (2016) The world's lowest nuclear state in thorium-229m. LDRL Annual Report 2015, Lawrence Livermore National Laboratory. https://ldrd-annual.llnl.gov/ldrd-annual-2015/nuclear/lowest. Accessed 12 Nov 2016
191. Ponce F (2017) High accuracy measurements of the nuclear decay of U-235m and search for the nuclear decay of Th-229m. Ph.D. thesis, University of California, US
192. Seiferle B et al (2017) Feasibility study of internal conversion electron spectroscopy of ^{229m}Th. Eur Phys J A 53:108
193. Borisyuk PV et al (2015) Preparation technique of thorium films by electrochemical deposition for nuclear optical frequency standard based on thorium-229. J Sol-Gel Sci Technol 73:580–585
194. Troyan VI et al (2015) Local electrochemical depostion of thorium on SiO_2/Si(111) surface. Phys Procedia 72:179–183
195. Borisyuk PV et al (2014) The band structure of submonolayer thorium coatings on a silicon oxide surface. Colloid J 76:645–650
196. Borisyuk PV et al (2014) The formation of submonolayer thorium coatings on a silicon oxide surface by electrochemical deposition. Colloid J 76:421–514
197. Troyan VI et al (2014) Experimental treatments for the investigation of nuclear optical transition in thorium-229. IEEE Eur Freq Time Forum (EFTF) 15615881
198. Tordoff B et al (2006) Investigation of the low-lying isomer in ^{229}Th by collinear laser spectroscopy. Hyperfine Interact. 171:197–201
199. Beloy K (2014) Hyperfine structure in $^{229g}Th^{3+}$ as a probe of the $^{229g}Th \rightarrow {}^{229m}Th$ nuclear excitation energy. Phys Rev Lett 112:062503
200. Tordoff B et al (2006) An ion guide for the production of a low energy ion beam of daughter products of α-emitters. Nucl Instrum Methods B 252:347
201. Sonnenschein V et al (2012) The search for the existence of ^{229m}Th at IGISOL. Eur Phys J A 48:52
202. Sonnenschein V et al (2012) Determination of the ground-state hyperfine structure in neutral ^{229}Th. J Phys B At Mol Opt Phys 45:165005
203. Sonnenschein V (2014) Laser developments and high resolution resonance ionization spectroscopy of actinide elements. Ph.D. thesis, University of Jyväskylä, Finland
204. Dembczynski J, Elantkowska M, Ruczkowski J (2015) Method for detecting the isomeric state $I = (3/2)^+$ in ^{229}Th with laser-induced fluorescence. Phys Rev A 92:012519
205. Thielking J et al (2017) Laser spectroscopic characterization of the nuclear clock isomer ^{229m}Th, submitted for publication. arXiv:1709.05325 [nucl-ex]
206. Brandau C et al (2010) Resonant recombination at ion storage rings: a conceptual alternative for isotope shift and hyperfine studies. Hyperfine Interact. 196:115–127
207. Brandau C et al (2013) Probing nuclear properties by resonant atomic collisions between electrons and ions. Phys Scr T 156:014050
208. Ma X et al (2015) Proposal for precision determination of 7.8 eV isomeric state in ^{229}Th at heavy ion storage ring. Phys Scr T166:014012
209. Pálffy A et al (2007) Isomer triggering via nuclear excitation by electron capture. Phys Rev Lett 99:172502
210. Pálffy A et al (2008) Nuclear excitation by electron capture followed by fast X-ray emission. Phys Lett B 661:330–334

Chapter 4
Experimental Setup

The experiments performed within the scope of this thesis were aiming for a first direct detection and unambiguous identification of the isomeric decay of ^{229m}Th. It was known from theory that several competing decay channels of ^{229m}Th exist [1, 2] (see Sect. 2.2). These include the photonic decay, decay via internal conversion (IC), decay via electronic bridge processes (EB) as well as α decay [1, 3]. The strengths of the individual decay channels were largely unknown and theory predicted a strong dependence of the individual branching ratios on the electronic environment (see e.g. Refs. [4, 5] ^{229m}Th half-life was also predicted to heavily depend on the electronic environment. The predictions span a half-life range of 9 orders of magnitude, reaching from tens of microseconds for neutral ^{229}Th under decay via internal conversion to up to several hours for a purely photonic decay in charged ^{229}Th [2, 6]. The multitude of decay channels and the large potential half-life range have led to a variety of experiments carried out in the search for the isomeric decay.

Two of the four potential ^{229m}Th decay channels were investigated by our group. Starting with the search for a potential photonic decay, it soon became clear that there was little hope for a successful observation of light emitted in the decay of ^{229m}Th. With these experiences, a second set of experiments was carried out in order to investigate the internal conversion decay channel. The latter experiments have led to the successful observation of the ^{229}Th isomeric decay [7].

Following an introduction of the experimental concept in Sect. 4.1, the experimental setup will be explained in detail. The experimental setup can be divided into three parts. The first part, namely the ion-beam formation system, provided the basis for all experiments and was used for the investigation of both decay channels. It will be described in Sect. 4.2. The vacuum ultra-violet (VUV) optical system, detailed in Sect. 4.3, was used for the investigation of the photonic decay channel only. The detection system was, again, the same for both decay channels and will be described in Sect. 4.4. Finally, in the last section, an efficiency estimation will be carried out.

L. Von der Wense, *On the Direct Detection of ^{229m}Th*,
Springer Theses, https://doi.org/10.1007/978-3-319-70461-6_4

4.1 Experimental Concept

The experimental concept has been developed for the purpose of a direct detection of the isomer-to-ground-state transition in ^{229}Th [8, 9]. Obviously, for the detection of any isomeric decay, it is a prerequisite to have ^{229}Th available in its isomeric state. In our experiments the natural population of ^{229m}Th via a 2% decay branch in the α decay of ^{233}U [10] is employed to populate the isomeric state. Several ways of isomer excitation from the ^{229}Th ground state have been discussed in literature, e.g. via direct radiative excitation [11] or the inverse electronic bridge process [12]. However, the advantage of the isomer population via the α decay of ^{233}U is that it is experimentally well under control, in this way reducing the experimental uncertainties. Following the production of ^{229m}Th from a ^{233}U source, ^{229}Th is spatially separated from this source in order to detect the subsequently occuring isomeric decay in a nearly background-free environment. The spatial separation of the population and depopulation of the first excited state of ^{229}Th is of major importance, as this concept allows for the suppression of significant amounts of background signals originating from the radioactive decays occuring in the ^{233}U source.

The spatial separation is experimentally implemented by producing a ^{229}Th ion beam (with a fractional content of ^{229}Th in its isomeric state) from ^{229}Th ions emitted in the α decay of ^{233}U. Having formed an ion beam allows for an efficient mass purification with the help of a quadrupole mass-spectrometer (QMS). This mass purification is required in order to purify the ^{229}Th ion beam from other (short-lived) daughter nuclides, that are contained in the ^{233}U decay chain and therefore also emitted from the source. In this way, a low-energy pure $^{229(m)}$Th ion beam is formed and guided to the spatially separated detection chamber.

A significant fraction of the ^{229}Th α-recoil ions remains charged during the process of ion-beam formation. This is of fundamental importance for the experimental concept, as otherwise the decay of the isomeric state, due to internal conversion, has to be expected to occur within microseconds of lifetime [2], significantly shorter than the time required for ion beam formation of a few ms, leading to a complete loss of thorium in the isomeric state. A conceptual overview of the ^{229}Th ion-beam formation system and the subsequent searches for a photonic as well as an internal-conversion decay of the isomeric state is given in the following.

4.1.1 Concept of the Ion-Beam Formation System

The ion-beam formation system, used for the production of a low-energy purified ^{229}Th ion beam with a fractional content of ^{229}Th in the isomeric state, provides the basis for all further investigations of the isomeric decay. A conceptual sketch of this system is shown in Fig. 4.1.

The system can be roughly divided into three segments: (i) a buffer-gas stopping cell used to thermalize the ^{229}Th α-recoil ions originating from the ^{233}U source,

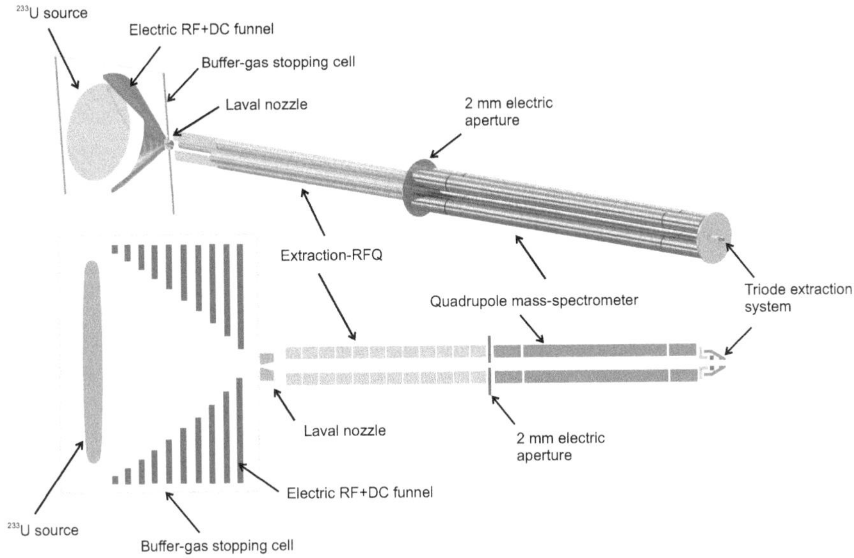

Fig. 4.1 Conceptual sketch of the ion-beam formation system used for the production of a low-energy, purified ^{229}Th ion beam. The availability of this ion beam was the basis for all further research aiming for a direct detection of the ^{229}Th isomeric decay. The upper part of the figure shows a 3-dimensional drawing, where all parts are shown on scale. In the lower part a 2-dimensional sketch is shown, where some of the electrodes were enlarged for better visibility. Further explanation is given in the text

(ii) a radio-frequency quadrupole (RFQ) to produce a low-energy ion beam and (iii) a quadrupole mass-spectrometer (QMS) for beam purification. The buffer-gas stopping cell together with the RFQ were developed and built by J. Neumayr between 2004 and 2006 [13–15]. They were designed for the general purpose of thermalization of high-energy nuclides produced in nuclear reactions. The QMS was built during the work of my PhD thesis to match the special purification requirements of the ^{229}Th ion beam [9, 16], thereby following design values found by E. Haettner [17].

The buffer-gas stopping cell houses the ^{233}U source used for ^{229}Th α-recoil isotope production. The ^{233}U source consists of a thin active layer, which allows the ^{229}Th isotopes, produced in the α decay of ^{233}U, to leave the source due to their kinetic recoil energy of about 84 keV [16, 18]. A significant fraction of the ^{229}Th α-recoil isotopes leaves the source as positively charged ions. The rather high kinetic energy, together with the divergent emission, does not directly allow for ion-beam formation by applying electric fields. Therefore, the recoil ions are first stopped in a buffer gas consisting of 40 mbar of ultra-pure helium. The collisions with the buffer gas do not only stop the ^{229}Th α-recoil nuclides, but also preserve thorium in the 2+ or 3+ charge state, respectively [16]. The small kinetic energy of the ions, stopped by the buffer gas, then allows to apply electric fields for ion guiding.

An RF+DC-funnel structure (Fig. 4.1, red) is used to guide the ions towards the exit of the buffer-gas stopping cell. The funnel structure consists of 50 ring electrodes, each with a different inner diameter, which are conically ordered. To each electrode a DC and an RF voltage is applied. The DC voltages lead to a field gradient, which guides the ions towards the exit of the buffer-gas stopping cell. The RF voltages differ by 180° in phase from electrode to electrode, leading to a repelling force, which prevents the ions from charge capture at the funnel electrodes. In this way the ions are fast and efficiently extracted from the stopping cell, even when being stopped significantly off-axis.

The exit of the stopping cell consists of a supersonic Laval nozzle with a 0.6 mm diameter nozzle throat, also itself acting as a last extraction electrode. Inside the nozzle, the helium buffer gas forms a supersonic gas jet. The α-recoil ions are extracted from the gas cell together with the helium gas. The large gas jet velocity leads to a fast extraction and prevents the ions from charge capture at the nozzle electrode. Together with the helium gas, the ions are entering an extraction chamber with a typical background pressure of 10^{-2} mbar. Here they are injected by the supersonic gas jet into a radio-frequency quadrupole (RFQ) structure (Fig. 4.1, green). The RFQ consists of four rods, to which electric RF fields are applied in order to stabilize the ions on the axis for ion-beam formation. While the ions are guided by the RFQ, the helium buffer gas is extracted. The remaining gas pressure leads to phase-space cooling of the ion beam, such that a sub-mm diameter ion beam is formed at the RFQ exit. Each RFQ rod is segmented into 12 parts, allowing for a DC voltage gradient. This voltage gradient is used to guide the ions through the remaining buffer-gas background.

At this point not only ^{229}Th is contained in the ion beam, but also all α-recoil daughter ions produced by the α decays in the ^{233}U decay chain. Further, a small impurity of ^{232}U is always contained in ^{233}U sources due to the production process. Therefore also the α-recoil ions originating from the decay chain of ^{232}U are found to play a role [16]. Both decay chains contain short-lived isotopes, which are sources of potential background. In order to allow for an unambiguous identification of the isomeric decay of ^{229}Th, a purification of the ion beam is required. For this purpose, a quadrupole mass-spectrometer (QMS) is used (Fig. 4.1, blue), aiming for an exclusive extraction of ^{229}Th. The QMS was built, following design values found by E. Haettner [17]. A mass resolving power of $m/\Delta m = 150$ was achieved with more than 70% transmission efficiency, sufficient to exclusively extract ^{229}Th in the 1+, 2+ or 3+ charge state [16]. Behind the QMS, the ions are guided towards the detection system by a triode extraction system, consisting of three ring electrodes.

The time required for ion extraction can be estimated to be in the range of up to 10 ms [14]. This is significantly below the expected lifetime of the isomeric state in charged ^{229}Th (expected to be in range of minutes to hours), which is important in order to allow for any isomer extraction. The maximum achieved ^{229}Th extraction rate was measured to be ~10 000 ions/s (individually in the 2+ and 3+ charge states), leading to about 200 extracted isomers per second (see Sect. 5.1.1).

4.1.2 Searching for a Photonic Decay

The availability of a low energy, pure ^{229}Th ion beam allowed searching for an isomeric decay signal. In a first attempt, a potential photonic decay of ^{229m}Th was investigated. Light, when emitted in the isomeric decay, has an expected wavelength of (159 ± 11) nm, corresponding to (7.8 ± 0.5) eV energy. This wavelength is in the vacuum ultra-violet (VUV) region and therefore heavily absorbed in air, which poses special requirements for the corresponding optical system. A vacuum-optical system was especially developed in order to match these requirements and will be described in the following. A sketch of the experimental concept is shown in Fig. 4.2 [9]. An important design criterion for the optical system was to provide a large signal-to-background ratio [9, 19]. For this purpose, the ^{229}Th ions, as extracted from the QMS with the triode extraction system, are collected on a 50 μm diameter collection surface. The collection of the ^{229}Th ions is supported by an attractive potential of −300 V and was experimentally investigated in preparatory measurements (Sect. 5.1.2). The surface itself consists of MgF_2-coated copper. The MgF_2 coating was foreseen in order to suppress the non-radiative isomeric decay via internal conversion. The thorium ions, when implanted into a MgF_2 crystal, should prefer the 4+ charge state in the ionic lattice, as is the case for CaF_2 [20]. Further, as the band gap of the material is larger than the isomeric excitation energy, also potential IC electrons are not expected to leave the valence band [4, 20]. Due to the expected suppression of the non-radiative decay, there is a chance to detect a photonic decay channel. This

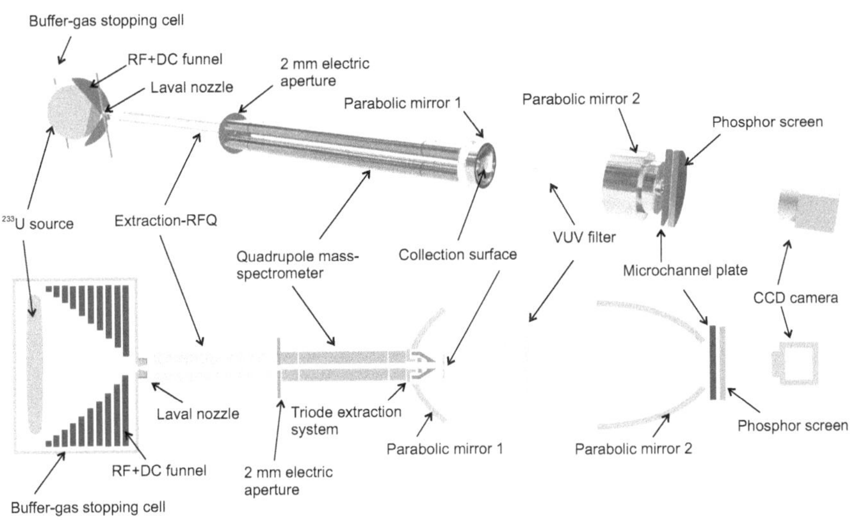

Fig. 4.2 Sketch of the experimental concept used during the search for a potential photonic decay channel of ^{229m}Th. After the setup used for ion-beam formation, a highly efficient focusing optical system is installed. This system allowed for the most sensitive search of a photonic decay channel that so far has been conducted

approach is comparable to other investigations, where a photonic decay of ^{229}Th in a crystal lattice environment is probed (see Sect. 3.6.3), however, with the advantage of a significantly reduced background and a several orders of magnitude improved achievable signal-to-background ratio.

Photons, as expected to be released in the isomeric-decay process, are emitted in all spatial directions. The collection surface is placed in the focus of an annular parabolic mirror (focal length $f_1 = 10$ mm, 39 mm aperture, 12 mm diameter center hole), which covers nearly one hemisphere in order to collect a significant amount of the emitted light. In this way, the light is collimated towards a second, deep annular parabolic mirror (focal length $f_2 = 2$ mm, 39 mm aperture, 12 mm diameter center hole), which acts as a focusing optics. This second mirror is designed in a way that its focal point is located behind the mirror exit (see Fig. 4.2). In preparatory measurements, a focal spot size of about 100 μm diameter (FWHM) was achieved (Sect. 5.1.3). The reflective mirror surfaces consist of MgF_2-coated aluminum, providing a reflectivity of about 70% in the VUV region [21].

The light is focused onto a CsI-coated microchannel-plate (MCP) detector [22] with an expected quantum efficiency of about 10% for light with a wavelength around 160 nm [23]. The MCP is combined with a phosphor screen, allowing for spatially resolved signal read-out by monitoring the screen with a CCD camera. The spatially resolved detection is required in order to make use of the highly efficient focusing optics in terms of an optimized signal-to-background ratio.

One of the design criterions of the VUV optical system was to allow for a wavelength determination of light emitted in the isomeric decay. The wavelength determination can theoretically be performed by introducing VUV filters with a steep absorption edge into the collimated part of the light path. A region of collimated light has to be used for that purpose, as otherwise the focal length of the system would be changed when introducing filter plates into the optical system. While the use of turnable interference filters would allow to determine the wavelength to better than 1 nm, an even significantly improved wavelength determination could be achieved by applying a VUV prism spectrometer.

However, in this first set of experiments, no γ-decay signal of ^{229m}Th could be detected (see Sect. 5.2.1). This observation was a strong indication for a significant internal-conversion decay channel, which was not efficiently suppressed by the MgF_2 coating of the collection surface.

4.1.3 Searching for an Internal-Conversion Decay

Motivated by the non-observation of a photonic decay-branch of ^{229m}Th, searches for an internal-conversion (IC) decay channel were started. For this purpose, the ^{229}Th ions were deposited with low kinetic energy (typically 50–75 eV, depending on the charge state) in "soft landing" directly on the surface of the MCP detector.

The same CsI-coated MCP detection system, as used for the optical measurements, was also employed in the search for an IC decay. ^{229}Th ions, when coming into

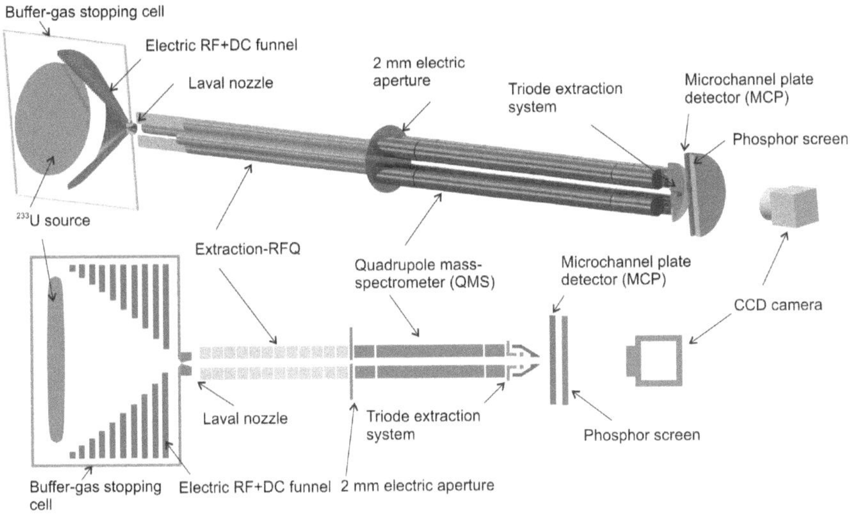

Fig. 4.3 Conceptual overview of the experimental setup used for the direct detection of the ^{229}Th isomeric decay via internal conversion. The upper panel of the figure shows a 3-dimensional drawing with electrodes shown on scale. The lower panel shows a 2-dimensional side view of the setup. See explanation in the text for details. With kind permission of Nature Research [7]

contact with the MCP surface, will immediately neutralize. This will trigger the internal conversion process, which is the favored decay channel of the isomeric state in neutral thorium atoms [2, 6]. The electron, emitted in this decay process, can be detected by the MCP detection system. A sketch of the experimental setup is shown in Fig. 4.3. This setup has finally led to the successful observation of the IC-decay signal of ^{229m}Th (see Sect. 5.2.2) [7].

4.2 The Ion-Beam Formation System

The ion-beam formation system is required for the production of a low-energy, pure ^{229}Th ion beam, which provides the basis for all investigations of a ^{229}Th isomeric decay. It consists of three main parts: (i) The buffer-gas stopping cell (housing an electric DC-cage, the ^{233}U source, the electric RF+DC funnel and the Laval nozzle) (ii) the extraction RFQ and (iii) the quadrupole mass-spectrometer (QMS). All contained elements will be described in this section individually in detail, starting with the ^{233}U source, being the origin of the ^{229}Th isotopes. A complete drawing of the ion-beam formation system is shown in Fig. 4.4.

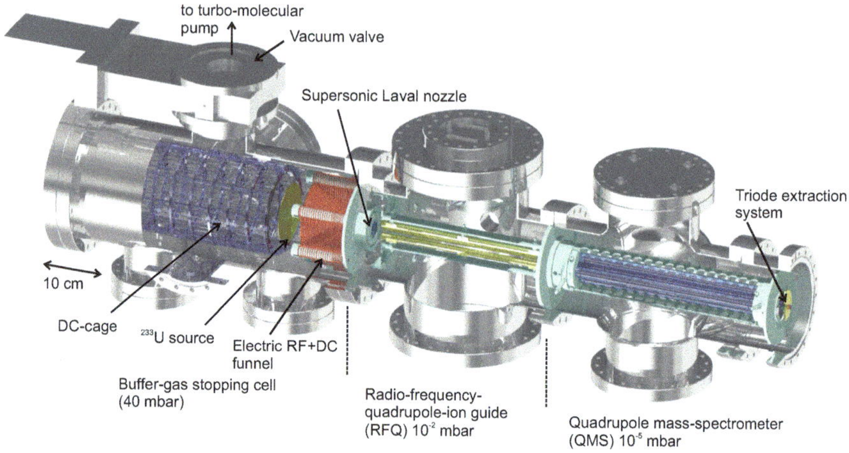

Fig. 4.4 Experimental setup used for the formation of a low-energy, pure ^{229}Th ion beam. The buffer-gas stopping cell (left) contains the ^{233}U source used for ^{229}Th α-recoil ion production. The ions are stopped by helium buffer-gas, guided by an electric RF+DC-funnel system towards the exit of the stopping cell and are extracted with the help of a supersonic Laval nozzle. Subsequently, the ions are injected into a radio-frequency quadrupole (RFQ) for ion beam formation and are further purified with the help of a quadrupole mass-spectrometer (QMS). With kind permission of Nature Research [7]

4.2.1 *The ^{233}U Source*

The experimental concept is based on the population of the isomeric state of ^{229}Th by a 2% decay branch in the α decay of ^{233}U [10]. Therefore the ^{233}U source plays a central role in the experimental setup. In total, three different sources were employed in the experiments, two of them are ^{233}U sources for ^{229}Th-recoil production. The third source is a ^{234}U source, used for ^{230}Th-recoil production as required for comparative measurements (see Sect. 5.3.4). All three sources will be discussed in detail in the following, an overview of their key parameters is shown in Table 4.1.

Table 4.1 Key parameters of uranium sources. See text for details

Number	Mat.	Compos.	Activity (kBq)	Area (cm^2)	Thickness (nm)	Recoil eff. (%)	Recoil rate (1/s)
Source 1	^{233}U	UF_4	200	3.14	360	5.0	10 000
Source 2	^{233}U	metallic[a]	290	62.5	6.9[b]	34.7	100 000
Source 3	^{234}U	metallic[a]	270	62.5	9.9[b]	28.9	78 000

[a]The exact chemical composition is unknown [24–26], however, simulations for metallic uranium are in good agreement with measured extraction rates, although oxidation of the thin uranium layer has to be expected

[b]These are calculated surface thicknesses, assuming a metallic uranium layer

The first source consists of UF_4, containing about 200 kBq of ^{233}U. The material was evaporated from a tantalum heater lined with a vitreous carbon crucible as a round surface with 20 mm diameter onto a stainless-steel plate of 22 mm total diameter and 2 mm thickness, as shown on the left-hand side of Fig. 4.6 [27]. This process was performed at the former hot-lab facility of the LMU Munich in Garching [28]. Unfortunately, this source was the last one to be produced before the facility was dismantled.

The ^{233}U source also contains a fraction of ^{232}U, unavoidably contained in the ^{233}U material due to the production process (see Appendix A.1). Via γ spectroscopy, this ^{232}U fraction was determined to about $6.1 \cdot 10^{-7}$ (at the time of material production) and the year of material production was inferred to be ~1969 (Appendix A.2). The ^{233}U material was not chemically purified before evaporation and in the time since 1969 a significant ingrowth of short-lived daughter isotopes had occurred, which will also be extracted by our setup (Appendix A.7) [16]. Although the fractional content of ^{232}U seems to be small, the daughter activity from radium downwards is comparable to that of the ^{233}U decay chain, as the ^{232}U and ^{228}Th half-lives are significantly shorter than the half-lives of ^{233}U and ^{229}Th, respectively. The decay chains of ^{233}U and ^{232}U are shown in Fig. 4.5 [29]. Note that only isotopes produced by α decay can efficiently leave the source material.

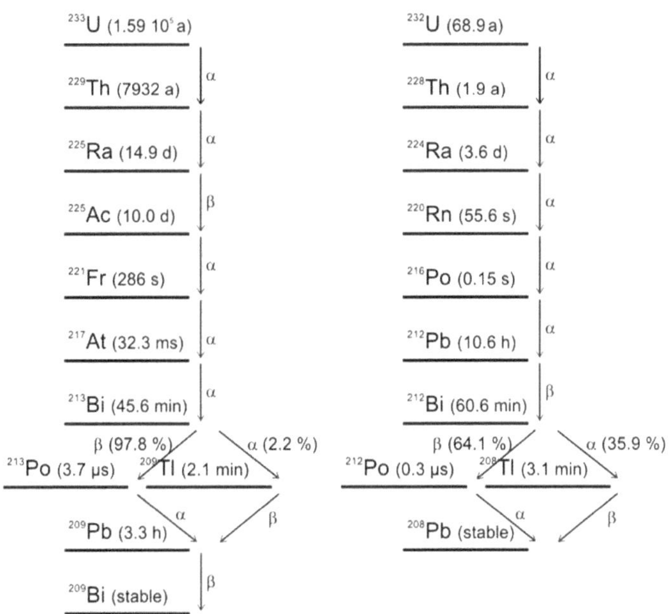

Fig. 4.5 Decay chains of ^{233}U and ^{232}U. Both decay chains are found to play a role as potential sources of background, as ^{232}U is always contained as a small impurity in ^{233}U source material due to the production process (see Appendix A.1). With kind permission of The European Physical Journal (EPJ) [16]

The α-recoil efficiency of the ^{233}U source was measured to be about 5.0% for ^{229}Th. This corresponds to an absolute number of ~10 000 ^{229}Th α-recoil ions leaving the source material per second (see Appendix A.5) [16]. The stopping range of 84 keV ^{229}Th α-recoil isotopes in amorphous $^{233}UF_4$ material was calculated by SRIM simulations (based on Ziegler-Biersack-Littmark stopping powers [30], SRIM version 2008 was used) to be about 23 nm. Given the source thickness of about 360 nm and taking into account the isotropic emission of α-recoils together with scattering in the uranium material, this leads to an expected recoil efficiency of ~1.6% (Appendix A.5), significantly smaller than the measured efficiency of 5.0% (Appendix A.4). The reason for this efficiency enhancement might be the production process via the evaporation technique. This production process may have led to the formation of a polycrystalline surface, opening the possibility for channeling and thus leading to a larger effective stopping range of ~76 nm. Corresponding simulations were performed with the MDrange program code [31], showing the potential to explain the measured recoil efficiency (Appendix A.5) [16].

Given the short ^{229}Th-recoil stopping range, it is obvious that it would be advantageous to employ a thinner ^{233}U source of larger surface area. For this reason, a new source was fabricated with a large active surface diameter of 90 mm (leaving a 12 mm diameter unplated region in the center). The source is shown on the right-hand side of Fig. 4.6. Also the ^{233}U activity was enhanced to 290 kBq, which is the maximum ^{233}U activity permitted to be handled in our laboratory. For this source the production process via the evaporation technique was not available anymore and instead the source was produced via electrodeposition at the Institute for Radiochemistry at the University of Mainz [24–26, 32]. This time, the ^{233}U material was chemically purified by ion-exchange chromatography before deposition in order to remove most

Fig. 4.6 Photographs of the small-area (20 mm diameter) ^{233}U source (source 1, left-hand side) and the large-area (90 mm diameter) newly available ^{233}U source (source 2, right-hand side). The new source was produced at the IRC of the University of Mainz [24]. Both sources are shown as they are mounted in the buffer-gas stopping cell. The large-area ^{233}U source provides about 10 times higher recoil efficiency, due to a smaller thickness of the uranium layer

of the short-lived daughter isotopes. A purification factor of better than 300 was determined by γ spectroscopy (Appendix A.8). In order to avoid back-implantation of the α-recoil isotopes into the surface, a mirror-polished silicon wafer of 0.5 mm thickness and 100 mm diameter was used as a carrier substrate. To allow for the electrodeposition, the wafer was sputtered with a 100 nm thick titanium layer. A center-hole of 8 mm diameter was foreseen in order to give free line of sight on the experimental axis. This might be of advantage for later Doppler-free collinear laser spectroscopy experiments.

Similar as for the small source, also for this large source, stopping-range calculations were performed to estimate the expected enhancement in recoil efficiency. Due to the fabrication method by electrodeposition, this time the SRIM simulations for an amorphous source structure are expected to give the correct result. As the exact chemical composition of the surface is unknown, for the simulations a surface consisting of metallic uranium was assumed, leading to an expected recoil efficiency of 34.7%. Given the 290 kBq source activity, this results in about $1 \cdot 10^5$ emitted ^{229}Th recoil ions per second, which is by a factor of 10 enhanced compared to the smaller source. This is in good agreement with the (by a factor of 10) measured improvement of the ion extraction rate (see also Fig. 5.4).

The last source, which was used for comparative measurements only, is a ^{234}U source, which was produced in the same way by electrodeposition, with the same geometries as the second, large-area ^{233}U source. The source activity is 270 kBq. As the half-life of ^{234}U ($t_{1/2} = 2.45 \cdot 10^5$ a) is comparable to that of ^{233}U ($t_{1/2} = 1.69 \cdot 10^5$ a), the ^{234}U layer thickness is expected to be of comparable range. This is important, as otherwise a strong reduction in recoil efficiency would have to be expected. Based on SRIM simulations, the α-recoil efficiency was calculated to 28.9%, leading to an absolute number of ~78 000 ^{230}Th recoil ions per second leaving the source material. This number is comparable to the extraction rate obtained for the large-area ^{233}U source, which is in agreement with experimental observations.

4.2.2 The Buffer-Gas Stopping Cell

The ^{229}Th α-recoil isotopes leave the ^{233}U source in all directions with kinetic energies of up to 84.3 keV. A significant amount of the ^{229}Th isotopes will leave the ^{233}U source as positively charged ions [33].[1] The kinetic energies are too high to easily guide and manipulate the ions with electric fields. As a prerequisite for ion-beam formation, the kinetic energy of the ions has to be reduced. This is done by collisions with helium atoms in an ultra-pure helium environment. The purity of the helium buffer-gas is an important requirement for the α-recoil isotopes to stay charged during the stopping process [35]. Even small amounts of impurities of the helium buffer-gas will lead to charge-state reduction and molecule formation of the ^{229}Th ions dur-

[1] In other studies, a large neutral fraction of α-recoil isotopes was found [34]. That the ionic fraction dominates in the considered case is most likely due to stripping of the α-recoil nuclides in the source material.

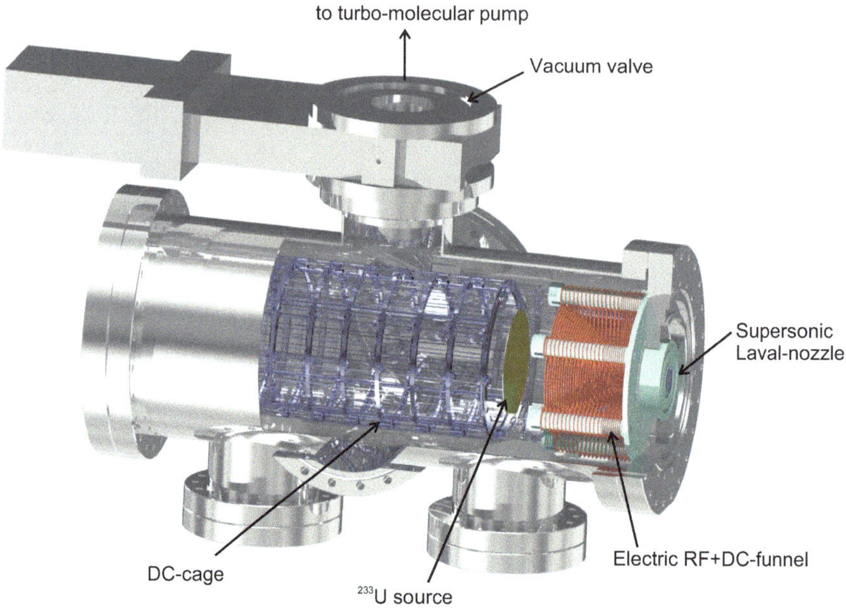

Fig. 4.7 The buffer-gas stopping cell contains the ^{233}U source (mounted to the front end of a DC-cage electrode system) and the electric RF+DC-funnel. The exit of the stopping cell is provided by a supersonic Laval nozzle for fast and efficient ion extraction

ing collisions, until they are neutralized and thus lost. Especially thorium is a highly reactive element, a reaction-rate constant of $6.0\cdot 10^{-10}$ cm^3/s is listed for the $Th^{+}+O_2$ reaction in Ref. [35], which is the largest listed value. Therefore, a buffer-gas stopping cell is used, which has been developed to provide the highest standards regarding cleanliness of the helium gas. For our experiments the Maier-Leibnitz-Laboratory (MLL) IonCatcher was used, which was developed and built by J. Neumayr between 2004 and 2006 [13]. It was designed based on the buffer-gas cell for SHIPTRAP at GSI, which was also developed and built at the MLL Tandem-accelerator facility in Garching [14, 15]. A helium-gas purity in the ppb-region and better can be achieved with this system [13]. A drawing of the buffer-gas stopping cell is shown in Fig. 4.7.

The MLL IonCatcher consists of a cylindrical vacuum chamber of about 480 mm length and 200 mm diameter (DN-200CF). It contains two electrode systems, the DC cage and the RF+DC-funnel system, both required for ion extraction from the gas cell. The electrode systems will be detailed in the following sections. The buffer-gas stopping cell also houses the ^{233}U source, which was described in the previous section. It is placed in a distance of about 80 mm from the exit of the stopping cell in a centered position in front of the electric RF+DC-funnel. A schematic overview of the stopping cell, together with its gas flow system and wiring scheme is given in Fig. 4.8.

The stopping cell has been designed with respect to strict UHV standards. It is made from UHV compatible stainless steel (type 316L) and bakeable up to 180 °C.

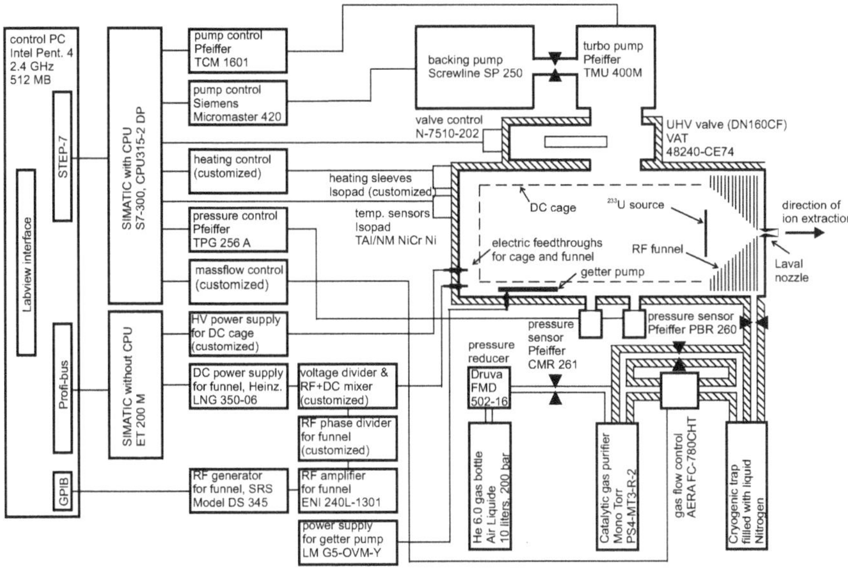

Fig. 4.8 Wiring scheme of the electronic environment of the buffer-gas stopping cell. Many of the components were built by the electronic workshop of the LMU Munich. See explanations in the text for a detailed description

Specially fabricated heating sleeves (Isopad) are used for bake out. They are controlled via a customized heating control system and the temperature is read-out by NiCrNi temperature sensors (Isopad), capable to read-out temperatures of up to 500 °C. Both, the heating control system and the temperature sensors, are connected to a programmable logic controller (Siemens SIMATIC SPS, type S7-300) with a CPU (CPU315-2 DP), which is controlled by a control PC via the STEP-7 control language. A Labview-based user interface allows for a full control of the bake-out procedure. After 48 h of bake out at a temperature of 130 °C, a typical pressure of $3 \cdot 10^{-10}$ mbar is reached. The system is pre-pumped by a Screwline SP 250 (Leybold, 250 m^3/h) and UHV pumped by a turbo-molecular pump of type TMU 400 M (Pfeiffer, 400 l/s). The pumps are controlled via a Siemens Micromaster 420 frequency converter and a Pfeiffer TCM 1601 controller, respectively. Both systems are connected to the SIMATIC and the Labview user interface is used for read-out and control.

In order to allow for the inlet of the helium buffer-gas, the turbo-molecular pump can be disconnected from the stopping cell volume with the help of an all-metal UHV valve (DN160 CF, VAT, type 48240-CE74). When the valve is closed, the cell can be vented with helium. In order to provide the highest possible cleanliness of the buffer-gas, helium with a purity of 99.9999% (He 6.0) is provided (Air Liquide, 10 liters, 200 bar). This gas is further purified by a catalytic purifier (SAES Getter Systems, Mono Torr, PS4-MT3-R-2) and a cryogenic trap attached to the 1/4" electropolished gas supply tubing (Dockweiler ultron), filled with liquid nitrogen. The inlet of the gas into the stopping cell is controlled by a gas-flow control (AERA, FC-780CHT). A typical helium gas pressure of 40 mbar was used in the experiments. The gas-flow control can be bypassed by a valve (Swagelok 4BG), which is required in order to also allow for a bake out of the gas tubing. The whole gas tubing system is made from stainless steel and was electropolished for highest cleanliness. To allow for a further gas purification during operation, a getter pump is installed (SAES, type St707-CTAM-30D) and can be activated by a power supply (Lambda, type LM G5-OVM-Y) providing an activation current of ca. 50 A. The chamber pressure is controlled via two pressure sensors, one for the low pressure region when no helium gas is entering the chamber (Pfeiffer, PBR 260) and a capacitance gauge for the operation with helium (Pfeiffer, CMR 261). All valves as well as the pressure sensors are controlled by the SIMATIC.

4.2.3 The DC Cage

Originally, the buffer-gas stopping cell was designed to stop high-energetic nuclides, as produced in nuclear fusion reactions at the Tandem-van de Graaff accelerator at the Maier-Leibnitz-Laboratory (MLL) in Garching, Germany [13]. The accelerator structure allows to accelerate ions from protons to uranium, starting with an electrostatic acceleration potential of up to 14 MV. When impinging onto a fixed target assembly with an energy above the Coulomb barrier, the resulting nuclear fusion products enter the buffer-gas stopping cell through a specially fabricated entrance window. Subsequently, they are stopped in the ultra-pure helium environment of the stopping cell. As the kinetic energies of the produced nuclides are typically several MeV, a stopping length of about 400 mm and a helium pressure of up to 150 mbar are foreseen. A DC cage was designed in order to provide an electric field gradient in the direction of extraction to drag the ions from the stopping region towards the chamber exit. The stopping range of ^{229}Th α-recoil ions with an energy of 84.3 keV in helium was estimated with the help of SRIM simulations to be about 10 mm. Therefore, for the stopping of the α-recoil ions, the full length of the DC cage is not required and the ^{233}U source is instead installed at the last segment of the DC cage, directly in front of the RF+DC-funnel system.

The DC cage consists of 10 cylindrical cage electrodes. An individual electric DC potential can be applied to each electrode (see Refs. [14, 15] for a similar device). Each cylindrical electrode has an outer diameter of 150 mm and a length of 35 mm, leading to a total length of the DC cage of about 350 mm. A photograph of the DC cage is shown in Fig. 4.9.

A voltage of up to 350 V can be applied to each electrode separately. The voltage is provided by an Iseg HV module (EHQ F025p), which is inbuilt into a customized voltage supply. The voltage is controlled by a second SIMATIC module without CPU (type ET 200M), which is connected to the control computer via a profibus connection (see Fig. 4.8). The user interface is provided by the same Labview program, which also controls the bake-out and gas-flow system.

Fig. 4.9 Photograph of the DC cage built for the extraction of high-energy nuclear-fusion reaction products. In our experiments, only the last DC-cage electrode was used to provide an electric offset for the ^{233}U source

For the ion extraction from the ^{233}U source only the electrode of the DC cage which is closest to the stopping cell exit is of importance. This electrode acts as a source holder, and allows to apply a voltage offset of typically 39 V to the ^{233}U source. By this voltage offset the ion extraction from the source can be controlled.

4.2.4 The RF+DC Funnel

A further electrode system is required in order to guide the α-recoil ions towards the extraction nozzle, that forms the exit of the stopping cell, even if they were stopped significantly off-axis. For this purpose, an RF+DC-funnel system was built [14, 15]. The funnel consists of 50 ring electrodes, which form a cone towards the extraction nozzle. The cone possesses a maximum inner diameter of about 115 mm and a minimum inner diameter of 5 mm. The first 20 ring electrodes (as counted from the direction of the cell exit) have a thickness of 0.5 mm and also a distance

Fig. 4.10 Photograph of the electric RF+DC-funnel system required to provide an efficient ion extraction from the buffer-gas cell volume [14]. All funnel electrodes were electro-polished in order to achieve a high surface smoothness to prevent electric discharges in the ion extraction region

of 0.5 mm from each other, while the thickness as well as the distance of the last 30 electrodes is 1 mm. Thus the funnel structure has a total length of 80 mm. The angle of the cone is 70°. A photograph of the funnel electrode structure is shown in Fig. 4.10. A combination of RF and DC voltages is applied to each electrode for an optimum ion guidance. A DC gradient of typically 4 V/cm (starting with 35 V at the funnel entrance and decreasing to 3 V at the last funnel electrode) drags the ions towards the chamber exit. An RF amplitude of about 220 V_{pp} at 850 kHz with an alternating phase of 180° between two consecutive funnel electrodes leads to a repelling force. This repelling force prevents the ions from charge capture at the funnel electrodes. The RF voltage is generated by an RF generator (SRS model DS 345) and then amplified by an RF amplifier of type ENI 240L-1301. The generated RF amplitude has to be phase divided in order to supply the funnel electrodes with voltages of alternating phase, which is done by a customized phase divider. The DC voltage is supplied by a power supply from Heinzinger of type LNG 350-06. Both, the RF voltage and the DC voltage, are coupled to a customized voltage-divider chain, where the 50 voltages are generated individually and then connected to the funnel structure.

4.2.5 *The Laval Nozzle*

Having guided the ions towards the exit of the buffer-gas cell with the RF funnel, they have to be extracted from the high-pressure region. This is done by a convergent-

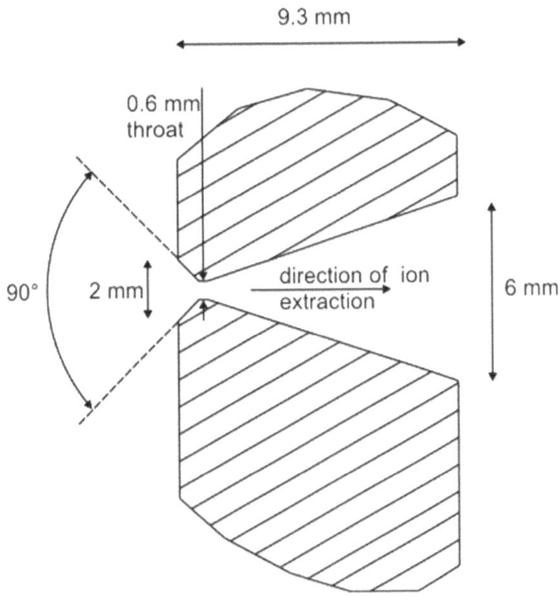

Fig. 4.11 Schematics of the supersonic Laval nozzle designed to provide a fast helium gas-flow from the high pressure region of the buffer-gas stopping cell to the next vacuum chamber [14, 15]. The ions are extracted together with the helium gas and subsequently injected into a radio-frequency quadrupole (RFQ)

divergent supersonic Laval nozzle [14, 15, 36]. The nozzle forms a supersonic helium gas jet from the buffer gas and the ions follow the gas flow. This system has the advantage of a fast extraction and it prevents the ions from charge capture at the chamber walls. Further, the previously thermalized ions gain enough kinetic energy to be injected into a subsequent RF-quadrupole system.

The shape of the nozzle is shown in Fig. 4.11. The entrance diameter is 2 mm with an opening angle of 90°. The nozzle throat has a diameter of 0.6 mm and a depth of 0.3 mm. The outlet is formed by a cone of 38° opening angle and a maximum diameter of 6 mm. This leads to a total length of 9.3 mm. The nozzle is made of V4A stainless steel and provides the possibility to apply a voltage offset (typically 2 V were applied). It is glued via metal-ceramic welding (Friatec) into a mounting flange.

4.2.6 *The Extraction RFQ*

Following the buffer-gas stopping cell, in a second step the helium-buffer gas has to be removed from the system, while the extracted ions have to be kept and radially confined in order to form a low-energy ion beam. For this reason the ions are injected into a radio-frequency quadrupole (RFQ), consisting of four rods of 11 mm diameter and 330 mm length. The inner rod distance is 10 mm. Each RFQ rod is segmented into 12 parts of about 25 mm length. Individual voltages can be applied to each segment

Fig. 4.12 Photograph of the front end (direction of ion injection) of the radio-frequency quadrupole (RFQ). The ions are radially confined by the provided RF voltages, while the helium background gas is reduced to 10^{-2} mbar. The remaining helium background leads to phase-space cooling of the ions. In this way a sub-mm diameter ion beam is provided at the RFQ exit

(see Refs. [14, 15] for a similar device). A photograph of the RFQ in the direction of ion injection is shown in Fig. 4.12.

An RF voltage amplitude of $200\,V_{pp}$ at a frequency of 880 kHz is applied to all RFQ rods. The RF voltages of neighboring rods are shifted in phase by 180°. This leads to a stabilizing force and a resulting radial confinement for the ions. The RF frequency is generated by an RF generator of type SRS model DS 345. The RF voltage is further amplified by an RF amplifier (RM Italy KL 500 HF) in combination with an air-coil transformer that generates the voltages of opposite phases. In a further step, 12 individual DC voltage components are added to the RF voltages in a DC+RF mixer. These voltages are then supplied to the individual RFQ segments.

An individual DC offset is applied to each RFQ segment. The DC voltages form a gradient along the beam axis, sufficiently large to guide the ions through the remaining helium buffer-gas background. Like for the DC cage, the voltage is provided by an Iseg HV module (EHQ F025p), which is controlled by the same control system used for the buffer-gas stopping cell (SIMATIC, connected to the control PC). The same Labview interface is used to apply the individual voltages. Typically, a low voltage gradient of −0.2 V/electrode is applied, starting at 1.8 V at the electrode closest to the Laval nozzle and reducing the voltage to 0 V at electrode 10. The last two electrodes are kept at 0 V offset.

While the ions are confined by the RFQ, the helium background pressure is reduced using a strong turbo-molecular pump (Pfeiffer, HiMag 2400) with a pumping speed of 2100 l/s, connected to the same pre-vacuum pump as the TMP of the stopping cell (Leybold, Screwline SP 250). For most measurements, the pumping speed is reduced to 50% during ion extraction, in order to achieve a pressure region of about 10^{-2} mbar. This ambient helium background pressure is required for phase-space cooling of the ions, leading to the formation of an ion beam of sub-mm diameter at the RFQ exit. The pressure can be monitored by a full-range gauge (Pfeiffer, PKR 261).

In order to provide the highest purity of the buffer-gas, the whole RFQ chamber was built to UHV standards and is bakeable up to 180 °C. The bake-out system is the same as used for the stopping cell. After two days of bake out, a typical pressure in the lower 10^{-10} mbar range is achieved.

4.2.7 The Quadrupole Mass-Spectrometer

At the point of ion-beam formation in the RFQ, still all α-recoil ions emitted from the uranium source are contained in the ion beam. As the ^{233}U decay chain contains several short-lived α and β^- emitters, these isotopes pose a significant source of background. In order to remove these short-lived isotopes and in this way allow for an unambiguous identification of a potential ^{229}Th isomeric decay, a quadrupole mass-spectrometer (QMS) is used. The QMS was built, following the design values of E. Haettner [17]. It consists of four stainless steel rods of $d = 18$ mm diameter with a total length of 400 mm. Each rod is segmented into three parts: a central part of 300 mm length, acting as the mass-separating region, and two segments (one on each side) of 50 mm length, acting as Brubaker lenses [37]. No mass-separating electric DC fields are applied to the Brubaker lenses and their purpose is to reduce any peripheral field distortions, which would otherwise reduce the QMS efficiency. An inner-rod distance of d_0 =15.96 mm was chosen, leading to a ratio $d/d_0 = 1.128$, which is an important design value for the QMS [17]. The tolerance for variations of d_0 is 40 μm only, posing strong requirements for the mechanical precision. A photograph of the QMS is shown in Fig. 4.13.

All QMS rods are supplied with RF and DC voltages, which have to match specific requirements in order to allow for the separation of a specified mass-to-charge ratio (see, e.g., Ref. [38]). The RF voltage is generated by a function generator (Tektronix, AFG 3022B) and is then further amplified by an RF amplifier (RM Italy, KL 500 HF). Similar to the amplification system of the RFQ, an air coil is used as a transformer for generation of high-voltage RF amplitudes (more than 1500 V_{pp} at a frequency of 825 kHz are achieved). The air coil generates two RF voltages, phase-shifted by 180°, that are applied to the QMS rods such that opposite rods are supplied with identical RF voltages. Two air coils are available, optimized for the generation of different RF-amplitude regions. An individual DC voltage can be applied to each QMS segment and is admixed to the RF voltages before they are applied to the QMS rods.

The DC voltages are generated by customized HV-voltage supplies, which allow to apply a well-defined voltage difference between neighboring rods, as required for mass-separation. Like for the mechanical precision, also strong requirements for the RF voltage precision exist. These have to be stabilized to the ppm region, which

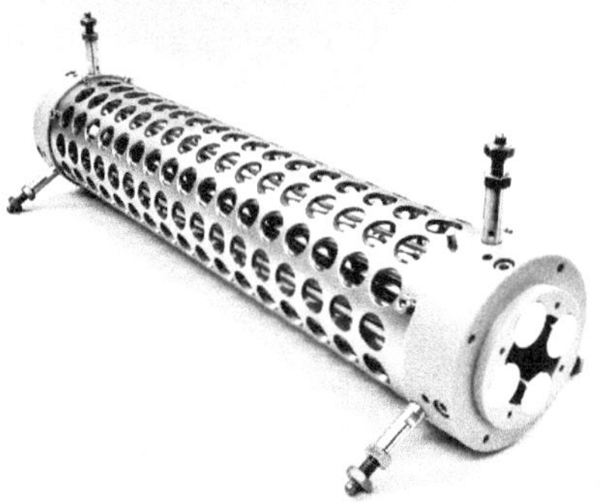

Fig. 4.13 Photograph of the quadrupole mass-spectrometer (QMS) required for ion-beam purification. The QMS was developed following design values given in Ref. [17]. The production of a mass-purified ^{229}Th ion beam is a prerequisite for the unambiguous identification of the isomeric decay. Otherwise, short-lived decay products of the ^{233}U and ^{232}U decay chains could not be excluded as signal origin. With kind permission of The European Physical Journal (EPJ) [16]

Table 4.2 Operational parameters of the QMS as used for the ^{229}Th ion extraction

q	m/q (u/e)	Δm/q (u/e)	m/Δm	Frequency (kHz)	RF amplitude (V_{pp})	DC voltage (V)
1+	229	1.5	~150	825	1434	119.7
2+	114.5	0.8	~150	925	901.2	75.23
3+	76.3	0.5	~150	925	600.5	50.15

is done by real-time monitoring of the voltages with a high-resolution oscilloscope (PicoScope 4227) and continuous voltage regulation if required. For this purpose a Labview program was developed, which allows for real-time read-out and stabilization of all QMS voltages [39]. The program also applies the voltages required to perform mass scans. The operational parameters of the QMS, used for the extraction of ^{229}Th ions of different charge states q, is shown in Table 4.2. For the extraction of ^{229}Th^{1+} a larger air-coil transformer was used in order to reduce the resonance frequency of the QMS. This led to a lower required RF amplitude, thus also allowing for the extraction of the 1+ charge state. Besides the RF and DC voltages required for mass separation, also a DC offset voltage can be applied to all QMS segments. The offset voltage was typically chosen to be −2 V for all segments.

A relative mass-resolving power m/Δm of about 150 has been experimentally obtained at a transmission efficiency of more than 70% [16]. This mass resolution is sufficient to securely separate all α-recoil ions of the ^{233}U decay chain with mass differences of 4 atomic mass units. In this way the QMS allows to form a (nearly) pure ^{229}Th ion beam. The only remaining contaminant is a fractional content of ^{228}Th ions originating from the ^{232}U decay chain (also present in the uranium source material). This cannot be avoided by the separation method, however, due to the comparably long ^{228}Th lifetime of 1.9 years, the contribution to the background is negligible.

The QMS is located in an individual vacuum chamber, which itself acts as a differential pumping stage. The vacuum chamber is separated from the RFQ by a metal orifice of 2 mm diameter opening, that is set to a voltage offset of usually −1 V. The chamber is pumped by a TMP (Pfeiffer, TMU 1601P, 1600 l/s), which is connected to the same pre-vacuum pump as the stopping cell and the RFQ (Leybold, Screwline SP250). A typical pressure range of 10^{-8} mbar is reached when no helium-gas is supplied to the system. No UHV conditions are required at this point and therefore no bake out system is installed. During operation the typical pressure is in the lower 10^{-5} mbar range. A pressure gauge (Pfeiffer, PKR 261) is installed for pressure read-out.

4.2.8 The Triode Extraction System

Following the QMS, the ^{229}Th ions are collected for the detection of any potential isomeric decay signal. In the first experimental attempt, a photonic decay of ^{229m}Th was investigated. In order to provide a nearly point-like source of light, as potentially emitted during the isomeric decay, the ^{229}Th ions were collected on a small collection surface of 50 μm diameter, which is described in Sect. 4.3.1. An electric potential was applied to this collection surface, in order to attract the ions. This nearly point-like source was then placed in the focus of an annular parabolic mirror, being the starting point of a highly efficient focusing optics (Sect. 4.3).

As the collection surface is small, the ions have to be close to this surface (within the range of a few mm) in order to feel the attractive electric potential. However, as the point of collection is located in the focus of a parabolic mirror (see Sect. 4.3.2), the ions first have to be guided through a central hole in the mirror towards the collection surface in order to allow for an efficient ion collection. A triode extraction system was designed in order to guide the ions towards the collection surface, thereby matching the special requirements originating from the spatial restriction of a 12-mm diameter mirror center-hole. A similar system is presented in Refs. [40, 41].

Although the triode extraction system in this special form was only required for the optical measurements, it was kept as a guiding system also during the search for an internal-conversion signal. Measurements of the extraction rate of thorium ions (Sect. 5.1.1), as well as extraction efficiency measurements (Appendix A.7), were performed behind the triode for individual charge states. For ^{229}Th in the 2+ and 3+ charge state, extraction rates of about 1000 ions/s for the small-area ^{233}U source (source 1) were obtained [9], while more than 10 000 ions/s could be extracted from the large-area source 2 [7]. The time required for ion extraction can be estimated to be below 10 ms [14].

The triode extraction system consists of three electric apertures of 4 mm, 2 mm and 2 mm diameter, respectively. The center electrode is typically set to an attractive

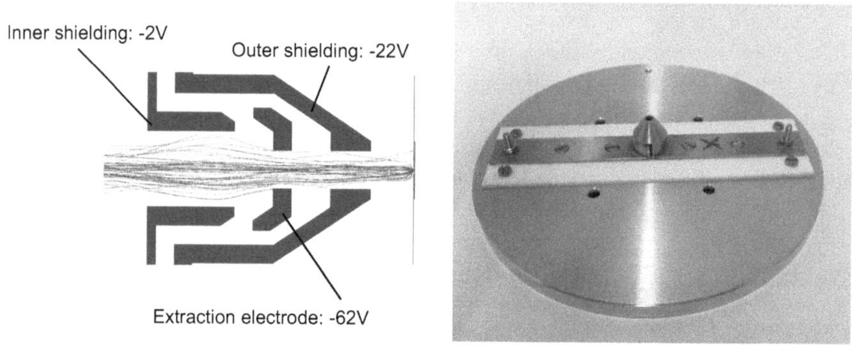

Fig. 4.14 Schematics of the triode extraction system (left) together with simulated ion trajectories. The system was developed in order to provide ion guiding through the central hole of the annular parabolic mirror described in Sect. 4.3.2. A photograph of the entire triode system is shown on the right

potential for the ions and is shielded to the outside by the other two electrodes. The applied voltages were optimized by SIMION simulations [42] (SIMION 8.1 was used) to be −2, −62 and −22 V and are applied by a Mesytec HV module (Mesytec, MHV-4). A cut through the triode system is shown next to a photograph in Fig. 4.14. The photograph shows a 100 mm diameter aluminum base-plate, which itself acts as the first electrode (inner shielding) of the triode. The remaining two electrodes are mounted on top of this base-plate, insulated by a ceramic plate. They are contacted from the sides to give space for the annular parabolic mirror in the center. The total height of the system is 16.7 mm. The electrodes form a nozzle-like shape of 10 mm outer diameter in order to fit through the 12 mm diameter center hole of the mirror. 1 mm space is left on each side, which allows for optical mirror-adjustment. The cone shape is required, as otherwise part of the mirror surface would be shadowed by the triode extraction system during operation. When introduced into the vacuum system, the base-plate is mounted directly in front of the QMS. Simulations have shown a combined triode extraction and collection efficiency of about 40%.

4.3 The Optical System

As soon as the ^{229}Th ion-beam formation and purification was experimentally under control (Sect. 5.1.1), the basis was obtained for experiments, searching for any direct isomeric decay signal of ^{229m}Th. The first attempt was to search for an isomeric decay via the photonic decay channel, as this observation would have allowed for an easy wavelength determination with the help of interference filters with sharp absorption edges. The photonic decay channel is expected to be significant for charged ^{229}Th. The reason is, that the second ionization potential of thorium is about 11.9 eV. As this energy value is above the expected isomeric excitation energy of 7.8 eV, the internal conversion (IC) decay via emission of an electron should be energetically forbidden. Of course the same argument is also valid for any higher thorium charge state. When the IC decay is suppressed, ^{229m}Th is expected to decay with a prolonged lifetime (in the range of minutes to hours) via a significant photonic decay channel [2, 6]. Further reduction of the direct photonic decay can occur due to the electronic bridge (EB) process, which is hard to be quantitatively discussed, but might easily amount to a reduction of the photonic decay rate by several orders of magnitude [12, 43, 44].

An experimental concept for the detection of a potential ^{229m}Th photonic decay branch was developed, for which the following requirements posed the main constraints for the design of the optical system [9]:

- The photon flux originating from the isomeric decay signal is expected to be extremely low (of unknown value), therefore an optical system with a high light yield is required.
- In order to obtain a good signal-to-background contrast, the expected light is emitted from a 50 μm diameter source and the total magnification factor of the optics system should be small.
- The exact energy of the isomeric transition is unknown, therefore the performance of the optical system should be largely independent of the exact wavelength of the light.
- The expected wavelength region of the light is in the range between 150 and 170 nm and thus in the vacuum ultra-violet (VUV) region. A vacuum-optical system is therefore required to allow for the detection of any isomeric decay signal.
- The optical system should allow for wavelength determination as soon as the emitted light has been successfully observed.

Numerical simulations were performed in order to optimize the optical system (see Sect. B.2). For this purpose, a multitude of different optical designs were numerically investigated. This includes all-diffractive two-component optical systems, consisting of spherical and aspherical lenses, as well as combinations of reflective and diffractive optical elements. The main optical concepts, that were theoretically studied, are disscussed in Ref. [9]. These studies have led to the development of a highly efficient all-reflective optical system, consisting of two parabolic mirrors (Fig. 4.15). Compared to that, the use of two spherical lenses is highly unfavored, leading to a factor of ~50 of reduction in signal contrast, due to spherical aberrations and an unfavourable numerical aperture [9]. While both problems could theoretically be solved by using aspherical instead of spherical lenses, in this case still chromatic aberrations would be present. In order to obtain the maximum optical efficiency of the system, the lenses would have to be adapted to the (unfortunately unknown) wavelength of the nuclear transition.

The use of parabolic mirrors has the advantage of no spherical and no chromatic aberrations, combined with a high light yield and a small factor of magnification, leading to a high achievable signal-to-background ratio. Further, a region of collimated light is obtained between the two mirrors, potentially allowing for wavelength determination of the emitted light with the help of interference filters. A conceptual overview of the optical setup is shown in Fig. 4.15.

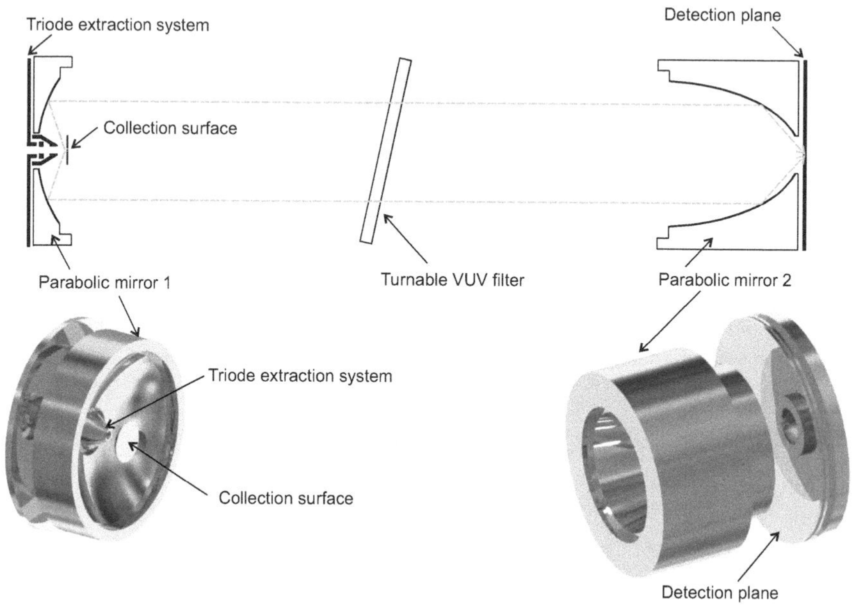

Fig. 4.15 Sketch of the VUV-optical detection concept. Two parabolic mirrors are used for collimation and focusing of the isomeric decay light, respectively. In this way the system does not suffer from any spherical or chromatic aberrations. The system provides a high light yield and a small magnification factor, as required for an optimum signal-to-background ratio. With kind permission of Elsevier [18]

4.3.1 The Collection Surface

Behind the quadrupole-mass-separator, the ions have to be collected on a small surface in order to provide the basis for the detection of a potential photonic isomeric decay with a high signal-to-background ratio. For this purpose, SIMION simulations (see Appendix B.1.6) were performed, resulting in a collection-electrode structure as shown in Fig. 4.16 [9].

The 50 μm diameter collection surface is placed in the center of a 10 mm diameter cover electrode, to which it is insulated by an insulator ring of 300 μm diameter. For ion collection, a voltage of −300 V is applied to the collection surface (CMTE, HV Supply 3160, 0–6 kV), while keeping the cover electrode on a potential of +3 V. In this way an electric field gradient between the collection surface and the cover electrode is provided, leading to an attractive force that guides the ions towards the center of the electrode structure. The concept of ion collection was experimentally proven to work (see Sect. 5.1.2). The high voltage, which is required in order to sufficiently attract the ions, has to be applied from the back side of the collection electrode, as otherwise the potential of the wire would dominate the attraction of the micro electrode. In this way, the cover electrode also acts as a shielding against

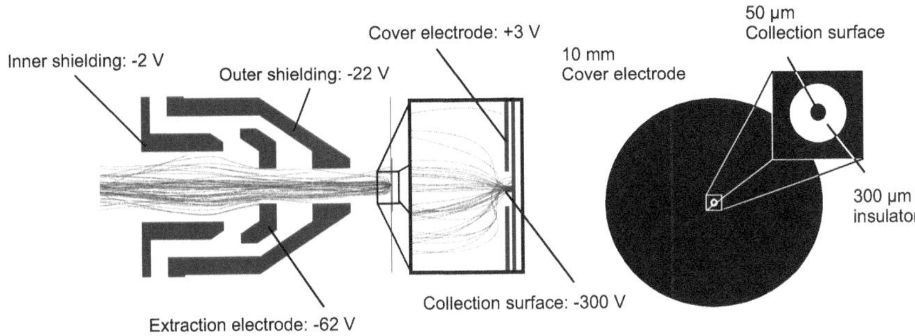

Fig. 4.16 Schematics of the collection surface together with simulated ion trajectories (left). The electrode structure is designed for the purpose of efficient ion collection on a small surface. For this purpose a voltage gradient of about 300 V is applied between a circular cover electrode and a centered collection electrode of 50 µm diameter. The electrodes are insulated by an insulator ring of 300 µm diameter (right). With kind permission of The Journal of Instrumentation (JINST) [9]

unwanted potential fields. In order to allow for an efficient ion collection, the whole electrode system is placed in about 3 mm distance in front of the 2 mm diameter orifice of the extraction triode (Sect. 4.2.8).

The collection surface was fabricated as a Kapton-based printed-circuit-board (produced by ILFA Feinstleitertechnik GmbH) of 10 mm diameter. Both front-surface electrodes are made of copper and were contacted to the back side of the printed-circuit board (using laser drilling and copper ingrowth), where they are solder-connected to copper wires. The insulator ring of 300 µm outer diameter was produced by etching the copper surface. A photograph and a microscopic image (microscope type Sensofar S neox, 20 times magnified) of the collection electrode-structure are shown in Fig. 4.17.

In order to avoid non-radiative decay via internal conversion, the whole surface was coated with MgF_2 of about 1 µm layer thickness. MgF_2 is a large band-gap material with a band gap of 10.8 eV, which is significantly above the expected isomeric energy of 7.8 eV. Further, thorium, when implanted into the MgF_2 material, is expected to prefer the 4+ charge state in the crystal lattice environment [20]. For these reasons the isomeric decay via electron emission is expected to be energetically forbidden [4]. As no intermediate electronic levels in the band-gap exist, also the electron bridge process should be suppressed [20, 45]. However, due to the low implantation energy of only a few keV, the implanted thorium ions will not reach the bulk material of the MgF_2 layer. As surface effects will influence the band structure of the material and also contaminants like, e.g., hydrocarbons have to be expected to be present on the surface, there is some uncertainty regarding the actual strength of a potential radiative decay branch [6, 46]. Note, that no space-charge effects have to be taken into account when collecting the ions onto the thin coating, as small currents in the material will occur as soon as the voltage gradient is large enough, leading to net charge neutralization [47].

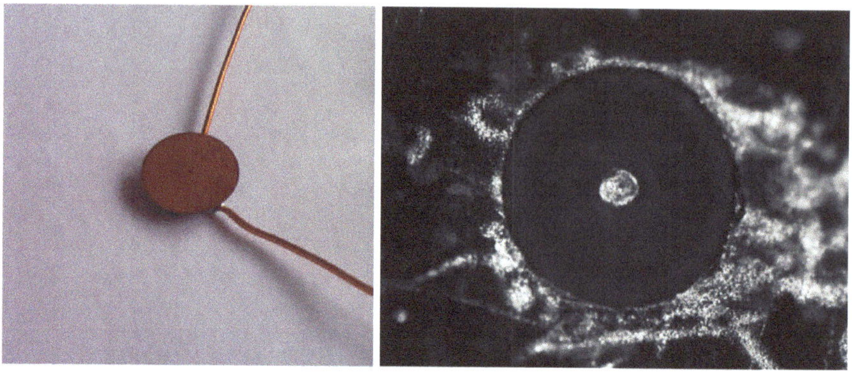

Fig. 4.17 Photograph of the collection-electrode structure consisting of the copper printed-circuit board of 10 mm diameter (left). The electrodes are connected by Kapton-isolated copper wires from the back side. A microscopic image of the 50 μm diameter collection surface surrounded by the 300 μm diameter insulator ring is shown on the right

Fig. 4.18 Photographs of the mirror mount for parabolic mirror 1, also carrying the collection electrode in a centered position. The motorization stage of the collection surface (Micromotion, Micro-Linear-Pusher) is seen on the left. The collection surface (white, phosphor coated as required for preparatory measurements described in Sect. 5.1.2) is seen on the right

The collection surface has to be placed in the focal point of the first parabolic mirror (denoted as mirror 1). The positioning precision should be in the range of 10 μm, which is a bit smaller than the size of the micro electrode, in order to avoid unwanted loss of light or signal contrast. For this purpose, the collection electrode is mounted adjustable to the mirror mount of parabolic mirror 1, as shown in Fig. 4.18. Adjustments in three axes are foreseen: two in the mirror plane (x and y direction) and one in the direction of the focus (z direction). The adjustment of the collection electrode in z direction is motorized, the smallest available linear pusher with a step size of 0.25 μm (6 mm maximum pitch) is used for this purpose (Micromotion, Micro-Linear-Pusher). The motorization is Labview controlled and allows for an easy adjustment and position changes also under vacuum. As the z axis has to be adjustable independently of position changes in the x and y direction, the micro electrode is directly mounted to the motorization stage and the combined system is placed on the central axis of parabolic mirror 1. For this purpose the printed-circuit board is glued to a ceramic insulator, which itself is fixed onto an aluminum adapter that can be mounted to the motorization stage. Three 1 mm thick aluminum struts are used to support the motorization system to the outside, where it is fixed to an aluminum ring, which itself is adjustable with respect to the mirror mount in x and y direction. The latter adjustment can only be done manually and is realized by three adjustment screws (OWIS, FGS 15-6-75).

4.3.2 *The First Parabolic Mirror*

The collection surface is placed in the focus of an annular parabolic mirror in order to collimate any light as potentially emitted in the ^{229m}Th isomeric decay. The mirror has a focal length of $f_1 = 10$ mm and an aperture of 39 mm diameter. A center hole of 12 mm diameter allows to collect the thorium ions head-on on the collection surface, which was shown to be the most advantageous collection method with the help of SIMION simulations. With this geometry, a large acceptance of 41.3% of all emitted light is obtained. A photograph of the parabolic mirror next to a schematic drawing is shown in Fig. 4.19.

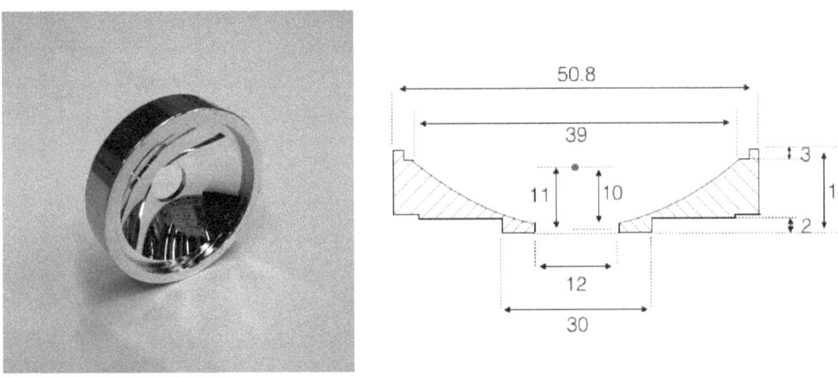

Fig. 4.19 Photograph of the first parabolic mirror, providing a large light yield for efficient collimation (left). The outer diameter of the mirror is 50.8 mm (2 in.). A dimensional drawing of the mirror surface with 39 mm aperture and 10 mm focal length is shown on the right [48]

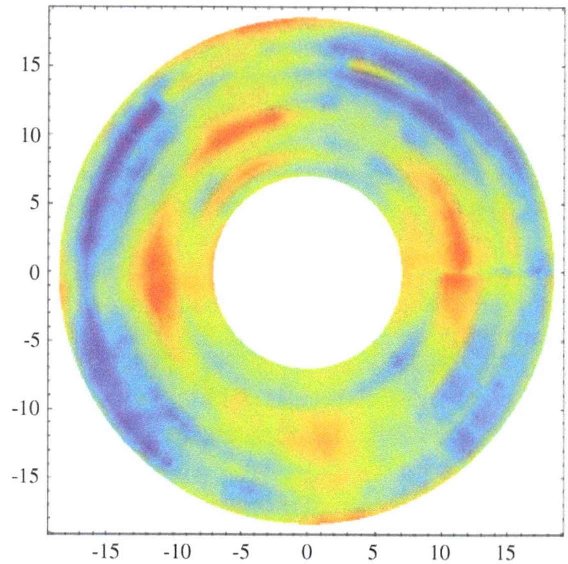

Fig. 4.20 White-light interferometric measurement of parabolic mirror 1 as performed by the Fraunhofer-Institut für Angewandte Optik und Feinmechanik (IOF). The determined surface roughness is below 1 nm rms

In order to provide a large reflectivity of more 70% in the VUV region around 160 nm, the reflective surface consists of MgF_2-coated aluminum. This coating provides high reflectivities down to 130 nm wavelength (Fig. 4.22) [21]. The mirror was fabricated at the Fraunhofer-Institut für Angewandte Optik und Feinmechanik (IOF) in Jena in a multi-step production process. In a first step, the special mirror-shape was produced by diamond turning from an aluminum block. In a second step, the surface was coated with a Ni-phosphorus layer, which allows for polishing to achieve surface roughnesses of less than 1 nm rms (root mean square) as measured by white-light interferometry shown in Fig. 4.20. In a third and last step, the surface is coated with aluminum and MgF_2 within the same manufacturing step under vacuum. Here the aluminum is the reflective coating, providing a high reflectivity in the VUV region, while the MgF_2 coating acts as a protective layer, preventing the aluminum from oxidation. The peak-to-valley shape deviation of the mirror was measured to be below 200 nm.

In order to provide the optimum optical efficiency, the mirror has to be turnable in two angular directions (θ and ϕ direction). For this purpose it is mounted to a two-inch mirror mount (Newport, U200-A) and two adjustment screws (Newport, AJS-0.5H V6) are used for mirror adjustment (Fig. 4.18). The mirror mount is placed on a linear translation stage (Newport, M-443-4 V6), allowing for up to 100 mm linear translation. This linear translation is required in order to shift the mirror in direction of the beam axis until it has reached its operational position, in which the collection surface (also fixed to the mirror mount) is placed in about 3 mm distance to the exit of the triode extraction system. The translational shift is motorized by a vacuum compatible motor (Faulhaber, AM2224) combined with a lead screw (Faulhaber, BS-1.5 4864). No high precision is required at this point, nevertheless the motorization is needed, as in this way the mirror can be shifted away from the triode extraction system under vacuum, allowing for ion-extraction efficiency measurements prior to the search for any isomeric decay.

4.3.3 The Second Parabolic Mirror

Collimated parallel light, as coming from the first mirror surface, has to be focused down to the detection plane. For this purpose, a second annular parabolic mirror was developed, which has a special shape, leading to a focal point that is located in the back side of the mirror. The focal length of this second parabolic mirror is $f_2 = 2$ mm at 39 mm diameter aperture. Like for the first mirror, a 12 mm diameter central hole is provided. In this way the focus is cut free, allowing for the positioning of the detection system in the focal plane. A photograph of the second parabolic mirror is shown in Fig. 4.21 together with a schematic drawing.

The magnification factor of the combined system consisting of mirror 1 and mirror 2 under paraxial approximation would be $m = f_2/f_1 = 1/5$, which suggests that the light, emitted from the 50 μm diameter source, would be focused down to a 10 μm diameter spot size. This calculation, however, is making use of simplifications that are only valid for paraxial optical systems, where the light is propagating close to the optical axis. As the described system is heavily non-paraxial, the approximation is not valid. Extensive numerical simulations were performed in order optimize the design of the optical system, using a self-written Matlab-based ray-tracing code (see Appendix B.2). Based on these simulations, an achievable focal spot size of

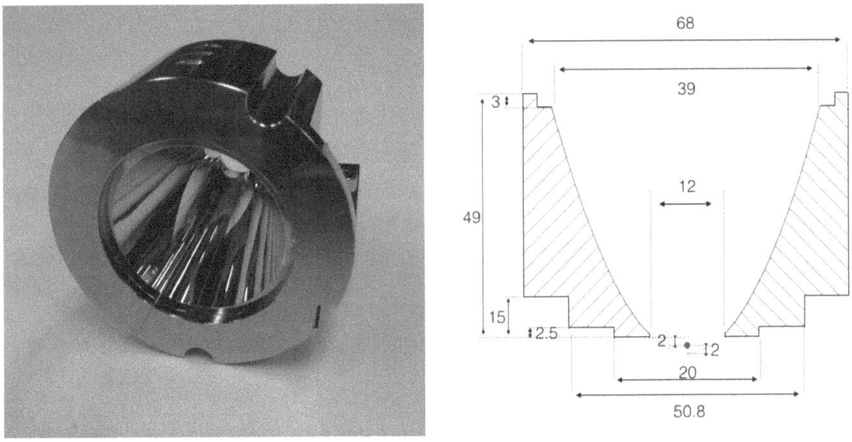

Fig. 4.21 Photograph of the second parabolic mirror used to focus the light onto the detection surface (left). A dimensional sketch of the deep parabola is shown on the right [48]. The aperture is 39 mm diameter at a focal length of 2 mm

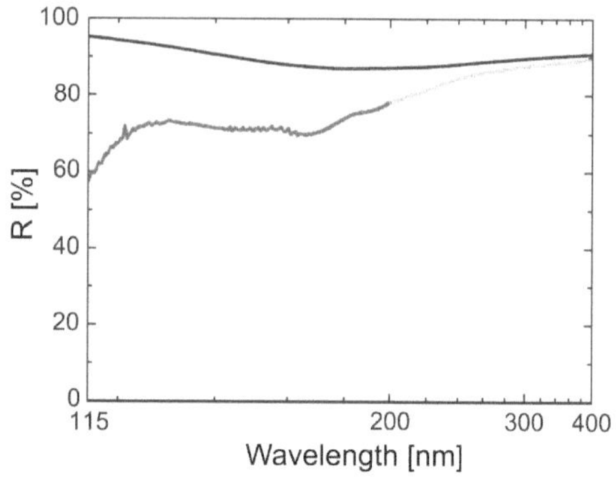

Fig. 4.22 Photograph of parabolic mirror 2 in its mirror mount (left). Mirror reflectivity for the MgF_2-coated aluminum surface as a function of wavelength (right). The reflectivity measurement was provided by the Fraunhofer-Institut für Angewandte Optik und Feinmechanik (IOF) in Jena [21]. A mirror reflectivity of about 70% around 160 nm is achieved. The measurement in the VUV below 200 nm was performed by Korth Kristalle GmbH, the measurement above 200 nm was performed by the IOF

44 μm (FWHM) is obtained, corresponding to a simulated factor of magnification of about 0.9 [9]. Experimentally, however, the best achieved focal spot size was 106 μm FWHM (Sect. 5.1.3) [19]. Most likely, this discrepancy originates from small misalignments, which are hard to be completely avoided.

With the help of the simulations it was found that an optical system of similar efficiency is obtained when replacing the second, deep parabolic mirror by a refractive optical component, namely an aspherical lens. In this case, however, the maximum efficiency is only obtained if the shape of the lens is adapted to the wavelength of the isomeric decay light. As this wavelength is unknown, the decision was to use a completely reflective optical system instead, providing a wavelength independent performance for wavelengths down to 130 nm (Fig. 4.22) [9, 21].

The reflective surface of the second mirror consists of MgF_2-coated aluminum, that was produced in the same multi-step process as for mirror 1. However, due to the depth of the mirror, the normal coating procedure could not be carried out and instead a new coating method was developed for this purpose. After diamond turning, the mirror was mounted to a special holder and cut into three pieces. Each mirror piece was coated individually and afterwards the pieces were mounted together in a well defined way, allowing for a precision of better than 10 μm. This fabrication procedure was especially developed by the IOF Jena for our purposes and is now frequently used for industrial fabrication of deep mirror surfaces.

In order to allow for an angular adjustment (θ and ϕ direction) of the mirror, it is mounted to a mirror mount (Newport, U200-A LH). The mirror mount had to be mechanically adapted for carrying the deep parabola (Fig. 4.22). Two large adjustment screws (Newport, AJS 100-02 V6) are used for the angular fine adjustment of the mirror surface. Besides the angular adjustment, also a precise adjustment of the focal distance to the detection plane is required (z direction). For this purpose, the mirror mount is fixed onto a linear translation stage (Newport, M-443-4 V6). This stage is motorized with a linear pusher (Pi-Micos, MP20-B), which allows for a linear translation of up to 76 mm at a step size of 0.54 μm. The long translation path is required, as the detector (MCP detection system described in Sect. 4.4.1) has always to be kept under vacuum. For this reason a vacuum valve is used (VAT, 10844-CE44-0005) that can be closed when the vacuum chamber is vented. The second parabolic mirror has to be moved under vacuum through the valve (of 70 mm length) to reach its final position in 2 mm distance to the detector surface.

All motorizations of the vacuum optical system are driven by Faulhaber motors, that are controlled via three Technosoft IPS210 modules. The modules are connected via CAN-bus and controlled by Labview (see Ref. [48] for a detailed description).

4.3.4 Optical Filters

The region of collimated light between the two mirrors conceptually allows for a wavelength determination to better than 1 nm, in case of a successful observation of light emitted in the isomeric decay. For this purpose, interference filters (Laseroptik, Long Wave Pass Coating) with a steep absorption edge around 160 nm were foreseen. The filters consist of coated CaF_2 windows of 50 mm diameter and 5 mm thickness, as shown exemplarily in Fig. 4.23. The wavelength of the absorption edge can be changed by turning the filter. Three different types of filters were provided by Laseroptik, one with an edge adjustable between 150 nm (45°) and 160 nm (0°) (L-02573), one with an edge between 160 nm (45°) and 170 nm (0°) (L-02574), and one also with an edge between 160 nm (45°) and 170 nm (0°), however on a smaller substrate (19114H2). The transmission curve of filter L-02574 is shown in Fig. 4.23 (measured and provided by Laseroptik). As no light originating from the isomeric decay was observed, the filters were not included in any of the performed experiments.

4.3.5 The Optical Chamber

The entire optical system is mounted in a vacuum chamber, which is of cubical shape with 230 mm inner dimension (Fig. 4.24). The flange types are DIN-160 CF and both parabolic mirrors with their support and positioning equipment have to fit movable through this flange size, posing strong limitations for the size of the optical mountings.

An aluminum optical base-plate of 15 mm thickness was fabricated to allow for the mounting of the optics in the chamber (Fig. 4.25). This base plate is adjustable with respect to the chamber walls in order to allow for adjusting the optical axis. Both linear translation stages are mounted on top of the base plate, together with their

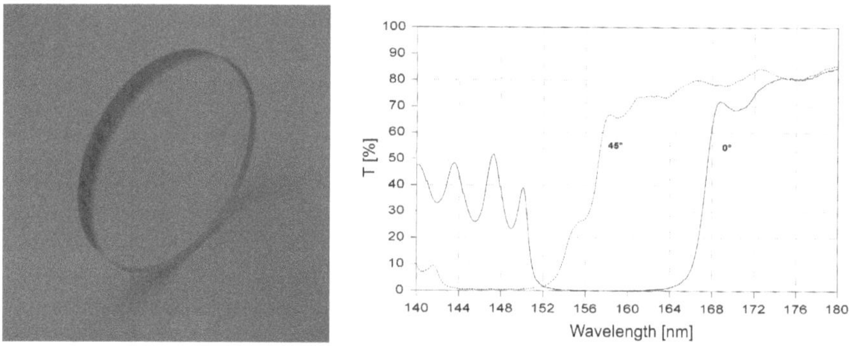

Fig. 4.23 Photograph of one of the turnable VUV filters with a steep absorption edge provided by Laseroptik (left). The corresponding measured transmission curves for two different angles are shown on the right

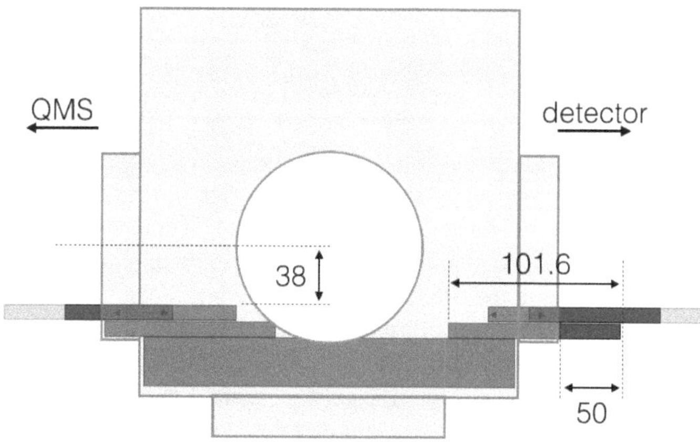

Fig. 4.24 Schematic drawing of the optical vacuum chamber together with the aluminum baseplate. The QMS is located to the left and the detection system to the right of the chamber. (drawing from Ref. [48])

motorizations. This is done in a way that the mirrors fit through the cube flanges of opposite sides. On one side, the vacuum chamber is mounted to the QMS, while on the other side the detector is mounted together with the vacuum valve (VAT, 10844-CE44-0005). The vacuum valve ensures that the detector is permanently kept under vacuum when the optical chamber is vented for adjustments.

During operation, the vacuum chamber is pumped by a turbo-molecular pumping station (Pfeiffer, TSU 521) with a pumping speed of 510 l/min. A pressure of typically about 10^{-6} mbar is reached, owing to the outgassing of some motor components, which are not designed for better vacuum conditions. The pressure is monitored by a full range gauge (Pfeiffer, PKR 261).

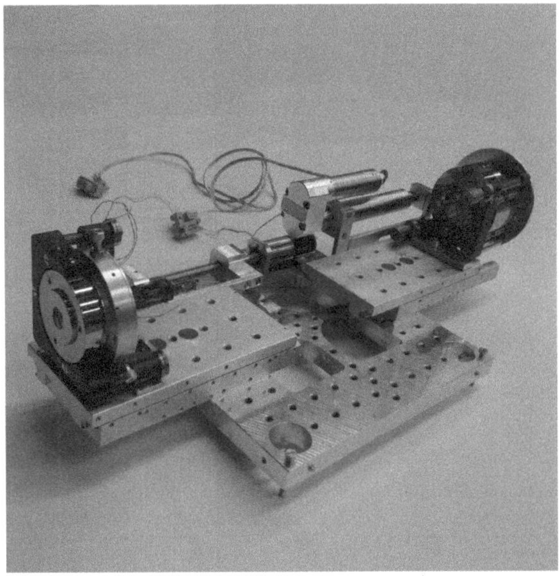

Fig. 4.25 Photograph of the optical system. The first linear translation stage is mounted on the left-hand side on top of the base plate together with parabolic mirror 1. The second linear translation stage is mounted on the right together with parabolic mirror 2. Both translation stages are motorized

4.4 The Detection System

For the detection of a potential photonic decay branch of ^{229m}Th, a detector with a high quantum efficiency in the VUV region around 160 nm is required. CsI-coated microchannel plate detectors (MCPs) provide quantum efficiencies of about 10% in this wavelength region and are among the most sensitive detectors in the VUV [22]. For this reason, a CsI-coated MCP was the detector of choice for our purpose. In order to make use of the small spot size provided by the optical focusing system, the detector has to allow for spatially resolved signal read-out. This was implemented by using an MCP detector combined with a phosphor screen. The phosphor screen is subsequently monitored by a CCD camera.

Although the CsI coating of the MCP surface was not required during the search for an internal conversion (IC) decay, the same detector system was used in these experiments. The MCP detector generates an electronic signal, which is then converted into visible light. In this way a high sensitivity for any potential decay is obtained. In the following, the MCP detector and the CCD camera will be described.

4.4.1 The MCP Detector

The MCP detector is of type BOS-75-FO (Beam Imaging Solutions) and consists of two imaging plates of 1 mm thickness (chevron geometry) with 75 mm outer diameter and 25 μm channel diameter. The two plates are placed in front of a flange-

Fig. 4.26 Photograph of the vacuum-flange mounted microchannel plate (MCP) detector (BOS-75-FO) used for the investigation of a potential photonic decay branch as well as for the detection of the internal-conversion decay of the ^{229}Th isomeric state

mounted optical fiber-glass window that is coated with a phosphor layer. The achieved resolution is about 50 μm and the dark count rate amounts to ~0.01 cts/(s mm)^2. Individual voltages can be applied to the front surface and the back surface of the stacked imaging plates, as well as to the phosphor screen. Typical voltages, as applied for the detection of the isomer's IC decay channel, are −25 V (front surface), +1900 V (back side) and +6000 V (phosphor screen), respectively. The voltages are applied by HV voltage supplies (MCP plates: Keithley Instruments, High Voltage Supply 246, −3− + 3 kV, phosphor screen: Fluke, High Voltage Power Supply 410 B, −10 to +10 kV). A photograph of the MCP detector mounted to a DN-160 CF flange is shown in Fig. 4.26.

The operational principle of the MCP detector is as follows: if an event occurs, which triggers the emission of one (or more) electrons in a microchannel of the MCP (e.g. radioactive decay, ion implantation, photo-electron emission), this electron feels the attractive potential of about 1000 V per channel plate through one of the microchannels. The electron will thus get accelerated along the channel, thereby hitting the channel walls. During the collisions, secondary electrons are generated, leading to the formation of an electron cascade. The typical factor of electron amplification is 10^6 for two plates. As soon as the electron cascade leaves the back side of the second channel plate, it will feel the +6000 V attractive potential of the phosphor screen. The phosphor screen is then hit by an electron cloud of about 10^6 electrons with a kinetic energy of about 4 keV, enough to generate a visible light signal. A schematic drawing of the operational principle is shown in Fig. 4.27.

The MCP front surface is coated with a CsI layer in order to improve the detection efficiency for VUV light around 160 nm to about 10% (Fig. 4.28) [23]. The detection

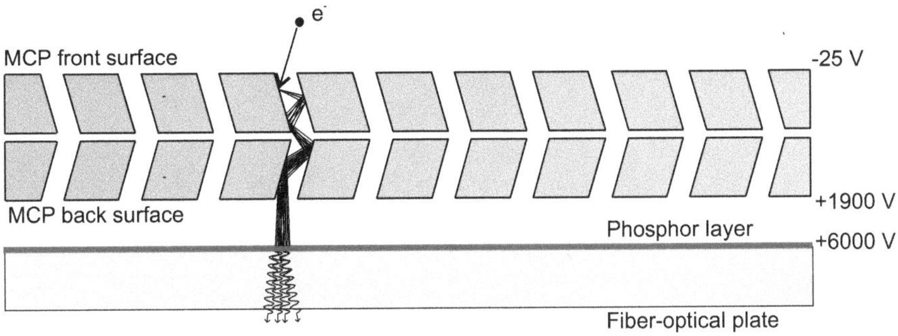

Fig. 4.27 Conceptual sketch of the MCP-detector operational principle. Any event that triggers the emission of an electron on the MCP surface can be detected. The electron is multiplied in the microchannels, leading to the production of an electron cloud. The corresponding electronic signal is converted into visible light with the help of a phosphor screen

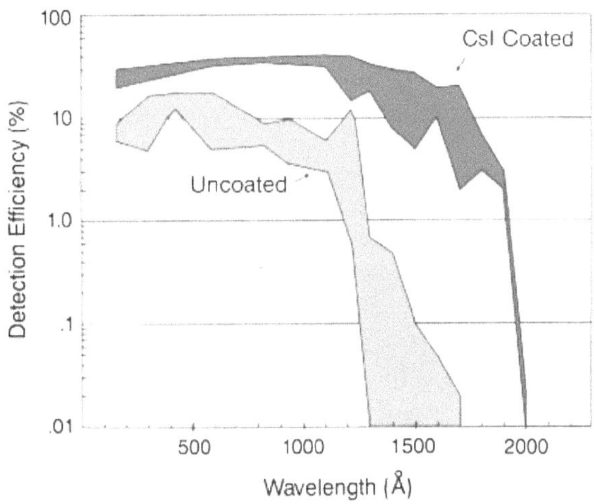

Fig. 4.28 MCP detection efficiencies for VUV light as a function of the wavelength taken from Ref. [50]. The efficiencies for CsI-coated MCP surfaces as well as for uncoated MCPs are shown. The detection efficiency of CsI-coated MCPs for light of about 160 nm wavelength is typically in the range of 10% [23]

efficiency of MCP detectors for low-energy electrons was discussed by Goruganthu and Wilson [49]. It was found that the detection efficiency starts to decrease below an optimum kinetic energy around 300 eV. However, even at very low kinetic electron energies around 0 eV there is a nonzero detection efficiency remaining. The reason is that the MCP detection principle is based on the generation of an electron cascade, which can be directly triggered by a free electron as soon as it is accelerated towards a microchannel of the MCP plate. This is a special feature of electron detection and was of significant help for the observation of the IC decay channel of ^{229m}Th. The kinetic energy of IC electrons, as emitted in the isomeric decay, is estimated to about $E = 1.5$ eV ((7.8–6.3) eV, where 7.8 eV is the expected isomeric energy and 6.3 eV is the thorium ionization potential).[2] The analytic expression obtained for the energy dependent relative electron detection efficiency $\eta(E)$ is [49]

$$\eta(E) = \frac{1 - e^{-k\delta(E)/\delta_{\max}}}{1 - e^{-k}}. \quad (4.1)$$

Here $k = 2.2$ is an empirical parameter, $\delta(E)$ is the secondary emission yield function and $\delta_{\max}$ equals 1 in case of normalization. For $\delta(E)$ the following theoretical equation holds [49]

$$\delta(E) = \delta_{\max}\left(\frac{E}{E_{\max}}\right)^{1-\alpha}\left(\frac{1 - e^{-T_{\max}(E/E_{\max})^{\alpha}}}{1 - e^{-T_{\max}}}\right), \quad (4.2)$$

where $\alpha = 1.35$ and $T_{\max} = 2.283$ are empirical parameters and $E_{\max}$ is the energy of the maximum detection efficiency of presumably 325 eV. Assuming a maximum absolute detection efficiency of 50% at 325 eV, the absolute electron detection efficiency for electrons of 1.5 eV is estimated to be about 1.5% based on Eq. (4.1).

Besides the detection efficiencies for photons and electrons emitted in the isomeric decay, also the MCP detection efficiencies for ions play an important role in the experiments for two reasons. Firstly, the ionic impact signal caused by ions of several keV kinetic energy is used to calibrate the QMS. Secondly, during IC detection, the remaining ionic impact signal caused by the low energy thorium ions has to be taken into account as a potential source of background. While the MCP detection efficiency for ions of several keV kinetic energy is relatively large (typically 50–60%, corresponding to the MCP's open area ratio) [51], MCP detection efficiencies for ions of low kinetic energy are harder to obtain and were reported in Refs. [52, 53]. These are very low, as most of the kinetic energy is transferred to phonons [54], and even decrease with increasing mass and ion charge state (Fig. 4.29). The very low detection efficiency for low energy ions was extremely helpful for the identification of the ^{229}Th isomeric decay, as it allowed for nearly background free detection of the IC electrons emitted in the isomeric decay.

4.4.2 The CCD Camera

A CCD camera (Point Grey, FL2-14S3M-C) combined with a zoom lens (Computar M2514MP2, 25 mm, C-mount) is used for monitoring the phosphor screen of the MCP detection system (Fig. 4.30). For the purpose of IC electron detection, the camera was positioned on an optical rail in about 300 mm distance to the screen. The corresponding field of view is 100 mm by 75 mm. For the optical measurements, the CCD camera was positioned combined with the zoom lens and a 5 mm spacer in a

[2]Note, that this is a simplification, as for a correct calculation of the electron energies the work function Φ of the surface material and the electon's binding energy E_b below the fermi-edge have to be taken into account: $E_e = E_\gamma - \Phi - E_b$. As CsI exhibits a work function of 6.2 eV, no significant correction for the maximum energy of the electrons is expected.

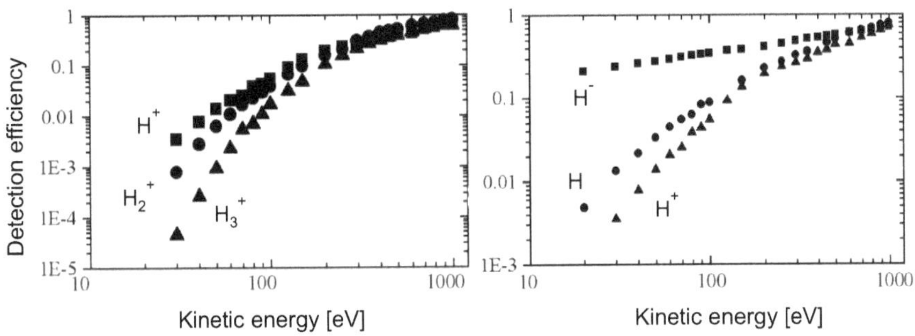

Fig. 4.29 MCP detection efficiencies for low-energy hydrogen ions. The detection efficiencies as function of the energy for hydrogen molecules of different masses are shown on the left. The detection efficiencies are reduced for increasing masses. The detection efficiencies for H^-, H and H^+ are shown on the right. A reduction of the detection efficiency occurs for increasing positive charge state. With kind permission of Elsevier [52]

Fig. 4.30 Photograph of the CCD camera (Point Grey, FL2-14S3M-C) combined with a zoom lens (Computar M2514MP2, 25 mm, C-mount) used for read-out of the MCP detection system (left). The right-hand side shows the CCD chip (Sony, ICX267 CCD) of an open camera. (Photograph from Ref. [48])

few cm distance to the phosphor screen, leading to 16 mm by 11.9 mm field of view. The whole camera system is contained in a light-tight housing, in order to reduce background light. The MCP fiber-glass window is covered by a 20 mm diameter aperture, in order to blind out field-emission effects originating from the detector's edge.

The CCD camera contains a CCD chip from Sony (ICX267 CCD, $4.64 \times 4.65\,\mu m$ pixel size, 1384×1032 pixels), which allowed for a maximum exposure time of four seconds for a single frame. The individual frames were automatically read out and saved by a Labview program. In individual frames, single events on the MCP detector can clearly be distinguished from noise and hot pixels of the CCD camera.

For this reason a Matlab program was written to automatically infer the position of MCP events from the individual frames (see Appendix B.3). A multitude of image frames (typically 500) are evaluated in this way and subsequently the event positions are added, leading to the acquisition of a single measurement (of typically 2000 s integration time). By applying this method, only a minor amount of CCD noise is not properly filtered and the background is dominated by the low MCP intrinsic dark-count rate of about 0.01 cts/(s mm)^2.

4.5 Efficiency Estimates

An estimate of the overall efficiency of the detection concept is an important requirement in order to evaluate if a direct detection of the ^{229}Th isomeric decay appears to be realistic. Such efficiency estimations are presented in the following separately for the determination of a photonic decay branch as well as for the internal conversion decay channel. The efficiency estimates are partly based on experimental observations and partly based on numerical simulations.

4.5.1 *Efficiency Estimate for the Ion Extraction*

Only the absolute number of extracted ^{229}Th ions per second is important for the observation of the isomeric decay. However, experiments were also performed in order to obtain relative extraction efficiencies. For this purpose the ^{233}U source (source 1) was investigated in detail in order to experimentally determine the absolute number of emitted ^{229}Th α-recoil nuclides per second to about 10 000 s^{-1} (see Appendix A.4). Given the α activity of the source to be about 200 kBq, this corresponds to a recoil efficiency of the ^{233}U source of $\sim$5.0% [16, 18]. The combined stopping, extraction and mass-purification efficiency for thorium could then be determined by measuring the absolute number of extracted ^{229}Th ions per second. Corresponding measurements were performed behind the triode extraction system (see Sect. 5.1.1). Results were obtained for individual charge states and revealed extraction efficiencies of (10$\pm$2.0)% for ^{229}Th^{3+}, (5.5$\pm$1.1)% for ^{229}Th^{2+} and (0.34$\pm$0.07)% for ^{229}Th^{1+} (Appendix A.7) [16]. The exact values are, however, subject to some changes, due to varying buffer-gas purity. Most of the measurements were performed with ^{229}Th in the 2+ or 3+ charge state, with an extraction efficiency in the range of 10%, leading to about 1000 extracted ^{229}Th ions per second in case of source 1. For the larger ^{233}U source 2, a factor of $\sim$10 increased absolute ion extraction rate was experimentally determined, leading to about 10 000 extracted ^{229}Th ions per second (in the 2+ and 3+ charge states). No experimental efficiency analysis was carried out with this second source. An overview of the efficiency budget is shown in Table 4.3.

Table 4.3 Efficiencies for the ion extraction

	Efficiency [%]	Source 1 [s^{-1}]	Source 2 [s^{-1}]
Source α activity		200 kBq	290 kBq
Recoil efficiency	5.3/35[a]	10 600	100 000
Extraction efficiency	~10	1000	10 000

[a] This value is estimated under the assumption of identical extraction efficiencies for source 1 and source 2 (see Sect. 5.1.1). It is in agreement with the calculated α-recoil efficiency assuming a metallic ^{233}U source (Sect. 4.2.1)

4.5.2 Efficiency Estimate for a Photonic Decay

For the optical system, the efficiencies are largely based on numerical simulations. The concept of an optical detection requires to collect the ^{229}Th ions on a small (50 μm diameter) collection surface. Experiments were carried out in order to prove that an ion collection does indeed take place (Sect. 5.1.2). However, these experiments, while suggesting that the collection is efficient, do not allow for a quantitative analysis of the collection efficiency. This efficiency could only be estimated based on SIMION simulations (Appendix B.1.6) to be 40%, which we consider to be a pessimistic estimation.

Detailed numerical ray-tracing simulations were performed in order to determine the efficiency of the optical system (see Appendix B.2). These include all geometrical aspects as well as the angular dependent reflectivity of the mirror surfaces [9]. The amount of photons that will arrive in the detection plane, relative to the total number of photons emitted from the detection surface, was determined to be about 23%. We refer to this number as the optical transmission efficiency. It consists of two parts, (i) the geometric efficiency, largely dominated by the acceptance of mirror 1 of about 41% and (ii) the reflectivity of the mirror surfaces, which is estimated to be 70% for each mirror. This simple approximation already results in an optical transmission efficiency of 20%. The minor deviation from the simulation-based value originates from the angular dependence of the reflectivity [55].

Not all photons that approach the detector surface will be located in the peak center. Simulations of the peak structure revealed, that about 62% of all photons that impinge in the detection plane will fall within the FWHM of the image peak, with a diameter of expectedly 44 μm (Appendix B.2). We refer to this value of 62% as the imaging efficiency of the optical system. Experimentally, an image diameter of 106 μm (FWHM) (corresponding to $7.8 \cdot 10^{-3}$ mm^2 detection surface) was the best value that could be achieved (Sect. 5.1.3) [19]. The deviation from the simulated value might occur due to minor misalignments of the optics, that are hard to be completely avoided.

The detection efficiency for photons around 160 nm of a CsI-coated MCP detector is typically in the range of 10% [23]. This value was taken from literature. In combination with the MCP dark-count rate of 0.01 cts/(s mm)^2, this allows to give a complete efficiency budget for the detection of a potential photonic decay channel of

Table 4.4 Efficiencies for a photonic decay detection

	Efficiency [%]	Source 1 [s^{-1}]	Source 2 [s^{-1}]
Extracted ions		1000	10 000
Isomeric fraction	2	20	200
Collection efficiency	40	8	80
Transmission efficiency	23	1.8	18
Imaging efficiency	62	1.1	11
Detection efficiency	10	0.1	1
Spot size	100 μm		
Signal intensity		13 s^{-1} mm^{-2}	130 s^{-1} mm^{-2}
Signal-to-background ratio		1300:1	13 000:1

^{229m}Th as shown in Table 4.4. The result is an expected signal-to-background ratio of 1300:1 when using the small ^{233}U source (source 1) and 13 000:1 when using the large-area ^{233}U source (source 2). Of course these numbers assume a purely radiative isomeric decay and do not account for non-radiative decay channels as have to be expected due to electronic-bridge processes even in case that internal conversion is successfully suppressed. The branching ratios of such non-radiative decay channels are hard to be quantified and can easily amount to several orders of magnitude [12, 43, 44].

4.5.3 *Efficiency Estimate for an Internal-Conversion Decay*

During the search for an internal conversion decay channel, the extracted ^{229}Th ions are directly accumulated on the MCP detector surface. No voltages are applied that could lead to a focusing of the ion beam and for this reason the surface area of ion accumulation corresponds to the size of the orifice of the triode extraction system of 2 mm diameter. The corresponding detection surface area of 3.1 mm^2 is large compared to the $7.8 \cdot 10^{-3}$ mm^2 obtained in the search for a photonic decay, however, the large rate of extracted ions and the low noise of the MCP detector can compensate for that.

Starting with the same number of extracted ^{229}Th ions of 1000 1/s and 10 000 1/s for source 1 and source 2, respectively, the only remaining relevant parameter is the MCP detection efficiency for low energy electrons. Based on studies performed in Ref. [49], this can be estimated to 1.5% for electrons of about 1.5 eV kinetic energy as expected to occur in the IC decay of ^{229m}Th (see Sect. 4.4.1). The resulting efficiency budget is shown in Table 4.5. An expected signal-to-background ratio of 10:1 is obtained for the small ^{233}U source 1 and 100:1 for the large-area source 2, respectively. While the expected signal-to-background ratios are smaller, the obvious advantage in the search for an IC-decay channel is, that in most chemical environ-

Table 4.5 Efficiencies for an internal-conversion decay detection

	Efficiency [%]	Source 1 [s^{-1}]	Source 2 [s^{-1}]
Extracted ions		1000	10 000
Isomeric fraction	2	20	200
Detection efficiency	1.5	0.3	3.0
Spot size	2 mm		
Signal intensity		0.1 s^{-1} mm^{-2}	1.0 s^{-1} mm^{-2}
Signal-to-background ratio		10:1	100:1

ments this has to be expected to be the dominant decay channel [6, 46]. Therefore no other branching ratios of unknown strength have to be taken into account.

References

1. Strizhov VF, Tkalya EV (1991) Decay channel of low-lying isomer state of the ^{229}Th nucleus. Possibilities of experimental investigation. Sov Phys JETP 72:387
2. Karpeshin FF, Trzhaskovskaya MB (2007) Impact of the electron environment on the lifetime of the $^{229}Th^m$ low-lying isomer. Phys Rev C 76:054313
3. Dykhne AM et al (1996) Alpha decay of the first excited state of the Th-229 nucleus. JETP Lett 64:345–349
4. Tkalya EV et al (2000) Decay of the low-energy nuclear isomer $^{229}Th^m(3/2^+, 3.5\pm1.0$ eV) in solids (dielectrics and metals): a new scheme of experimental research. Phys Rev C 61:064308
5. Tkalya EV (1999) Nonradiative decay of the low-lying nuclear isomer ^{229m}Th (3.5eV) in a metal. JETP Lett 70:371–374
6. Tkalya EV et al (2015) Radiative lifetime and energy of the low-energy isomeric level in ^{229}Th. Phys Rev C 92:054324
7. von der Wense L et al (2016) Direct detection of the ^{229}Th nuclear clock transition. Nature 533:47–51
8. Thirolf PG et al (2007) Optical access to the lowest nuclear transition in ^{229m}Th. Annual Report of the Maier-Leibnitz Laboratory, Garching, p 18
9. von der Wense L et al (2013) Towards a direct transition energy measurement of the lowest nuclear excitation in ^{229}Th. JINST 8:P03005
10. Barci V et al (2003) Nuclear structure of ^{229}Th from γ-ray spectroscopy study of ^{233}U α-particle decay. Phys Rev C 68:034329
11. Rellergert WG et al (2010) Constraining the evolution of the fundamental constants with a solid-state optical frequency reference based on the ^{229}Th nucleus. Phys Rev Lett 104:200802
12. Porsev SG et al (2010) Excitation of the isomeric ^{229m}Th nuclear state via an electronic bridge process in $^{229}Th^+$. Phys Rev Lett 105:182501
13. Neumayr JB et al (2006) Performance of the MLL-Ion catcher. Rev Sci Instrum 77:065109
14. Neumayr JB (2004) The buffer-gas cell and the extraction RFQ for SHIPTRAP. Ph.D. thesis, LMU Munich, Germany
15. Neumayr JB et al (2006) The ion-catcher device for SHIPTRAP. Nucl Instrum Methods B 244:489–500
16. von der Wense L et al (2015) Determination of the extraction efficiency for ^{233}U source α-recoil ions from the MLL buffer-gas stopping cell. Eur Phys J A 51:29

17. Haettner E (2011) A novel radio frequency quadrupole system for SHIPTRAP & New mass measurements of rp nuclides. Ph.D. thesis, University of Giessen, Germany
18. von der Wense L et al (2016) The extraction of $^{229}Th^{3+}$ from a buffer-gas stopping cell. Nucl Instrum Methods B 376:260–264
19. Seiferle B et al (2016) A VUV detection system for the direct photonic identification of the first excited isomeric state of ^{229}Th. Eur Phys J D 70:58
20. Dessovic P et al (2014) ^{229}Thorium-doped calcium fluoride for nuclear laser spectroscopy. J Phys Condens Matter 26:105402
21. Wilbrandt S et al (2016) Aluminum reflectors for the DUV and VUV, IOF annual report (2011). http://www.iof.fraunhofer.de/content/dam/iof/en/documents/publications/annual-report/2011/reflectors-duv-vuv-2011-wilbrandt.pdf. Accessed 16 Aug 2016
22. Wiza JL (1979) Microchannel plate detectors. Nucl Instrum. Methods 162:587–601
23. Carruthers GR (1988) Further investigation of CsI-coated microchannel plate quantum efficiencies. Appl Opt 27:5157–5159
24. Vascon A et al (2012) Elucidation of constant current density molecular plating. Nucl Instrum Methods A 696:180–191
25. Vascon A et al (2013) Smooth crack-free targets for nuclear applications produced by molecular plating. Nucl Instrum Methods A 714:163–175
26. Vascon A et al (2013) The performance of thin layers produced by molecular plating as α-particle sources. Nucl Instrum Methods A 721:35–44
27. Maier HJ, Grossmann R, Friebel HU (1991) Radioactive targets for nuclear accelerator experiments. Nucl Instrum Methods B 56–57:926–932
28. Grossmann R, Maier HJ, Friebel HU (1997) The new hot-lab facility for radioactive target preparation at the University of Munich. Nucl Instrum Methods A 397:39–45
29. NNDC Interactive Chart of Nuclides (2016) Brookhaven National Laboratory, Brookhaven. http://www.nndc.bnl.gov/chart. Accessed 22 Mar 2016
30. Ziegler JF, Biersack JP, Littmark U (1985) The stopping and range of ions in matter. Pergamon, New York
31. Nordlund K (1995) Molecular dynamics simulation of ion ranges in the 1–100 keV energy range. Comput Mater Sci 3:448–456
32. Henderson RA et al (2011) Electrodeposition of U and Pu on thin C and Ti substrates. Nucl Instrum Methods A 655:66–71
33. Wakeling MA (2014) Charge states of ^{229m}Th: Path to finding the half-life, Thesis performed at LLNL, Livermore (2014)
34. Wandkowsky N et al (2013) Modeling of electron emission processes accompanying radon-α-decays within electrostatic spectrometers. New J Phys 15:083040
35. Yu K et al (2001) A gas cell for thermalizing, storing and transporting radioactive ions and atoms. Part 1: Off-line studies with a laser ion source. Nucl Instrum Methods B 179:412–435
36. Taylor SE et al (2008) Pulsed Laval nozzle study of the kinetics of OH with unsaturated hydrocarbons at very low temperatures. Phys Chem Chem Phys 10:422–437
37. Brubaker WM (1968) An improved quadrupole mass analyser. Adv Mass Spectrom 4:293–299
38. Dawson PH (1976) Quadrupole mass spectrometry and its applications. Elsevier Scientific Pub. Co., Amsterdam
39. Kalb D (2012) Entwicklung eines Regelsystems zum Przisionsabgleich der RF-Amplituden eines Quadrupole-Massenspektrometers, Bachelor thesis, LMU Munich, Germany (in German)
40. Abdelrahman MM et al (2012) SIMION calculations for the triode extraction system. Int J Theor Math Phys 2:122–129
41. Soliman BA et al (2011) Simulation of ion beam extraction and focusing system. Chin Phys C 35:83–87
42. Dahl DA et al (1990) SIMION PC/PS2 electrostatic lens design program. Rev Sci Imstrum 61:607–609
43. Porsev SG, Flambaum VV (2010) Effect of atomic electrons on the 7.6 eV nuclear transition in $^{229m}Th^{3+}$. Phys Rev A 81:032504
44. Porsev SG, Flambaum VV (2010) Electronic bridge process in $^{229}Th^{+}$. Phys Rev A 81:042516

45. Jeet J et al (2015) Results of a direct search using synchrotron radiation for the low-energy ^{229}Th nuclear isomeric transition. Phys Rev Lett 114:253001
46. Borisyuk PV et al (2015) Band structure and decay channels of thorium-229 low lying isomeric state for ensemble of thorium atoms adsorbed on calcium fluoride. Phys Status Solidi C 12:1333–1337
47. Budenstein PP et al (1969) Destructive breakdown in thin films of SiO, MgF_2, CaF_2, CeF_3, CeO_2, and teflon. J Vac Sci Tech 6:289–303
48. Seiferle B (2015) Setup of a VUV detection system for the direct identification of the fluorescence radiation of ^{229m}Th, Master thesis, LMU Munich, Germany
49. Goruganthu RR, Wilson WG (1984) Relative electron detection efficiency of microchannel plates from 0–3 keV. Rev Sci Instrum 55:2030–2033
50. Barnstedt J (2016) Advanced practical course microchannel plate detectors University of Tübingen (2016). http://www.uni-tuebingen.de/fileadmin/Uni_Tuebingen/Fakultaeten/MathePhysik/Institute/IAAT/AIT/Lehrveranstaltungen/F-Praktikum/Dokumente/VersuchsAnleitungMCP_english.pdf. Accessed 16 Aug 2016
51. Oberheide J et al (1997) New results on the absolute ion detection efficiencies of a microchannel plate. Meas Sci Technol 8:351–354
52. Peko BL et al (2000) Absolute detection efficiencies of low energy H, H^-, H^+, H_2^+, H_3^+ incident on a multichannel plate detector. Nucl Instrum Methods B 171:597–604
53. Stephen TM et al (2000) Absolute calibration of a multichannel plate detector for low energy O, O^-, and O^+. Rev Sci Instrum 71:1355–1359
54. Bay HL, Winters HF, Coufal HJ, Eckstein W (1992) Energy transfer to a copper surface by low energy noble gas ion bombardment. Appl Phys A 55:174–278
55. Maier-Komor P et al (2002) VUV reflective coatings on thin concave float glass substrates with a perimeter of 86 cm to be used as provisional HADES RICH mirror segments. Nucl Instrum Methods A 480:65

Chapter 5
Measurements

Different sorts of measurements were performed, aiming for the direct detection of ^{229m}Th. The first focus was on preparatory measurements, required to verify (i) a high extraction rate of ^{229}Th ions from the buffer-gas stopping cell, (ii) the possibility to collect the ions on a 50-μm diameter micro electrode and (iii) the performance of the optical system. These measurements will be detailed in Sect. 5.1. The preparatory measurements were the basis for the next part of experiments, searching for a direct decay of the isomeric state, as will be detailed in Sect. 5.2. This includes the search for a photonic as well as an internal conversion (IC) decay. The successful direct detection of the isomeric state in the IC decay channel has led to a third part of experiments, performed in order to unambiguously show that the detected signal originates from the isomeric decay of ^{229m}Th. These experiments will be detailed in Sect. 5.3.

5.1 Preparatory Measurements

Three different types of preparatory measurements were carried out. (i) Following a first extraction of ^{229}Th ions from the ^{233}U source, the extraction voltages of the setup were optimized and a high absolute ^{229}Th extraction rate was verified. Also extraction efficiency measurements were performed. These are detailed in Appendix A.7. (ii) As a basis for the investigation of a potential photonic decay channel, the possibility to collect the extracted thorium ions on a microscopic collection surface was probed. (iii) Further, the achievable focal spot size of the optical system was experimentally investigated, being the most important parameter for the performance of the optics.

L. Von der Wense, *On the Direct Detection of ^{229m}Th*,
Springer Theses, https://doi.org/10.1007/978-3-319-70461-6_5

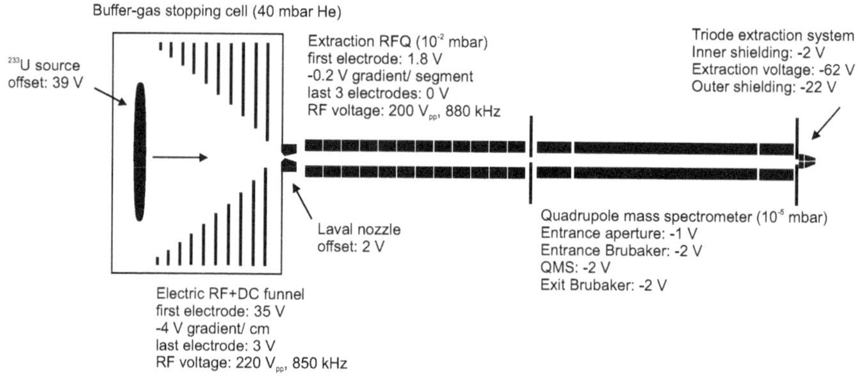

Fig. 5.1 Schematic drawing of the ion-beam formation system, used for the production and purification of a low-energy ^{229}Th ion beam. The optimized extraction voltages are indicated for each electrode. With kind permission of The European Physical Journal [1]

5.1.1 Extraction Rate for ^{229}Th Ions

Following a successful extraction of ^{229}Th ions from the buffer-gas stopping cell, a detailed study and optimization of the extraction rate was performed, leading to a complete optimization of all extraction voltages. These extraction conditions were fixed for all further investigations if not specified differently. They are detailed in Fig. 5.1.

With these extraction voltages, a complete mass scan from 10 u/e to 290 u/e was performed, using the small-area ^{233}U source (source 1). For this purpose a single anode MCP detector (Hamamatsu, F2223) with 27 mm surface diameter was placed in about 5 mm distance behind the triode extraction system. Voltages were applied as required for single-ion detection (−2000 V front surface, back surface connected to ground via a 5-MΩ resistor, anode on ground). The MCP detector was read out by a preamplifier (Ortec, VT120 fast timing preamplifier) connected to a shaping amplifier (Ortec, type 571) and a constant fraction discriminator (Canberra, type 1428A). Signal acquisition was performed by a DAQ card (National Instruments, NI-DAQmx) connected to a control PC. To perform the scan between 10 u/e and 200 u/e, the QMS was operated at a frequency of 925 kHz with RF amplitudes varying between 78.71 V_{pp} and 1574 V_{pp}. For the mass-over-charge region between 200 u/e and 290 u/e the operating frequency was lowered to 825 kHz using a larger air coil. The maximum operational voltage required for the extraction of 290 u/e was 1816 V_{pp}. The resulting mass scan is shown in Fig. 5.2.

It can be inferred that ^{229}Th is extracted in the 1+, 2+ and 3+ charge state. Besides ^{229}Th, also ^{233}U is detected, originating from sputtering of source material via the α decay. Some molecule formation can be observed, leading to the extraction of thorium oxides, hydrides as well as thorium with an argon adduct. The whole ^{233}U decay chain becomes visible in the 2+ charge state [1, 2].

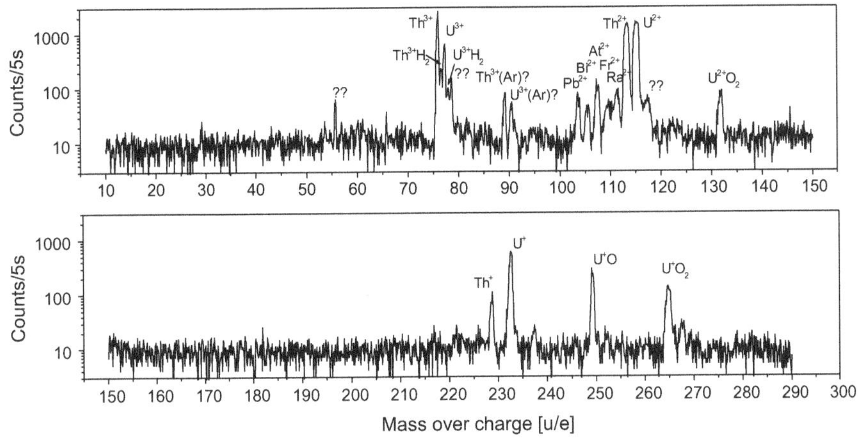

Fig. 5.2 Mass scan of the extracted α-recoil ions from 10 u/e to 290 u/e [1]. Ion extraction from the small-area ^{233}U source 1 was performed for this scan. The 1+, 2+ and 3+ charged species of ^{229}Th are clearly visible. Besides some hydrogen and oxygen-molecule formation, also the whole ^{233}U decay chain becomes visible in the 2+ charge state. ^{233}U is extracted due to sputtering of the source material via the α decay. With kind permission of Elsevier [2]

Typically, only 1+ and 2+ ions can be extracted from buffer-gas stopping cells to a significant amount [3]. Thorium poses an exception, due to its very low 3rd ionization potential of only 18.3 eV [4]. As the ionization potential of helium of 24.6 eV is above this value, it is energetically favorable for the electron to stay attached to the helium atom instead of reducing the 3+ charge state of thorium during collisions of the thorium ions with the helium buffer-gas background. The result is a significant Th^{3+} ion extraction rate [1, 2]. That was the first time that a considerable extraction rate for triply-charged ions from a buffer-gas stopping cell was reported.

A $^{229}Th^{3+}$ count rate of about 2500 cts in 5 s is determined. The detection efficiency for ions with a kinetic energy of several keV typically amounts to 50–60% [5], depending on the open-area ratio of the MCP detector. Assuming for the following an MCP detection efficiency of about 50%, the extraction rate for $^{229}Th^{3+}$ ions from the small-area ^{233}U source (source 1) can be estimated to be 1000 ions/s. The corresponding values for $^{229}Th^{2+}$ and $^{229}Th^{1+}$ are ~600 ions/s and ~40 ions/s, respectively. These numbers are, however, subject to some fluctuations, depending on the purity of the helium buffer gas.

A second, completely different, detection technique was applied in order to further validate the high extraction rate for $^{229}Th^{3+}$ ions. For this purpose, the thorium ion beam was accumulated directly onto the surface of a silicon detector (Ametek, BU-016-300-100) with 100 μm thickness and 300 mm^2 active area. A voltage offset of −1250 V was applied to the silicon detector surface in order to attract the ions. The QMS was set to extract $^{229}Th^{3+}$ with a mass resolution of $\Delta m/q = 1$ u/e. Due to the long ^{229}Th half-life of 7932 years, a continuous accumulation time of 5 days was chosen. Following the accumulation of $^{229}Th^{3+}$, subsequently α decays were

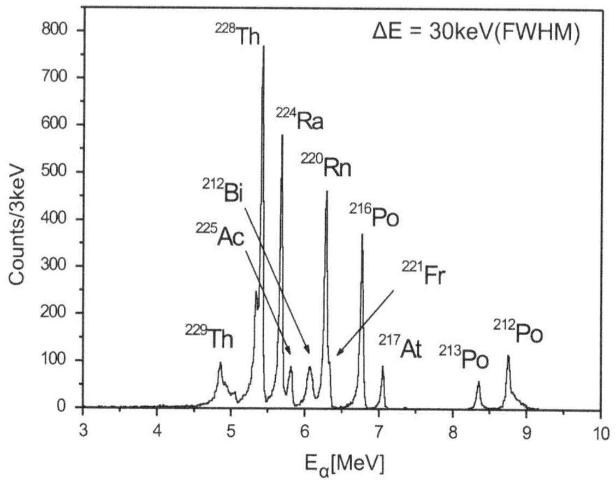

Fig. 5.3 α-energy spectrum as obtained after 5 days of continuous ^{229}Th (and ^{228}Th) α-recoil ion accumulation on the surface of a silicon detector and further 100 days of decay measurement [1]. The energy resolution is about 30 keV (FWHM). The small-area ^{233}U source 1 was used for ion extraction. A line broadening of the ^{229}Th α decay arises from the fact that only 56% of all decays occur with the dominant energy of 4.845 MeV (Table A.5). With kind permission of Elsevier [2]

registered in a remote setup for 100 days. Signal acquisition was performed with a charge sensitive preamplifier (CSTA) connected to a shaping amplifier (Ortec, model 571) and a portable multi-channel analyzer (Amptek, MCA-8000 A). The resulting α-energy spectrum is shown in Fig. 5.3.

The α decay of ^{229}Th is clearly visible on the left of the spectrum. The width of the corresponding line results from a superposition of several α-decay energies (Table A.5). Besides ^{229}Th, also ^{228}Th is contained in the spectrum and even dominates the α activity. ^{228}Th is extracted, as also ^{232}U is contained in the source as an unavoidable contaminant of the ^{233}U source material (Appendix A.1) and the mass resolving power of the QMS is not sufficient to separate ^{229}Th and ^{228}Th in the 3+ charge state. ^{228}Th dominates the spectrum, due to its shorter half-life of 1.9 years compared to 7932 years of ^{229}Th. The decay chains of ^{229}Th and ^{228}Th are also visible, as all daughter nuclei possess significantly shorter lifetimes.

The number of detected ^{229}Th α decays amounts to 5400 ± 500 events in 100 days of data registration. The error was estimated and is dominated by the integration limits. Under the realistic assumption that 50% of all α particles are detected as being emitted into the hemisphere of the silicon detector, the total decay rate is about 100 per day. This corresponds to $(4.5 \pm 0.3) \cdot 10^8$ ^{229}Th ions collected during 5 days of accumulation. The $^{229}Th^{3+}$ extraction rate is therefore estimated to be (1000 ± 100) ions per second, which is well in agreement with the MCP-based measurement [2].

Using the large-area ^{233}U source (source 2), this extraction rate could be increased by a factor of more than 10, as it was found by mass scans as well as by a comparison of isomeric decay signals (Sect. 5.2.2) [6]. The absolute extraction rate can be estimated to be 10 000 thorium ions per second in the 2+ and 3+ charge state, respectively. The

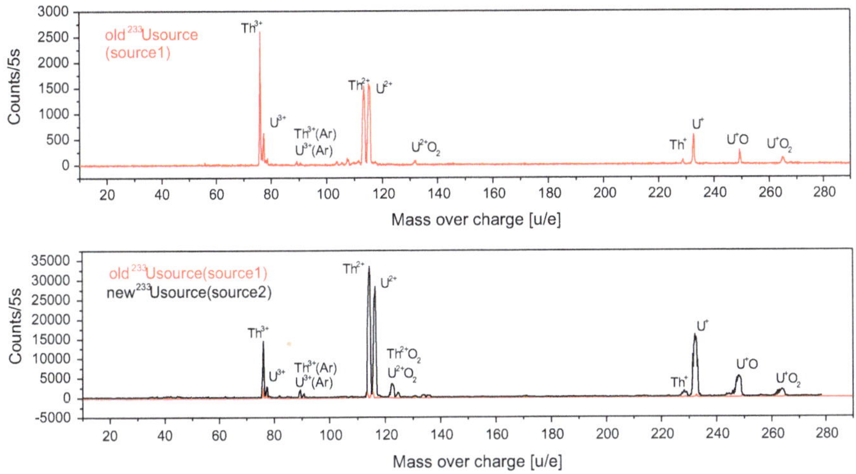

Fig. 5.4 Comparison of mass scans performed with the small-area ^{233}U source 1 (upper panel) and the large-area ^{233}U source 2 (lower panel, black). For better comparison, also the scan performed with source 1 is included (lower panel, red). About a factor of 10 of improvement of the extraction rate is obtained for source 2. Relative differences of the extraction rate between ^{229}Th^{2+} and ^{229}Th^{3+} presumably arise from changing levels of buffer-gas purity

same extraction rate for both charge states is assumed as the 2+ charge state can be populated on cost of the 3+ charge state, e.g., by increasing the buffer-gas pressure or reducing the extraction velocity. A comparison of two mass scans performed with the old, small-area ^{233}U source 1 and the new large-area ^{233}U source 2 is shown in Fig. 5.4. The upper panel shows a mass scan performed with source 1 (the same as shown in Fig. 5.2), the lower panel shows a mass scan performed with source 2 together with the previous scan. About an order of magnitude increased counting rate can be inferred. Also a relative intensity decrease of the Th^{3+} compared to the Th^{2+} signal is visible. This might result from a different helium buffer-gas purity achieved in the two measurements performed with a time distance of about three years.

5.1.2 *Verification of the Ion Collection*

As a prerequisite for optical measurements of a potential photonic ^{229m}Th decay channel, the ^{229}Th ions have to be collected on a micro-electrode collection surface of 50 μm diameter (described in Sect. 4.3.1). The design of the electrode structure was optimized based on SIMION simulations (Appendix B.1.6). Measurements were performed in order to verify that the collection is actually working as theoretically expected. For this purpose, the quadrupole mass-spectrometer was set to extract ^{221}Fr^{2+} instead of thorium ions. ^{221}Fr is also contained in the ^{233}U decay chain and for this reason it is emitted from the chemically unpurified ^{233}U source 1. It is an α emitter of 286 s half-life. The ^{221}Fr ions were collected on a phosphor-coated collection electrode (Fig. 4.18), in this way each α decay leads to the emission of a multitude of photons on the phosphor layer. ZnS:Ag (P11, 2–3 μm grain size) of 15 μm thickness was chosen as a phosphor, emitting light in the wavelength range around 460 nm. The coating of the surface was performed by ProxiVision GmbH.

A microscopy optical system was used in order to localize the point of ion collection on the collection electrode for different applied voltages. For this purpose, the second parabolic mirror of the optical system described in Sect. 4.3 was exchanged by a lens (CaF_2, $f = 132$ mm at 460 nm wavelength). The light was focused onto an MCP-based single-photon imaging system (Quantar, Model 2601B Mepsicron-II Single-Photon Imaging Detector System [7]), which allows for a spatially-resolved electronic read out [8]. In this way, the collection surface was imaged onto the imaging system with a magnification factor $m \approx 13$. The CsI-coated MCP detector, which is used for the investigation of the isomeric decay, could not be used for this purpose, as it does not provide a significant quantum efficiency at 460 nm wavelength. A conceptual sketch of the setup is shown in Fig. 5.5a.

In a first measurement it was verified that the ions react to the voltages applied to the collection surface. For this purpose, a voltage of −1000 V was applied to the collection electrode and the cover electrode (see Fig. 5.5b). In this way the ions were attracted, however not exclusively collected on the 50 μm diameter collection surface and also not guided to the center of the optical system. The resulting image is shown on the left-hand side of Fig. 5.6. For a single measurement, 150 frames with an exposure time of 4 s per frame were recorded. A high Quantar-MCP voltage of −2500 V had to be applied in order to visualize the uncollimated light. Afterwards, a collection voltage of −1000 V was applied to the collection electrode, while the cover voltage was set to +3 V. In this way, the ions were collected on the collection surface, leading to a centered spot position and a high signal intensity as shown on

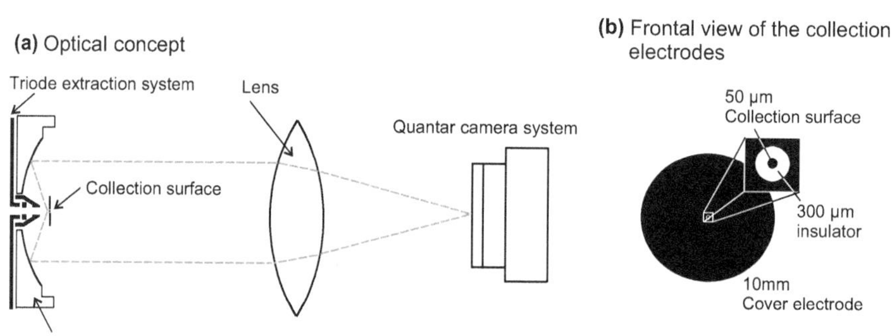

Fig. 5.5 **a** Conceptual sketch of the optical setup that was used for the verification of the ion collection on the micro electrode. The collection surface is phosphor coated and $^{221}Fr^{2+}$ ions ($t_{1/2} \approx 4.8$ min) were accumulated in order to visualize the α decays. The luminescence light is collimated by the first parabolic mirror and further imaged onto a Quantar MCP surface [8]. **b** Schematic drawing of the electrode structure used for ion collection on a 50 μm-diameter collection surface [3]

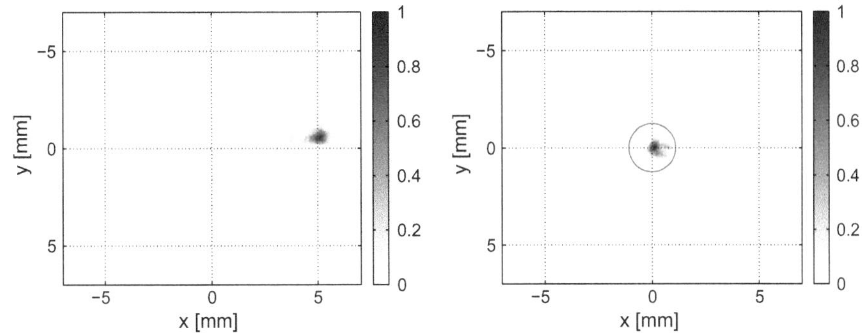

Fig. 5.6 Photon distributions recorded by the Quantar camera with the setup shown in Fig. 5.5 [8]. The x- and y-axis correspond to the length scale on the image plane of the Quantar camera, which is by a factor of about 10 magnified compared to the scale on the collection surface. The left panel shows a measurement (10 min acquisition time) during accumulation of $^{221}Fr^{2+}$ with voltages of −1000 V applied to the collection electrode and the cover electrode (−2500 V MCP voltage). The right panel shows the same measurement, however, with −1000 V on the collection electrode and +3 V applied to the cover electrode (−2010 V MCP voltage). The collected ions react on the collection voltage in the intended way. With kind permission of The European Physical Journal (EPJ) [8]

the right-hand side of Fig. 5.6. The measured spot size is about 0.6 mm diameter, corresponding to the magnified 50 µm diameter collection surface. The Quantar-camera voltage had to be reduced to −2010 V for image read out, as otherwise saturation of the images occurred, for this reason the intensities of the two images are not comparable. No signal could be obtained when collecting ions with −1000 V applied to both electrodes in case of −2010 V camera voltage. This proves that the ions react on the collection voltage in the intended way. The shift of the intensity maximum occurs as the ions are following the symmetry axis of the triode extraction system when no collection voltage is applied. This symmetry axis is not completely aligned with the axis of the optical system and a shift on the sub-mm scale becomes visible.

In a second set of experiments the signal intensity was probed as a function of the collection voltage. For this purpose, the voltage was varied between 0 and −2000 V, leaving the cover electrode at +3 V. The number of α decays occurring in a circle of 1.2 mm radius around the center on the detector (shown in Fig. 5.6) were counted. 150 image frames with 4 s exposure time per frame were recorded for each data point. Prior to each measurement, $^{221}Fr^{2+}$ was accumulated for 10 min. The resulting normalized count rate is shown in Fig. 5.7. As expected, the curve shows a saturation behavior [8]. −2000 V was the maximum collection voltage that could be applied to the electrode during preparatory measurements. At higher voltages field emissions started to occur. −300 V was the collection voltage applied during optical measurements for thorium ion collection (Sect. 5.2.1) resulting in a collection

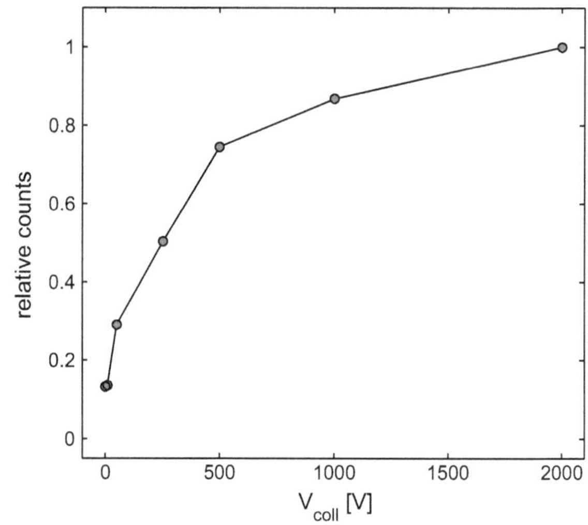

Fig. 5.7 Relative counts (normalized to the maximum) measured with the Quantar camera for different collection voltages (V_{coll}). The collection voltages were varied between 0 and −2000 V at a constant cover electrode voltage of +3 V. Each measurement was performed with 10 min acquisition time. With kind permission of The European Physical Journal (EPJ) [8]

efficiency of estimatedly 40–50%. At higher voltages field emissions started already to occur. The reduced maximum available collection voltage might be caused by the MgF_2 coating of the collection surface.

5.1.3 *Verification of the Optical Performance*

A dedicated alignment method of the optical system was developed and is described in detail in Refs. [8, 9]. Following the alignment of the optics, it was a prerequisite for the search of a potential ^{229m}Th photonic decay signal to probe the achievable focal spot-size on the MCP detector as described in Sect. 4.4. Prior to any measurement, this size was investigated by numerical ray tracing simulations to be 44 μm FWHM (Appendix B.2).

In order to experimentally probe the optical performance, a 2π light source of ~50 μm diameter (corresponding to the size of the collection surface) was produced in the focal point of parabolic mirror 1. This was achieved by focusing the light of a deuterium lamp (Heraeus, V05 30 W with 1 mm diameter aperture) down to a small spot size with the help of an optical system consisting of two CaF_2 lenses (Korth Kristalle GmbH, $f \approx 52$ mm at 160 nm). A deuterium lamp had to be used for this purpose, as it offers a high light intensity in the deep ultra-violet wavelength region, which provides some overlap with the quantum efficiency of the CsI-coated MCP detector. The setup used to produce the focal spot on the collection surface is shown in Fig. 5.8.

The calculated factor of image magnification of this optical system (at 160 nm wavelength) is $m \approx 1/13 \cdot 1/5 \approx 1/65$, leading to a calculated focal spot size

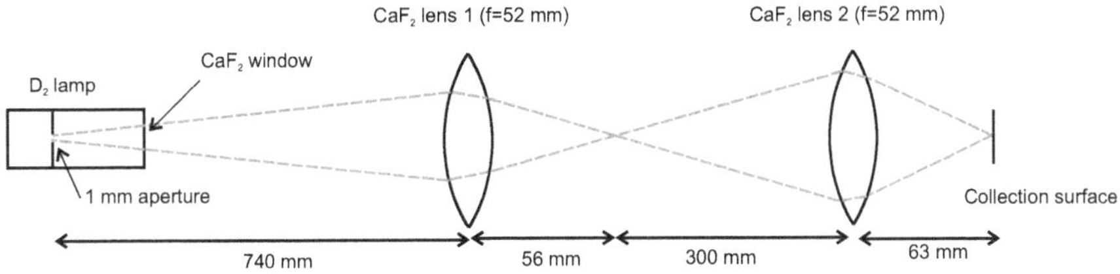

Fig. 5.8 Optical focusing system used to generate a small spot of light on the collection surface. Light, emitted by a D_2 lamp, is focused with the help of two CaF_2 lenses onto the collection surface. Subsequently, the light scattered off the collection surface is used to experimentally probe the achievable magnification of the VUV optical system

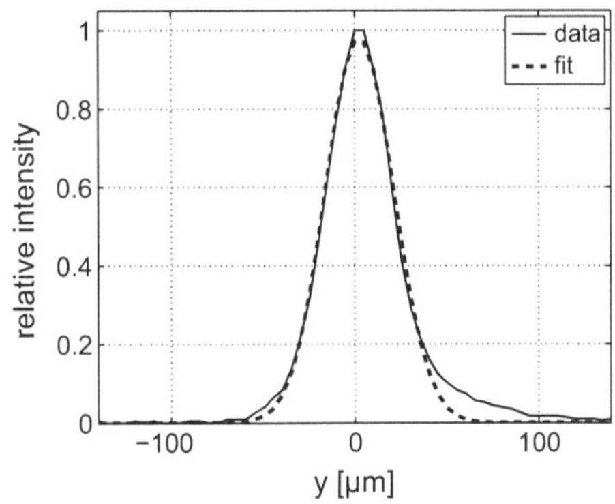

Fig. 5.9 Image spot, obtained by focusing the light of a D_2 lamp onto a bare CCD chip. The corresponding optical setup is shown in Fig. 5.8 with the collection surface exchanged by the CCD chip. About 44 μm (FWHM) spot size are obtained. With kind permission of The European Physical Journal (EPJ) [8]

of ~15 μm diameter. In reality, however, some broadening occurs due to spherical and chromatic aberrations and also the short-wavelength part of the spectrum will be considerably absorbed as the setup is exposed to air. In order to obtain a better impression of the actually achieved focal spot size, the collection surface was exchanged by a CCD sensor (Sony ICX267 CCD). In this way the spot size of the D_2-lamp image on the CCD sensor was measured to be 44 μm FWHM, with a 80 μm diameter tail (measured at 10% of the maximum intensity, 1/16 s exposure time). The corresponding measurement is shown in Fig. 5.9 [8].

Following the production of a small spot size of deep UV light, this light was scattered off the collection surface in order to probe the performance of the optical system. For this purpose, the collection surface was placed under vacuum in the double-parabolic mirror system (see Sect. 4.3), as shown in Fig. 5.10.

The system has to be placed under vacuum of 10^{-6} mbar in order to allow for the operation of the MCP detector. In this setup, the light, as emitted from the D_2 lamp, travels a long way of about 1 m in air, leading to a strong absorption of light emitted in the VUV region. However, the remaining deep-UV component is still sufficiently large to be detected by the CsI-coated MCP detector. As the performance of the all-reflective optical system is wavelength independent, this deep-UV component can be used in order to probe the achieved focal spot-size on the MCP detector.

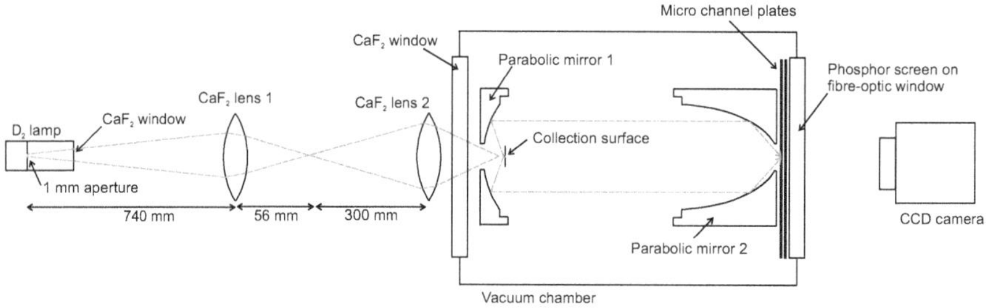

Fig. 5.10 Experimental setup used to probe the focusing properties of the VUV-optical system [8]. The light emitted by a D_2 lamp is focused by two CaF_2 lenses onto the collection surface in order to generate a small spot size. Subsequently, the scattered light is focused with the help of two parabolic mirrors onto the MCP detection surface. A CCD camera is used for MCP image read out

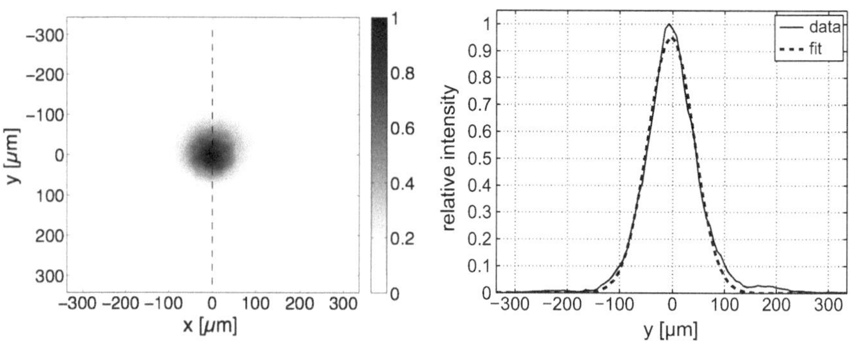

Fig. 5.11 Left-hand side: Focal spot detected on the MCP detector with the setup shown in Fig. 5.10 after 50 s acquisition time [8]. The analysis was performed as described in the text. Right-hand side: Projection of the photon distribution onto the y axis along the dashed line shown in the previous image. A FWHM of 106 μm of the image spot is obtained. With kind permission of The European Physical Journal (EPJ) [8]

For this purpose, the MCP was operated in single-photon counting mode at voltages of 0, +2000 and +5000 V on the front surface, back surface and phosphor screen, respectively. The CCD camera (Pointgrey, Flea 2 14S3M-C) was placed in a few cm distance to the fiber-optical base-plate of the phosphor screen. The phosphor screen was imaged to the CCD chip (Sony, ICX267 CCD) with the help of a zoom lens (Computar, M2514 MP2, 25 mm, C-mount) and a 10 mm spacer ring, leading to a field of view of $10 \times 7.4\,\text{mm}^2$. 800 frames with an exposure time of 1/16 s per frame were recorded with the CCD camera. The image frames were evaluated with the help of a Matlab program as described in Sect. 4.4.2. The result is shown in Fig. 5.11 [8].

The resulting spot size, which could be obtained in this way, is 106 μm diameter FWHM and thus larger than the value obtained by simulations (44 μm FWHM) [3, 8]. The reason might be some small misalignments, which are hard to be experimentally avoided, or the 80 μm diameter tail, measured for the spot on the collection surface (Fig. 5.9). When plotting the center positions of single MCP events, it is inferred that

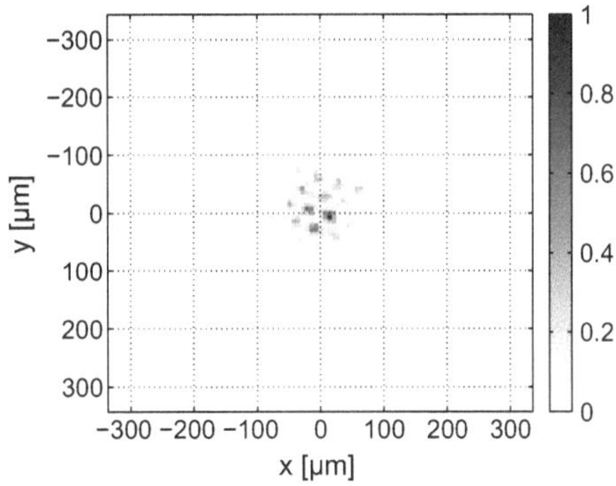

Fig. 5.12 Plot of the peak-center positions of the photon distribution shown in Fig. 5.11 [8]. A pixelated distribution is visible, which results from the resolved MCP channels. With kind permission of The European Physical Journal (EPJ) [8]

the performed measurements are already close to the MCP resolving power, as the individual MCP channels can be resolved, leading to a pixelated image shown in Fig. 5.12 [8].

5.2 Searching for the ^{229}Th Isomeric Decay

Following the preparatory measurements, the search for the direct decay detection of ^{229m}Th was started. In first measurements the photonic decay channel was investigated. The corresponding experiments will be described in the first part of this section. As the investigation of the photonic decay channel did not lead to the observation of a decay signal, in a further set of experiments the internal conversion (IC) decay branch was probed. These measurements led to the first successful direct detection of ^{229m}Th and will be described in the second part of this section.

5.2.1 Investigation of the Photonic Decay Channel

The experimental setup used for the investigation of a potential photonic decay signal of ^{229m}Th was described in Sect. 4.1.2. It is comparable with the test setup used for the investigation of the optical performance (Sect. 5.1.3). However, this time the lens-based optical part was replaced by the ion-beam formation system (Sect. 4.1.1). Ion extraction was performed under experimental conditions described in Sect. 5.1.1. A conceptual sketch of the experimental setup is shown in Fig. 5.13.

It is a prerequisite for the detection of a photonic decay channel of ^{229m}Th that the internal conversion decay is suppressed [10, 11]. Theoretically, this could be achieved by appropriate choice of the electronic surrounding [12]. The IC decay channel of isomeric ^{229}Th is expected to be energetically forbidden when thorium is embedded into a MgF_2 crystal [13]. The reason is that thorium prefers the 4+ charge

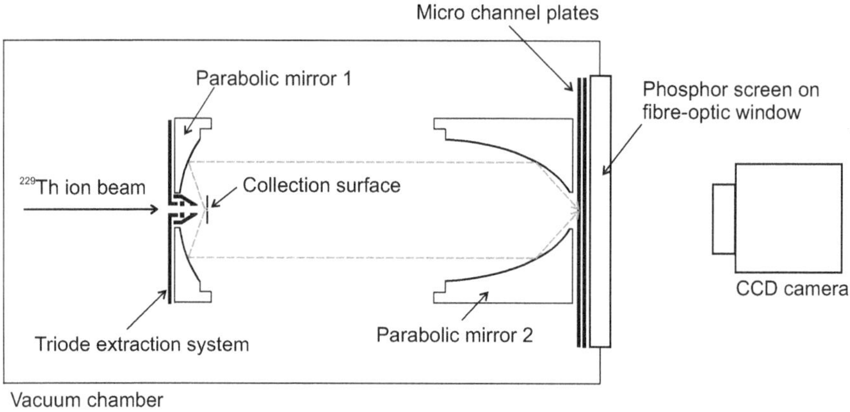

Fig. 5.13 Simplified sketch of the experimental setup used for the investigation of a potential photonic decay channel of ^{229m}Th. ^{229}Th ions, with a fractional content of ^{229m}Th, are collected on the microscopic collection surface. Light, when emitted in the isomeric decay, is collimated by the parabolic mirror 1 and then focused onto the MCP detector by the parabolic mirror 2. A spatially resolved signal read-out is realized with the help of a phosphor screen and a CCD camera

state when being embedded into the crystal-lattice environment and the band gap of the material is larger than the isomeric energy. For this purpose, a MgF_2-coated collection surface (~1 µm thickness) was used in our experiments. When applying a collection voltage of −300 V, the thorium ions are implanted into the MgF_2 layer with a kinetic energy of 0.6–0.9 keV, depending on the charge state. Before implantation, the surface was ozone cleaned at air pressure for a few minutes with the help of a deuterium lamp [14].

Prior to any measurement, the optical system was adjusted following a dedicated procedure described in Refs. [8, 9]. Voltages of 0, +1800 and +6000 V were applied to the MCP detector and the phosphor screen (Beam Imaging Solutions, BOS-75-FO). For read out, the CCD camera (Pointgrey, Flea 2 14S3M-C) was positioned in the same way as for the test measurements (Sect. 5.1.3) in a few cm distance to the fiber-optical plate. Events were imaged onto the CCD chip (Sony, ICX267 CCD) via a zoom lens (Computar, M2514 MP2, 25 mm, C-mount) mounted to the CCD camera together with a 5 mm spacer. The corresponding field of view is $16 \times 11.9\,\mathrm{mm}^2$.

The isomeric lifetime under the described conditions was expected to be in the range of minutes to hours [10, 11]. The uncertainty originates partly from an unknown exact isomeric energy value and partly from non-radiative decay channels of unknown strength. Such non-radiative decay channels can be expected to be present even in case that the IC decay is efficiently suppressed and occur in the form of electronic bridge processes, which could easily dominate the radiative decay by a factor of 10^2–10^3 [15]. In our experiments, the decay detection was performed simultaneously to ion accumulation. In this way, isomeric lifetimes down to practically 0 s could be probed, provided that no isomeric decay during the time of ion extraction (a few ms) occurs. The upper limit of the photonic lifetimes probed in our experiments was only limited by the measurement times. These were chosen to 20 h, in this way probing isomeric lifetimes of up to about 10 h.

Individual image frames of 4 s exposure time were recorded for 20 h (resulting in 18 000 image frames for one measurement) with the help of a Labview program. Following the image acquisition, a Matlab program was used to infer the positions of individual MCP events for all 18 000 image frames (see Appendix B.3). Afterwards, all recorded events were added to form one single measurement. This way of image read-out has the advantage that, in individual frames, CCD intrinsic noise and hot pixels can be easily distinguished from MCP events by size and intensity. Therefore, the resulting measurement is dominated by the low MCP-intrinsic dark-count rate of 0.01 cts/(s mm)^2.

Measurements were performed, extracting $^{229}Th^{2+}$ and $^{229}Th^{3+}$ ions for 20 h from the large-area ^{233}U source 2, providing more than 10 000 thorium ions per second in each charge state. Voltages for ion extraction were applied as described in Sect. 5.1.1. The accumulated ^{229}Th α-decay activity after 20 h corresponds to one decay every 500 s. However, potential light emission caused by the α decays on the collection surface occurs in a spectral region that is not detected by the MCP-detection system. The functionality of the ion extraction was probed by a second MCP detector (Hamamatsu, F2223), that was located sideways under an angle of 90° to the beam axis. Prior to every measurement, the parabolic mirror 1 was moved under vacuum by a motorized linear translation stage in order to give free sight onto the MCP detector. When applying -2000 V onto the detector surface, mass scans could be performed despite its off-axis position, in this way allowing to probe the functionality of the ion extraction and the correct calibration of the QMS.

The absolute number of dark counts occurring in the focal spot of the optical system of $\sim$100 µm diameter ($7.9 \cdot 10^{-3}$ mm^2) in 20 h of measurement time is approximated to be 5.6. Even in case of a three orders of magnitude reduced photonic decay rate of the isomer due to non-radiative decay processes like electronic bridge, the expected signal-to-background ratio would be 13:1 (see Sect 4.5.2). The absolute number of detectable isomeric decay events would thus amount to more than 70, which is sufficient to allow for a secure detection of any isomeric decay.

Despite such high sensitivity, no isomeric decay signal was observed. This result is interpreted as being caused by an unsuccessful suppression of the internal conversion decay channel. Although a MgF_2-coating of the collection surface was applied, surface effects like band-gap reduction, hydrocarbon contamination or unbound surface electrons can easily change the expected decay behavior [11, 16]. If the IC decay is not successfully suppressed, the intensity of the photonic decay channel has to be expected to be reduced by a factor of about 10^9 [10]. Such a reduction in signal intensity cannot be compensated by the applied optical system.

In a further experiment, a different technique was tested in order to suppress the IC decay channel. This time, a piezo-actuated valve (MaxTec Inc., MV112) was mounted to the RFQ vacuum chamber in order to allow for a controlled inlet of air into the system. Due to the high reactivities of thorium and oxygen, this led to $^{229}ThO^{2+}$-molecule formation on the fly. As the third ionization potential of ^{229}ThO is above the isomeric energy of 7.8 eV, no internal conversion should occur during

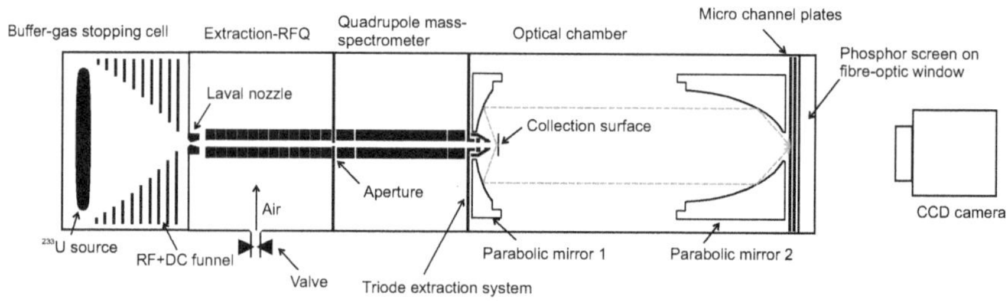

Fig. 5.14 Complete schematic drawing of the experimental setup used for the investigation of a ^{229m}Th photonic decay channel. A piezo-controlled valve was mounted to the RFQ vacuum chamber in order to allow for a controlled air inlet. In this way a strong $^{229}ThO^{2+}$ ion beam was produced by on-the-fly molecule formation. This allowed for the investigation of the isomeric decay under different chemical conditions

the time of flight. Further, oxygen provides a high electronegativity, leading to an ionic bonding. For this reason, thorium is expected to be ionically bound even after neutralization of the molecule on the collection surface. It was hoped for that this ionic bonding would lead to a suppression of the IC decay channel. A schematic sketch of the experimental setup is shown in Fig. 5.14

Also in this second type of experiment no isomeric decay signal could be observed within 20 h of measurement time. The experimental techniques were the same as already described above. The failure of this experiment to detect an isomeric decay signal is not completely unexpected, as the ionization potential of neutral thorium-oxide is below the isomeric energy value. Therefore, even when thorium stays charged in the ionic bonding, one has to expect internal conversion to occur as soon as $^{229}ThO^{2+}$ neutralizes on the collection surface. If this was the case, it would be very interesting to perform the same experiment with on-the-fly $^{229}ThF^{2+}$ molecule formation, as thorium fluoride exhibits a larger ionization potential. While experiments with $^{229}ThF^{2+}$ have not yet been performed, experiments searching for an IC decay during extraction of $^{229}ThO^{2+}$ will be described in the following section. Interestingly, no such decay could be observed, pointing at an isomeric decay process that occurs on time scales shorter than the extraction time (of below 10 ms).

5.2.2 *Investigation of the Internal-Conversion Decay Channel*

The failure of the optical system to observe any isomeric decay was a strong indication for the presence of a dominant internal conversion (IC) decay channel. In order to also investigate IC, a different set of experiments was performed, which has led to the successful observation of the ^{229}Th isomeric decay [6]. For this purpose, ^{229}Th ions were directly collected with low kinetic energy on the surface of the CsI-coated MCP detector (Beam Imaging Solutions, BOS-75-FO). The corresponding experi-

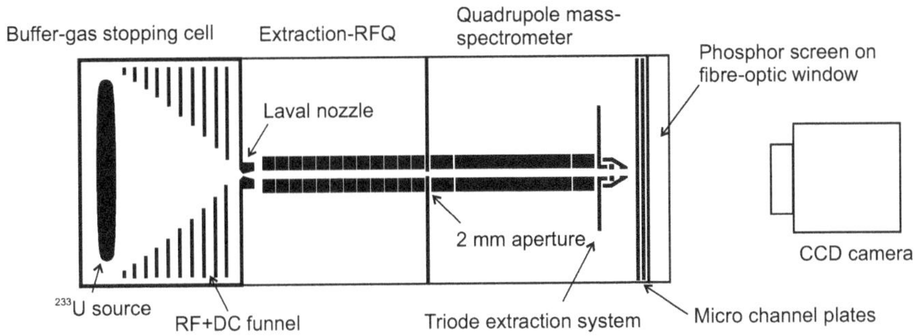

Fig. 5.15 Schematic drawing of the experimental setup used for the investigation of an internal-conversion decay channel of ^{229m}Th. ^{229}Th ions, with a fractional content of thorium in the isomeric state, are accumulated directly on the surface of the MCP detector. The detector is placed in 5 mm distance to the orifice of the triode extraction system. Signals were acquired simultaneously to low-energy ion collection

mental setup was described in Sect. 4.1. It is the same as used for the investigation of a photonic decay, however with the optical system removed. Instead, the MCP detection surface is placed directly in 5 mm distance to the orifice of the triode extraction system. Extraction voltages were applied as described in Sect. 5.1.1. A simplified sketch of the experiment is shown in Fig. 5.15.

The isomeric lifetime in thorium ions is expected to be in the range of minutes to hours. For this reason, no isomeric decay is expected to occur during the time of ion extraction. However, as soon as the ions come into contact with the MCP-detector surface, charge exchange occurs, forming neutral ^{229}Th, for which the rapid (a few μs lifetime) internal conversion decay channel expectedly dominates the isomeric decay [10, 11]. The electrons emitted in the IC decay can be detected by the MCP detector. These experiments are making use of the non-vanishing MCP detection efficiency for low-energy electrons (see Sect. 4.4.1). Similar experiments are reported for the detection of metastable molecular states [17].

Prior to any decay measurement, the functionality of ion extraction was tested and the QMS was set to extract the desired ion species. For this purpose, the MCP detector was operated at voltages of −900, +1000 and +5000 V applied to the front surface, back surface and phosphor screen, respectively. In this way, ions are accelerated onto the MCP and the ionic impact signals become visible. Subsequently, a mass scan can be performed, allowing for a QMS calibration as shown in Fig. 5.16. The exposure time of the CCD chip was 1 s for image acquisition. The shown image sections correspond to about $10 \times 10\,\mathrm{mm}^2$ on the phosphor screen. They were taken during extraction from the small-area ^{233}U source 1.

After QMS calibration, the MCP operation voltages were set to −25, +1900 and +6000 V, respectively. In this way the kinetic energy of collected ions is reduced (to 50 or 75 eV, depending on the charge state) and no ionic impact signal is detected, in agreement with the very low MCP detection efficiency for low energy ions (see Sect. 4.4.1) [18, 19]. ^{229}Th ions were continuously collected in the 2+ and 3+ charge

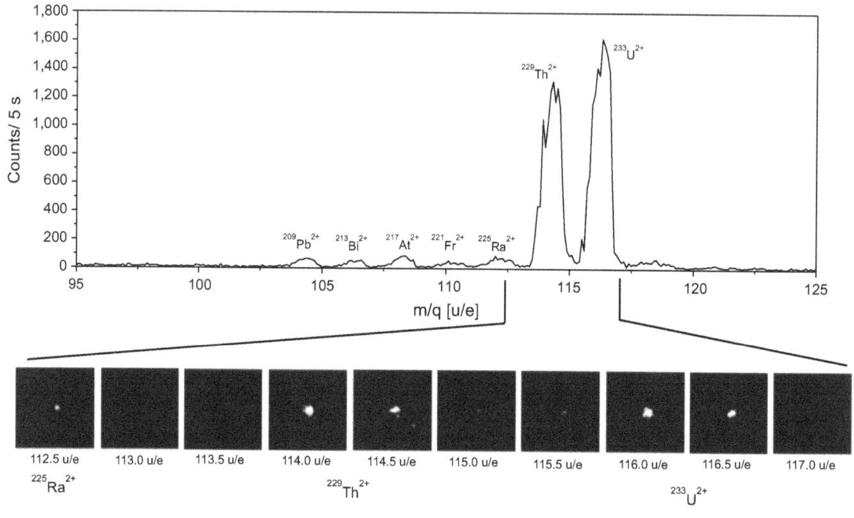

Fig. 5.16 Upper panel: Mass scan in the range of the 2+ ion species performed when using the chemically unpurified ^{233}U source 1. Ion counting was performed with a single-anode MCP detector (Hamamatsu, type F2223). Lower-panel: Ion impact profile measurements performed with the phosphor-screen mounted MCP detection system (Beam Imaging Solutions, BOS-75-FO) allowing for spatially resolved signal read-out. The measurement was performed at −900 V surface voltage and 1 s CCD exposure time when using the ^{233}U source 1. The ^{229}Th and ^{233}U mass peaks can clearly be separated. With kind permission of Nature Research [6]

state for 2000 s on the MCP surface. Read out of the phosphor screen was performed with the CCD camera (Pointgrey, Flea 2 14S3M-C) combined with a zoom lens (Computar, M2514 MP2, 25 mm, C-mount) placed in about 30 cm distance to the fiber-optical plate (see also Sect. 4.4.2). The resulting field of view is about $100 \times 75\,\text{mm}^2$. As described in Sect. 4.4.2, individual image frames of 4 s exposure time were recorded, corresponding to 500 frames per measurement. A Matlab program (Appendix B.3) was used to infer the event positions of individual MCP events, which were added to form one image. As CCD intrinsic noise and hot pixels can clearly be distinguished from MCP events in individual image frames, this read-out technique leads to a significant noise reduction and the background is dominated by the MCP dark count rate of 0.01 cts/(s mm^2).

Corresponding measurements were carried out during the extraction of ^{229}Th in the 2+ and 3+ charge states, originating from the small-area ^{233}U source 1, as well as from the large-area ^{233}U source 2. Clear signals were observed in all cases and are shown in Fig. 5.17 [6]. Measurements were also performed during the extraction of ^{233}U in the 2+ and 3+ charge states from both ^{233}U sources in order to exclude any remaining ionic impact as signal origin. In order to also exclude that the detected signal originates from an atomic shell effect, a ^{234}U source (source 3, see Sect. 4.2.1) for ^{230}Th α-recoil isotope production was employed for a different set of measurements. No signal could be detected in all exclusion measurements. The

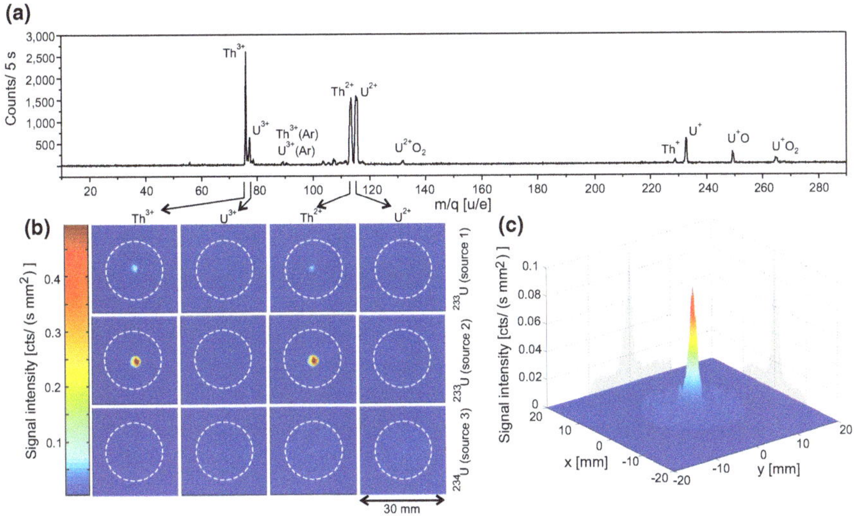

Fig. 5.17 **a** Mass scan performed with the small-area ^{233}U source 1 (see also Sect. 5.1.1). **b** Comparison of MCP signals obtained during accumulation of thorium and uranium in the 2+ and 3+ charge states. Each row corresponds to an individual ^{233}U source as indicated on the right-hand side. Each column corresponds to a different extracted ion species as indicated by the arrows from the mass scan. Images correspond to individual measurements of 2000 s integration time (20 mm diameter aperture indicated by dashed circles). **c** Signal of the ^{229}Th isomeric decay obtained during $^{229}Th^{3+}$ extraction using the ^{233}U source 1 (corresponding to the upper left image of panel b). With kind permission of Nature Research [6]

measurements were performed under the same conditions as described above. The results are presented in Fig. 5.17, together with a mass scan of the extracted ions.

The decays occur on a surface area of about 2 mm diameter, corresponding to the aperture size of the triode extraction system. The background corrected, spatially integrated decay count rate, as observed for the extraction of $^{229}Th^{3+}$ from the ^{233}U source 1, is ~0.25 cts/s, leading to a signal-to-background ratio of 8:1. This is in good agreement with the expectations, when taking the MCP detection efficiency for low energy electrons of expectedly 1.5% into account (see Sect. 4.5.3). For the large-area ^{233}U source 2, the detection rate was measured to ~3.4 cts/s, suggesting a factor of about 13.5 enhanced ^{229}Th extraction rate compared to the old source. Also the peak signal-to-background rate was increased by more than a factor of 10 to about 100:1.

Besides the 2+ and 3+ charge states, also the 1+ charge state was probed for an internal conversion decay. No IC decay signal was observed during these searches. When using the small-area ^{233}U source 1, this was not surprising, as the $^{229}Th^{1+}$ extraction efficiency is extremely low (only 0.3%, compared to up to 10% for the 2+ and 3+ charge states) [6]. In the meantime, however, detailed investigations of the 1+ charge state with the large-area ^{233}U source 2 could be carried out. While the relative extraction efficiency for $^{229}Th^{1+}$ is still small (comparable to the old source), the absolute extraction rate could be increased by more than a factor of 10, resulting in about 500 extracted $^{229}Th^{1+}$ ions per second. This number is comparable to the extraction rate for $^{229}Th^{2+}$ from the old ^{233}U source 1 (see Fig. 5.4), for which an IC decay signal could easily be detected. Surprisingly, also when using the new ^{233}U source 2, no IC decay signal was detected when extracting $^{229}Th^{1+}$. This observation is a strong hint towards a short lifetime (significantly below the ion extraction time of a few ms) of the isomeric state in singly charged ^{229}Th.

Like for the investigation of the photonic decay branch, also during the search for an internal conversion decay signal, experiments with thorium molecules were performed. For this purpose, air was introduced into the system at the point of the RFQ (see Fig. 5.14), using a piezo-actuated valve (MaxTec Inc., MV112). $^{229}ThO^{2+}$ as well as $^{229}ThAr^{3+}$ could be produced in this way in significant amounts. None of the molecule measurements led to the detection of an IC decay signal. In combination with the negative result of the searches for a photonic decay signal in $^{229}ThO^{2+}$, this is a strong indication for a fast isomeric decay in thorium molecules. Such a decay can be expected to occur, as di-atomic molecules possess a rich spectrum of rotational and vibrational modes in the VUV region. One might assume that the isomeric energy is transferred to such a molecular mode.

The detection of the ^{229m}Th internal conversion decay signal also allowed to probe for surface influences of the isomeric decay. The case of a dielectric material with a large band gap (significantly above the expected isomeric energy of 7.8 eV) is most interesting, as in this case one may hope that the IC decay channel is energetically forbidden [12], in which the IC decay channel was investigated, while collecting $^{229}Th^{2+}$ ($^{229}Th^{2+}$) ions on a MgF_2-coated copper surface (MgF_2 exhibits a band gap of about 10.8 eV). The experimental setup is shown in Fig. 5.18.

During the experiments, ions were extracted from the large-area ^{233}U source 2. Instead of collecting them directly on the MCP detector, a hemispherical reflection grid (OCI Vacuum Microengineering, type BDL450, 26 mm radius, gold plated) was mounted in front of the opening and supplied with a voltage of +75 V. In this way, the extracted ions were turned and collected onto a MgF_2-coated copper grid (TEM grids, Sciences Services, copper type 300 Mesh, 3 mm diameter), which was mounted directly to the last electrode of the triode extraction system set to −22 V extraction voltage. Electrons, potentially emitted in the isomeric decay, are attracted by the hemispherical grid and detected by an MCP detector (Hamamatsu, type F2223), placed at the back side of the reflection grid and supplied with voltages of +300, +2100 and +2250 on the front surface, center plate and anode, respectively. In this way it was possible to detect conversion electrons also from the MgF_2-coated surface, which is further evidence for the assumption that no radiative decay of the isomeric state could be detected due to a significant IC decay branch (Sect. 5.2.1). This result might be interpreted in terms of surface contamination or unbound surface electrons [16].

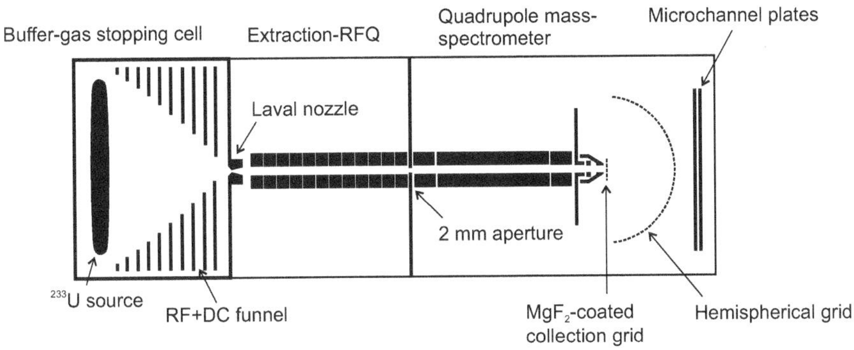

Fig. 5.18 Experimental setup used for the investigation of the isomeric decay via internal conversion on a MgF_2-coated surface. Despite the large band gap of the material, the experiments show that IC remains to be the favored decay channel

5.2.3 *Investigation of the Isomeric Properties*

The successful direct detection of the ^{229}Th isomeric state allows to draw some conclusions on the isomeric lifetime and energy. In order to probe the isomeric lifetime of ^{229}Th, $^{229}Th^{2+}$ ions were stored in the RFQ for a well defined time period. Such ion storage was achieved by applying a gate voltage to the last RFQ segment and reducing the RFQ helium buffer-gas background to the lower 10^{-4} mbar range (using the full pumping speed of the TMP). When the liquid-nitrogen based gas purifier is operated, storage times of more than 1 min for $^{229}Th^{2+}$ could be achieved. The functionality of the ion storage was probed prior to any measurement by operating the MCP detector at voltages of −900, +1000 and +5000 V, respectively, thus allowing for ion-impact detection.

In order to infer a lower limit for the isomeric lifetime in $^{229}Th^{2+}$, the ions are stored in the RFQ by applying a gate voltage of 5 V. Ions are loaded into the RFQ for 10 s. Afterwards, the ^{233}U source offset is reduced from 39 to 0 V, in this way hindering any further ions from leaving the buffer-gas stopping cell. The ions in the RFQ are stored for 1 min, waiting for the isomeric decay to occur. In the last step, the gate is opened by applying 0 V offset to the last RFQ electrode. In this way the ions are extracted and accumulated on the MCP detector, where the remaining isomeric decay signals are detected when applying voltages of −25, +1900 and +6000 V, as required for IC-signal detection. In order to reduce the dark-count rate, the CCD camera is triggered to only acquire frames during ion accumulation.

With this procedure, 200 pulses were evaluated with 3 acquired frames per pulse and 4 s exposure time per frame. A clear signal could be detected, from which a half-life of greater than 1 min in $^{229}Th^{2+}$ can be inferred [6]. No significant reduction in IC signal intensity could be inferred during this time, for this reason the storage time of 1 min in the RFQ is found not to be sufficient to determine a value for the isomer's lifetime.

In a further experiment, an upper limit for the isomeric lifetime in neutral ^{229}Th could be obtained. For this purpose, a gate voltage of 0.5 V is applied to the last RFQ electrode and $^{229}Th^{2+}$ ions are loaded into the RFQ for 1700 ms before releasing the ion cloud by reducing the gate voltage to 0 V for 500 ms. The CCD camera is triggered in a way that frames are only acquired at times when no ions are approaching the MCP surface. For this purpose, the camera is always started 500 ms after gate closing and stopped after 1200 ms, in parallel to gate opening, allowing to acquire one image frame of 1 s exposure time per bunch.

By this sequence, 1200 frames were acquired, corresponding to 1200 s total acquisition time. No signals were detected during this measurement, showing that the isomeric half-life for neutral ^{229}Th must be significantly shorter than 1 s [6]. In the meantime, the implementation of an improved ion bunching scheme [20] and an electronic read out of the IC decay signals has allowed for the first half-life measurement of ^{229m}Th in the neutral thorium atom, which is determined to be about 7 μs [21].

The isomeric lifetime limits allow to draw some preliminary conclusions on the isomeric energy. The long lifetime of the isomer in $^{229}Th^{2+}$ is a clear indication that the IC decay is energetically forbidden. Thus the isomeric energy must be below the 3rd ionization potential of thorium of 18.3 eV. Further, the IC decay is energetically allowed for neutral ^{229}Th, indicating that the isomeric energy is above the 1st ionization potential of thorium of 6.3 eV. Therefore, based on this direct detection, the isomeric energy can be inferred to range between 6.3 and 18.3 eV. This is a simplified picture, as surface influences have to be expected to play a role during the IC decay on the MCP detector. In this way, the work function of the collection material might dominate the lower limit of the energy range. In our experiments, however, collection takes place on a CsI-coated surface with a work function of 6.2 eV, which is comparable to the thorium 1st ionization potential. For this reason no significant change to the presented energy range is inferred.

The presented energy range is a further support of today's most accepted value of (7.8 ± 0.5) eV [22, 23]. Based on the presented direct detection, a significantly more precise energy value (exceeding 0.1 eV in precision) could be inferred by performing high-precision electron spectroscopy, using a magnetic-bottle electron spectrometer [24]. Experiments along this line are currently performed by our group.

Besides an improved energy value, also the determination of the isomeric lifetime in $^{229}Th^{2+}$ and $^{229}Th^{3+}$ is of interest. This value can be expected to be in the range of up to several hours and the current experimental limitation is the achieved ion storage time in our RFQ of only about 1 min. For this reason, a new, cryogenic linear Paul trap is being developed, following the Cryptex design obtained at MPIK in Heidelberg [25, 26]. This device will allow for the storage of highly charged ions for up to several hours, thus providing the basis for first ^{229m}Th lifetime measurements in $^{229}Th^{2+}$ and $^{229}Th^{3+}$. Full optical access will also be provided, as required for laser cooling and manipulation. In this way, the Paul trap will lay the foundation for the development of a ^{229}Th-based nuclear clock.

5.3 Confirmation Measurements

The measurements that have led to the direct detection of the isomeric decay signal of ^{229}Th were presented in Sect. 5.2.2. Also first measurements were presented that allow to exclude other sources of signal origin (comparisons with ^{233}U and ^{230}Th). Such exclusion measurements provide the basis for a confirmation of an unambiguous signal identification and are a central aspect of this work. In the following, it will be presented which exclusion measurements were performed and how they allow to exclude all potential sources of background, thus proving that the detected signal originates from the decay of ^{229m}Th. For this purpose, all potential sources of background are classified into four categories as follows:

- Background signals originating from setup components (Sect. 5.3.1)
- Background signals caused by ionic impact (Sect. 5.3.2)
- Nuclear decay signals other than from ^{229m}Th (Sect. 5.3.3)
- Signals caused by excited shell states (Sect. 5.3.4)

Most of the sources of background could be excluded in multiple ways. In the last section, also the problem of detector contamination due to continuous ^{229}Th (^{228}Th) accumulation is discussed.

5.3.1 *Background Signals Originating from Setup Components*

Various types of background signals could potentially occur due to components of the setup. These include background originating from the ^{233}U source, the ion transport system and the detection system. As these types of background would be constant throughout the measurements, they were easily excluded as signal origin by means of dark-count measurements. All measurements were dominated by the uniform MCP-intrinsic dark-count rate of 0.01 cts/(s mm^2), caused by random field emission and cosmic rays. In the following, only the contribution of the ^{233}U source to the background will be estimated.

The buffer-gas stopping cell houses the open ^{233}U source. During extraction, the stopping cell is filled with 40 mbar of ultra pure helium to provide the stopping and successive extraction of the α-recoil ions. The ^{233}U source is producing a significant amount of high-energy radiation. While ^{233}U itself is a pure α emitter, within the decay chain also some strong β^- emitters are produced (^{225}Ra, ^{213}Bi, ^{209}Tl, ^{209}Pb). Further, the decay chain of ^{232}U, which is unavoidably contained as a small impurity due to the radiochemical production process (Appendix A.1), also contains β^- emitters (^{212}Pb, ^{212}Bi, ^{208}Tl). Within both decay chains various γ rays are emitted, ranging up to energies of 2.6 MeV (^{208}Tl). As the ^{233}U source is open, also the α-recoil ions, as emitted within both decay chains, have to be considered. The emitted radiation is distributed isotropically in 4π. Only the γ rays can be expected to penetrate the walls of the surrounding vacuum chamber and thus possibly lead to an isotropic background signal in the MCP detector. However, the distance between the source and the MCP is about 85 cm, therefore only a fraction of $4.9 \cdot 10^{-4}$ of the emitted γ rays can be expected to reach the MCP detector surface of 75 mm diameter. Further, the γ -ray detection efficiency of an MCP detector is rather low (at the 1% level and below). The most dominant γ-ray line can be expected to be the 13.0 keV line of ^{233}U itself with 5.2% relative intensity. Assuming the maximum source activity of 290 kBq for the large source, the expected background rate of the MCP detector caused by the γ emission of the ^{233}U source is $1.7 \cdot 10^{-5}\,\text{s}^{-1}\text{mm}^{-2}$. However, as the intrinsic dark count rate of the MCP detector is in the range of $1 \cdot 10^{-2}\,\text{s}^{-1}\text{mm}^{-2}$, the contribution caused by the γ rays turns out to be negligible. This is in agreement with the observed dark count rates.

Most of the remaining high-energy radiation as originating from the ^{233}U source (α particles as well as α-recoil isotopes) will directly be implanted into the walls of the vacuum chamber and does not approach the detector. However, there is a small remaining direct line of sight between the ^{233}U source and the MCP detection surface through the 0.6 mm diameter extraction nozzle throat, which could allow for an enhanced signal in a central position at least if no buffer gas is contained in the stopping cell. The fraction of emitted particles, potentially able to reach the detector, strongly depends on the source geometry. For the smaller source this fraction is larger and by numerical ray tracing calculations a value of $1.8 \cdot 10^{-9}$ is obtained. Comparing this value with the α activity of 200 kBq of the small ^{233}U source, an absolute detection rate of $3.6 \cdot 10^{-4}\,\text{s}^{-1}$ can be expected, which corresponds to about one α particle every 45 min. This is also far below the intrinsic background rate of the MCP detector, however, it was made visible with a silicon detector within about 20 days of continuous detection. For the β^- activity a calculated value of $4.9 \cdot 10^{-6}$ s^{-1} is obtained. For the α-recoil ions the estimated rate of impact is $1.9 \cdot 10^{-5}\,\text{s}^{-1}$ and therefore negligible. As expected, none of these effects was detected with the MCP detector.

As soon as helium buffer-gas enters the vacuum chamber, Cherenkov radiation may emerge from β decays, giving rise to background signals of unknown strength and wavelength distribution. This radiation could potentially be detected by the MCP detector through the 0.6 mm diameter extraction nozzle throat. As the resulting signal is hard to be predicted, a measurement was performed instead in order to estimate the corresponding background rate. No Cherenkov light was observed within 20 h of measurement time.

5.3.2 Background Effects Caused by Ionic Impact

The impact of the ions on the MCP detection surface is a conceptually unavoidable source of potential background. This includes signals that might be caused by the ionic kinetic energy, as well as by the ionic charge state.

The detected ions possess a kinetic energy, which can be controlled by changing the MCP surface voltage. When operating the MCP for ion counting, typically −900 V offset are applied. The resulting kinetic energy of the impinging ions varies between 900 and 2700 keV, depending on the charge state, and is large enough to drive electrons off the MCP surface, leading to ion detection (Fig. 5.16). Experimentally we found, that the detection efficiency for the ionic impact can be reduced to practically zero by lowering the kinetic energy of the ions when applying an MCP surface voltage between 0 and −40 V. This is in agreement with a very low MCP detection efficiency for low energy ions, as reported in literature (see Sect. 4.4.1) [18, 19].

The fact that this technique is indeed applicable was shown in a comparative measurement between the accumulation of $^{229}Th^{2+}$ and $^{233}U^{2+}$. These ions were used for comparison, as they are extracted with nearly equal count rates from the buffer-gas stopping cell (see Fig. 5.4). Furthermore, they possess the same charge state and similar mass, leading to nearly the same velocities. The number of the MCP events was measured as a function of MCP surface voltage during separate accumulation of $^{229}Th^{2+}$ and $^{233}U^{2+}$, while varying the MCP surface between 0 and −100 V. The measurement was performed with the small-area ^{233}U source 1 with 1200 s measurement time per data point. Ion extraction and data acquisition were performed under the same conditions as described in Sect. 5.2.2. During the measurement, all MCP voltages were adjusted accordingly, in order to exclude any MCP intrinsic effects. The resulting measurement is shown in Fig. 5.19.

For the $^{233}U^{2+}$ signal, the result is as expected for a pure ionic impact signal: a moderate signal intensity is still detected at −100 V MCP offset, continuously decreasing towards higher offset voltages, as the kinetic energy of the ions is reduced. At about −40 V, there is practically no remaining signal (within measuring times of

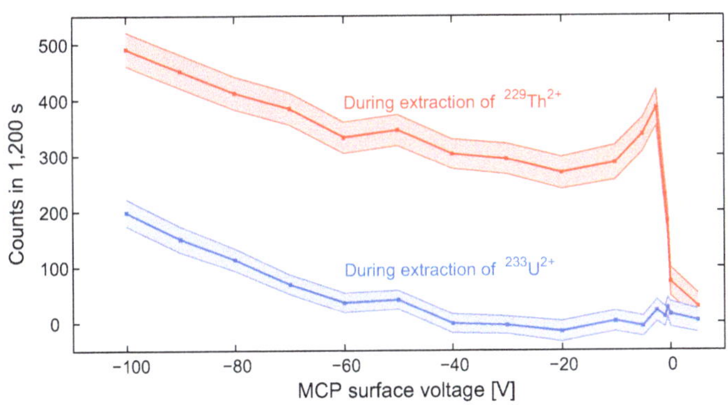

Fig. 5.19 Detected signals as a function of the MCP surface voltage during extraction of $^{233}U^{2+}$ (blue) and $^{229}Th^{2+}$ (red). The measurement was performed using the small-area ^{233}U source 1 with 1200 s acquisition time for each individual data point. Both curves are background corrected and the error bands show the statistical uncertainty. The curves exhibit systematically different behavior (see text). With kind permission of Nature Research [6]

20 min). The $^{229}Th^{2+}$ signal, however, exhibits a significantly different behavior: even at voltages as low as -2 V a remaining signal is detected. A steep signal decrease occurs between -2 and 0 V. A minor increase in signal intensity is observed before the cut off, which originates from electrons that are repelled by the last electrode of the triode extraction system due to the voltage gradient between the triode and the MCP surface. This signal behavior is not expected for a pure ionic impact signal, while a decay would allow instead for an explanation. The cut off occurs exactly at the point at which the ions start to be repelled by the MCP surface voltage, which indicates that the signal occurs even at zero kinetic energy of the accumulated ions. For this reason it can be excluded that the detected signal originates from an ionic impact of the thorium ions. This measurement also excludes a signal origin from background of the first type, i.e. originating from the setup components.

Besides the kinetic energy, the ions are known to also carry energy corresponding to their charge states. The energies involved are 11.9 and 18.3 eV for the 2nd and 3rd ionization potentials of thorium, respectively. Both energies lie significantly above the work function of CsI (6.2 eV), in this way providing enough energy to generate a signal on the MCP surface. Experimentally, however, one finds that no signal is generated corresponding to the ions charge state. The detection efficiency for low-energy ions even tends to decrease with increasing charge state (see Fig. 4.29) [18]. The finding that the neutralization energy does not lead to a background signal is also supported by the comparative measurements with $^{233}U^{2+}$. The 2nd ionization potential of uranium is 11.6 eV and thus close to the one of thorium. Therefore, one should expect to obtain the same signal behavior during extraction of $^{223}U^{2+}$ and $^{229}Th^{2+}$. As shown in Fig. 5.19, the two ion species show a systematically different behavior.

5.3.3 Nuclear Decay Signals Other Than ^{229m}Th

In previous experiments, background signals caused by short-lived nuclides of the ^{233}U and ^{232}U decay chains have prevented the unambiguous direct identification of the ^{229m}Th isomer. Our experimental setup was especially designed to exclude this type of potential background. A detailed list of nuclides that could potentially play a role in the radiochemistry of the ^{233}U source material is given in Fig. A.2. All nuclides except for ^{229m}Th could be excluded as signal origin.

In order to suppress radioactive decays not originating from ^{229}Th, a customized quadrupole mass separator (QMS, see Sect. 4.2.7) was build, following design values of E. Haettner [27]. This device was shown to possess a mass resolution of $m/\Delta m \approx 150$ at 70% of transmission efficiency [1]. This resolution is sufficient for a complete separation of α-recoil ions with a difference of four atomic mass units. Corresponding mass spectra are shown in Fig. 5.20 for two different MCP surface voltages. At -2000 V surface voltage (blue curve), impact signals are detectable due to the kinetic energy of several keV of the impinging ions. At -25 V MCP surface voltage (red curve) no ionic impact is detectable and the remaining signal originates

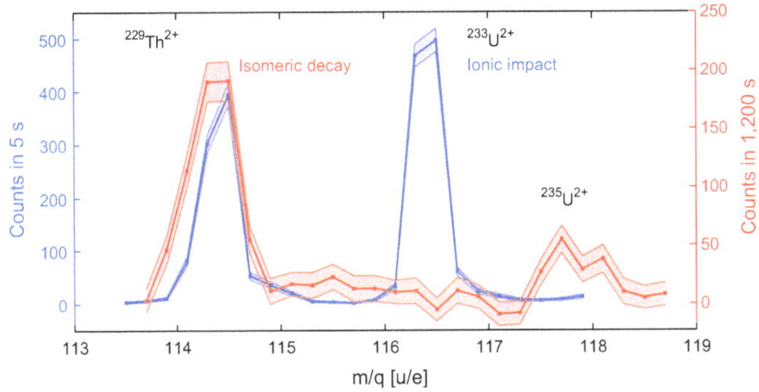

Fig. 5.20 Mass scans of the 2+ ion species for different detection conditions [6]. The blue curve shows the ionic impact signal acquired with 5 s integration time with a single-anode MCP detector (Hamamatsu, F2223) operated at −2000 V surface voltage. The red curve shows the background corrected isomeric decay signal obtained when operating the phosphor-screen mounted MCP detector (Beam Imaging Solutions, BOS-75-FO) at −25 V surface voltage. Each data point corresponds to 1200 s integration time. Error bars show the estimated statistical error. Besides the $^{229m}Th^{2+}$ isomeric decay signal at 114.5 u/e, a further signal at 117.5 u/e is visible. This signal corresponds to the isomeric decay of ^{235m}U. ^{239}Pu was shown to be present in the source material by α spectroscopy (see Fig. A.5) and the isomer is populated by a 70% decay branch. With kind permission of Nature Research [6]

from the ^{229m}Th isomeric decay. A clear signal restriction to the m/q value of $^{229}Th^{2+}$ (114.5 u/e) can be inferred. Both mass spectra were measured with the small-area ^{233}U source 1. Note the different acquisition times of 5 s for the ionic impact signal (blue) and 1200 s for the isomer detection (red). Besides the ^{229}Th isomeric decay at 114.5 u/e, a further signal at 117.5 u/e occurs. This signal originates from the isomeric decay of ^{235}U. ^{239}Pu was shown to be contained in the source by α spectroscopy (Fig. A.5). ^{235m}U is populated in the α decay of ^{239}Pu by a decay branch of 70% and is the second lowest isomeric state with an excitation energy of 76 eV. It decays with a half-life of about 26 min to its ground state via internal conversion.

However, operating the QMS at high mass resolving power is not sufficient to exclude short-lived nuclides as signal origin, as molecular sidebands may be populated (e.g. $^{213}Bi^{16}O$ reveals the same mass as ^{229}Th). The fact that this is not the case and that the registered signal does indeed originate from ^{229}Th was shown in four independent ways that will be detailed in the following [6].

The first way of exclusion is based on the parallel observation of a decay signal in the 2+ and 3+ charge states (Fig. 5.17). In preparatory measurements of the ion extraction rates from the buffer-gas stopping cell, ^{229}Th was shown to be extracted to a significant amount in the 3+ charge state (see Sect. 5.1.1). The reason is the exceptionally low 3rd ionization potential of thorium of just 18.3 eV, which is even below the ionization potential of the helium buffer gas (24.6 eV) [1]. During ion stopping in the helium environment, it is energetically favorable for the electrons to stay attached to the helium atoms, instead of reducing the thorium 3+ charge state [28]. None of the short-lived daughter nuclides that could potentially be extracted

Table 5.1 List of ionization energies [4]

Element	Atomic no.	1st [eV]	2nd [eV]	3rd [eV]
Curium	96	5.99	12.4	20.1
Americium	95	5.97	11.7	21.7
Plutonium	94	6.03	11.5	21.1
Neptunium	93	6.27	11.5	19.7
Uranium	92	6.19	11.6	19.8
Protactinium	91	5.89	11.9	18.6
Thorium	90	6.31	11.9	18.3
Actinium	89	5.38	11.8	17.4
Radium	88	5.28	10.1	31.0
Francium	87	4.07	22.4	33.5
Radon	86	10.75	21.4	29.4
Astatine	85	9.32	17.9	26.6
Polonium	84	8.41	19.3	27.3
Bismuth	83	7.29	16.7	25.6
Lead	82	7.42	15.0	31.9
Thallium	81	6.11	20.4	29.9
Mercury	80	10.44	18.7	34.5

Ionization energies of the first 3 charge states of elements potentially contained in the source material are listed. From radium downwards, all elements reveal 3rd ionization potentials which are above the 1st ionization potential of helium (24.6 eV)

from the ^{233}U source has a 3rd ionization potential which is below 24.6 eV and would thus allow for an extraction in the 3+ charge state (see Table 5.1). Experimentally, a suppression of the extraction efficiency of three to four orders of magnitude was observed for the 3+ compared to the 2+ charge state for all short-lived daughter nuclides of the ^{233}U and ^{232}U decay chains (of atomic number $Z = 88$ or below, see Appendix A.7) [1].

Table 5.1 reveals that also the relatively long-lived nuclides from curium to actinium provide 3rd ionization potentials below 24.6 eV. Their lifetimes are, however, too long to explain the detected signal (see Appendix A.1). In addition all of them have larger mass values than ^{229}Th, except actinium and a few thorium isotopes. In this way they can be excluded as signal origin with the help of the mass scans, as no mass shifts from heavier to lighter masses can be explained by molecular sidebands.

A second way of exclusion of short-lived daughter isotopes as signal origin is provided by the comparison of signals originating from the two different ^{233}U sources (see Fig. 5.17). The small-area ^{233}U source 1 was produced from ^{233}U material which was not chemically purified since about 1969. For this reason, a significant ingrowth of short-lived daughter isotopes has occurred. The large-area ^{233}U source 2 was produced from chemically purified ^{233}U material. The factor of chemical purification was shown to be larger than 300, based on comparisons of γ-ray spectra (see Appendix A.8). In case that the detected signal would originate from any of the short-lived daughter isotopes, a strong reduction of the signal intensity would be expected when using ^{233}U source 2. However, no such reduction occurs and instead an increase of the signal intensity by a factor of 13.5 is observed, attributed to a higher α-recoil efficiency of the source.

A third way of exclusion of nuclear background is provided by the lifetime measurements. In preliminary measurements, a half-life of significantly shorter than 1 s was obtained for the observed decay, when ions are collected on the detector surface (Sect. 5.2.3). This information already allows for an exclusion of all nuclides other than ^{229m}Th as signal origin [6]. While the long-lived nuclides are trivially excluded, all potentially extracted short-lived isotopes (with a half-life of 1 s or below) are α emitters and can be excluded by means of the signal shape and intensity of α decays on the MCP detector (see Fig. 5.21). There is no β emitter extracted from the source which provides a sufficiently short half-life to explain the detected signal (see Appendix A.1). The same argument also serves for the exclusion of potential isomeric states other than ^{229m}Th (see Table A.1).

Recently, the half-life of the detected IC decay could be measured to be about 7 μs [21]. This is even faster than the time required for ion extraction from the buffer-gas stopping cell, thus immediately excluding any radioactive decay as signal origina except for ^{229m}Th.

The fourth and last way of nuclear background exclusion is based on the direct search for α and β decays. For the exclusion of α decays, ^{229}Th^{2+} and ^{229}Th^{3+} ions were directly accumulated (with extraction conditions described in Sect. 5.1.1) onto a silicon detector (Ametek, BU-014-150-100) set to -10 V offset and 20 V bias voltage. α decays were detected in parallel to accumulation. The detector was mounted directly behind the triode extraction system in about 5 mm distance, replacing the CsI-coated MCP detector. A charge sensitive preamplifier (CSTA) and a shaping amplifier (Ortec, model 571) were used for signal read-out. The spectra were acquired with the help of a portable multi-channel analyzer (Amptek, MCA-8000 A). Mass scans

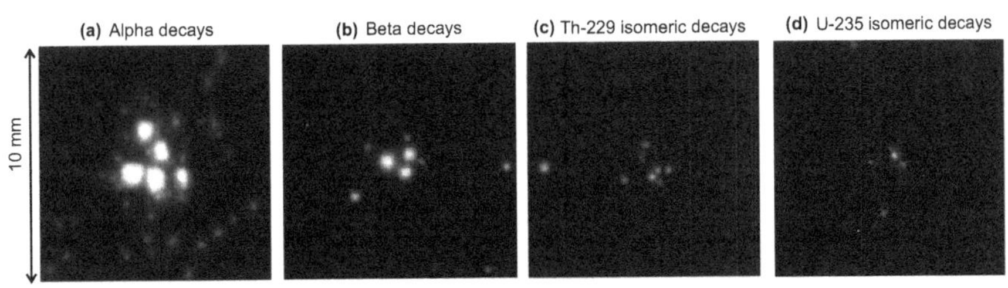

Fig. 5.21 Different classes of decay events as observed during ion accumulation on the MCP surface (Beam Imaging Solutions, BOS-75-FO) [6]. Individual frames of 4 s exposure time at an MCP surface voltage of -25 V are shown. The extracted ion species was chosen by mass-to-charge separation with the QMS. **a** α decays originating from ^{221}Fr ($t_{1/2} = 4.8$ min). **b** β decays originating from ^{209}Pb ($t_{1/2} = 3.25$ h). **c** ^{229}Th isomeric decay ($t_{1/2} \approx 7$ μs). **d** ^{235}U isomeric decay ($t_{1/2} = 26$ min). With kind permission of Nature Research [6]

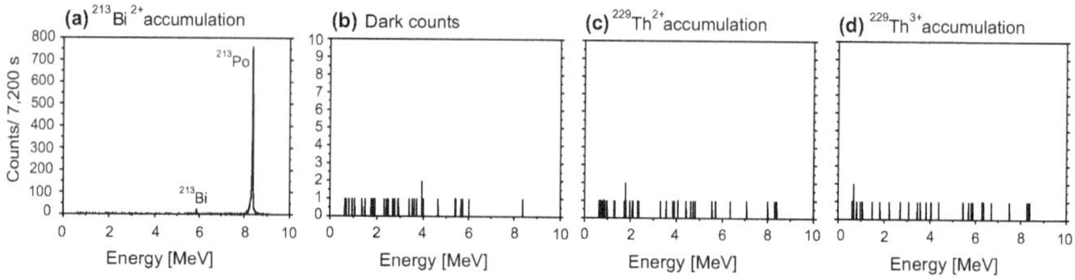

Fig. 5.22 α-energy spectra of different Si-detector-based measurements, each accumulated for 7200 s. The extracted ion species was chosen by a mass-to-charge separation with the QMS. Measurements were performed using the chemically unpurified ^{233}U source 1. **a** Accumulation of $^{213}Bi^{2+}$. **b** Dark count measurement. **c** Accumulation of $^{229}Th^{2+}$. **d** Accumulation of $^{229}Th^{3+}$. With kind permission of Nature Research [6]

could be performed for QMS calibration using an MCP (Hamamatsu, F2223) that was mounted sideways at 90° to the triode extraction system.

Four different measurements were performed with this setup, using the small-area ^{233}U source 1, each with 2 h detection time: The functionality of the detector system was tested by accumulation of $^{213}Bi^{2+}$ (Fig. 5.22a, 2.0 cts/s), a dark count measurement (Fig. 5.22 b, $5.7 \cdot 10^{-3}$ cts/s) was performed prior to the extraction of $^{229}Th^{2+}$ (Fig. 5.22c, $6.0 \cdot 10^{-3}$ cts/s) and $^{229}Th^{3+}$ (Fig. 5.22d, $5.3 \cdot 10^{-3}$ cts/s). No signals above the background were detectable, leading to a clear exclusion of α decays as signal origin. The count rate of $\sim$0.25 cts/s (see Sect. 5.2.2) would otherwise have been easily detectable. The α decay of ^{229}Th cannot be expected to be detected on such short accumulation times due to its long lifetime (see Sect. 5.1.1 for comparison).

For the exclusion of β decays or high-energy internal conversion electrons as signal origin, a lithium-drifted silicon (Si(Li)) detector was used (Canberra, type ESLB-3000-300). The detector was cryogenically cooled with liquid nitrogen and replaced the previously used silicon detector. The read out was performed with a preamplifier (Eurisys Measures, PSC 761) and a shaping amplifier (Ortec, type 572). For operation, a bias voltage of −400 V was applied to the front surface, such that no further voltage offset was required. Four different measurements were performed using the ^{233}U source 1 with 10 h acquisition time per measurement: The functionality of the detector was verified during accumulation of $^{209}Pb^{2+}$ (2.13 cts/s), one dark count measurement (0.47 cts/s) was carried out prior to the accumulation of $^{229}Th^{2+}$ (0.44 cts/s) and $^{229}Th^{3+}$ (0.48 cts/s). If the detected signals were β decays or high-energy internal-conversion electrons, an enhancement of the integrated signal rate by 0.25 cts/s above the level of the dark count rate would have been easily detectable.

5.3.4 Signals Caused by Excited Shell-States

Atomic shell effects originating from atoms other than thorium itself can be excluded in a similar way as nuclear background. The parallel observation of the signal in the

2+ and 3+ charge states, in combination with a clear signal restriction to the m/q-value of ^{229}Th, serves as one way of exclusion. Further evidence is provided by the measurements with the chemically purified ^{233}U source 2. For this reason, the only remaining type of potential background originates from thorium atomic shell effects.

Thorium atomic shell effects, like a long-lived atomic excitation or a chemical reaction on the detector surface could, lead to MCP signals. Such shell effects could be special to thorium and cannot be excluded by comparison with other elements. In order to exclude such effects as signal origin, measurements were performed with ^{230}Th instead of ^{229}Th. A chemically purified ^{234}U source (source 3, see Sect. 4.2.1) was produced for this purpose by the Institute for Radiochemistry (IRC) at the University of Mainz via electrodeposition. The ^{234}U source possesses an activity of 270 kBq and has the same geometry as the large-area ^{233}U source 2. ^{230}Th ions were shown to be extracted from this source with a similar rate as ^{229}Th ions in case of the ^{233}U source.

$^{230}Th^{2+}$ as well as $^{230}Th^{3+}$ ions were extracted for 2000s directly onto the MCP surface, and images were acquired under the same conditions as described for ^{229}Th in Sect. 5.2.2. As shown in Fig. 5.17, no signal was detected when extracting ^{230}Th for both charge states. Therefore any kind of thorium shell effects can be excluded as signal origin.

This measurement also serves for further exclusion of background signals originating from the setup components and ionic impact. In this way it allows to exclude most of the systematic background effects.

5.3.5 Estimation of the ^{229,228}Th Intrinsic Activity

The α decay of ^{229}Th or ^{228}Th as signal origin has already been excluded in multiple ways, namely by (i) their long lifetimes, (ii) the different signal shapes and intensities of α decays and (iii) by direct silicon-detector measurements. However, after long times of continuous accumulation, a contamination of the MCP detector has to be taken into account.

The QMS is not designed to separate ^{229}Th and ^{228}Th ions in the 2+ and 3+ charge states. Both isotopes are extracted from the ^{233}U source and ^{228}Th has a half-life of 1.9 years, significantly shorter than the ^{229}Th half-life of 7932 years, therefore the accumulated ^{228}Th activity even exceeds the ^{229}Th activity by a factor of 3.6 (Fig. 5.3) [1]. An MCP-detector contamination is assumed to be measurable as soon as the number of decays reaches the order of the dark counts of the MCP detector (about 0.01 cts/(s mm^2)). The accumulation takes place on a surface area of about 2 mm diameter, leading to a maximum allowed decay rate of 0.03 1/s. This activity corresponds to an absolute number of ^{228}Th atoms of $2.6 \cdot 10^6$. Based on the relative α-decay rates in the spectrum shown in Fig. 5.3, the number of $^{228}Th^{3+}$ isotopes, extracted from the large-area ^{233}U source 2, can be estimated to 8.7 per second (compared to 10 000 extracted $^{229}Th^{3+}$ ions per second for ^{229}Th). Thus, after $3 \cdot 10^5$ s of continuous extraction from the ^{233}U source 2, an MCP detector contamination should be detectable. This corresponds to 3.5 days. Moreover, after this time the activity on the detector surface will soon grow due to daughter ingrowth.

Fortunately, the detected signal rate is significantly above the background level (by a factor of about 100 for source 2) and therefore no dilution of the signal by accumulated activity has to be expected on short time scales.

References

1. von der Wense L et al (2015) Determination of the extraction efficiency for ^{233}U source α-recoil ions from the MLL buffer-gas stopping cell. Eur Phys J A 51:29
2. von der Wense L et al (2016) The extraction of $^{229}Th^{3+}$ from a buffer-gas stopping cell. Nucl Instrum Methods B 376:260–264
3. von der Wense L et al (2013) Towards a direct transition energy measurement of the lowest nuclear excitation in ^{229}Th. JINST 8:P03005
4. Kramida A., Ralchenko Yu., Reader J (2015) NIST Atomic Spectra Database (ver. 5.2), National Institute of Standards and Technology, Gaithersburg, MD. http://physics.nist.gov/asd. Accessed 16 June 2015
5. Oberheide J et al (1997) New results on the absolute ion detection efficiencies of a microchannel plate. Meas Sci Technol 8:351–354
6. von der Wense L et al (2016) Direct detection of the ^{229}Th nuclear clock transition. Nature 533:47–51
7. Quantar Technology (2016) Mepsicron-II^{tm} Series Single-Photon Imaging Detector System, Model 2601B Technical Description (2007). http://www.quantar.com/pdfpages/2601desc.pdf. Accessed 29 Nov 2016
8. Seiferle B et al (2016) A VUV detection system for the direct photonic identification of the first excited isomeric state of ^{229}Th. Eur Phys J D 70:58
9. Seiferle B (2015) Setup of a VUV detection system for the direct identification of the fluorescence radiation of ^{229m}Th, Master Thesis, LMU Munich, Germany
10. Karpeshin FF, Trzhaskovskaya MB (2007) Impact of the electron environment on the lifetime of the $^{229}Th^{m}$ low-lying isomer. Phys Rev C 76:054313
11. Tkalya EV et al (2015) Radiative lifetime and energy of the low-energy isomeric level in ^{229}Th. Phys Rev C 92:054324
12. Tkalya EV et al (2000) Decay of the low-energy nuclear isomer $^{229}Th^{m}(3/2^{+}, 3.5\pm1.0$ eV) in solids (dielectrics and metals): a new scheme of experimental research. Phys Rev C 61:064308
13. Dessovic P et al (2014) ^{229}Thorium-doped calcium fluoride for nuclear laser spectroscopy. J Phys Condens Matter 26:105402
14. Vig JR (1985) UV/ozone cleaning of surfaces. J Vac Sci Technol A 3:1027–1034
15. Porsev SG, Flambaum VV (2010) Electronic bridge process in $^{229}Th^{+}$. Phys Rev A 81:042516
16. Borisyuk PV et al (2015) Band structure and decay channels of thorium-229 low lying isomeric state for ensemble of thorium atoms adsorbed on calcium fluoride. Phys Status Solidi C 12:1333–1337
17. Jongma RT, Rasing T, Meijer G (1995) Two-dimensional imaging of metastable CO molecules. J Chem Phys 102:1925–1933
18. Peko BL et al (2000) Absolute detection efficiencies of low energy H, H^{-}, H^{+}, H_2^{+}, H_3^{+} incident on a multichannel plate detector. Nucl Instrum Methods B 171:597–604
19. Stephen TM et al (2000) Absolute calibration of a multichannel plate detector for low energy O, O^{-}, and O^{+}. Rev Sci Instrum 71:1355–1359
20. Plaß WR et al (2008) Isobar separation by time-of-flight mass spectrometry for low-energy radioactive ion beam facilities. Nucl Instrum Methods B 266:4560–4564

21. Seiferle B et al (2017) Lifetime measurement of the ^{229}Th nuclear isomer. Phys Rev Lett 118:042501
22. Beck BR et al (2007) Energy splitting of the ground-state doublet in the nucleus ^{229}Th. Phys Rev Lett 109:142501
23. Beck BR et al (2009) Improved value for the energy splitting of the ground-state doublet in the nucleus ^{229m}Th, LLNL-PROC-415170
24. Yamakita Y et al (2000) A highly sensitive electron spectrometer for crossed-beam collisional ionization: a retarding-type magnetic bottle analyzer and its application to collision-energy resolved Penning ionization electron spectroscopy. Rev Sci Instrum 71:3042–3049
25. Schwarz M et al (2012) Cryogenic linear paul trap for cold highly charged ion experiments. Rev Sci Instrum 83:083115
26. Schmöger L et al (2015) Coulomb crystallization of highly charged ions. Science 347:1233–1236
27. Haettner E (2011) A novel radio frequency quadrupole system for SHIPTRAP and new mass measurements of rp nuclides, Ph.D. thesis, Univ. Giessen, Germany
28. Maier WB (1978) Electron transfer in collisions between atomic ions and rare-gas atoms for primary-ion energies below 200 eV. J Chem Phys 69:3077–3092

Chapter 6
Conclusion and Outlook

In the following, the results of the presented work will be shortly summarized and interpreted with respect to ongoing experimental efforts on ^{229}Th in different groups. Also perspectives for future experiments are given, that would provide a step towards the development of a ^{229}Th-based nuclear frequency standard.

6.1 Conclusion

The main result of the presented work is the first direct detection of $^{229\text{m}}$Th, which was achieved by investigating the internal conversion decay channel (Sect. 5.2.2) [1]. Corresponding experiments were performed, after it became evident that the previous investigation of a potential radiative decay channel would not lead to a successful observation of the isomeric decay (Sect. 5.2.1). The non-observation of photons, despite a high sensitivity for a radiative decay channel of $^{229\text{m}}$Th, was a clear hint towards a significant non-radiative decay branch.

The concept of isomer detection is visualized in Fig. 6.1. $^{229\text{m}}$Th ions in the 2+ or 3+ charge state are collected with low kinetic energy directly on the surface of an MCP detector. The internal conversion decay channel in thorium ions is energetically forbidden. Therefore the isomeric state is long-lived (in the range of minutes to hours), while thorium remains charged and no decay occurs during time of flight. However, as soon as the ions come into contact with the MCP detector, charge exchange occurs, forming neutral thorium atoms. In this case, the fast internal conversion decay channel of the isomeric state dominates and leads to the ejection of an electron within the timescale of a few microseconds. The internal conversion electron is accelerated into a microchannel of the MCP detector, thereby multiplied and finally detected. It is remarkable that the isomer detection is practically background free, as neither the ionic impact of the low-energy ions (with a kinetic energy of a few ten eV) nor the charge exchange on the MCP detector lead to a detectable signal. In both processes, the released energy is predominantly transformed to phonons [2, 3].

L. Von der Wense, *On the Direct Detection of* ^{229m}*Th*,
Springer Theses, https://doi.org/10.1007/978-3-319-70461-6_6

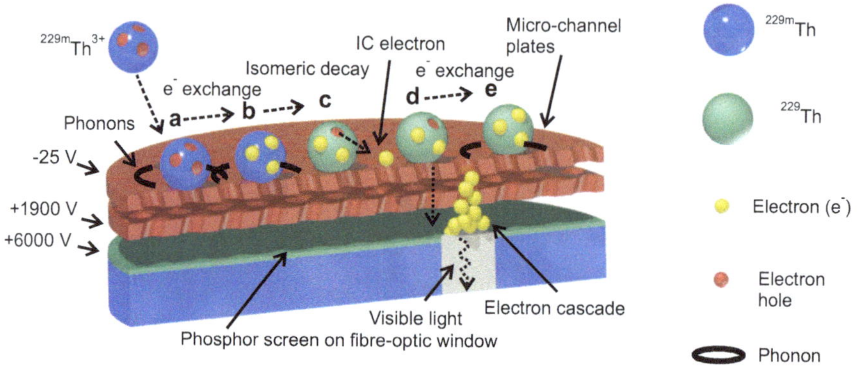

Fig. 6.1 Schematic drawing of the isomer detection process. **a**: A $^{229m}Th^{3+}$ ion impinges on the MCP surface. The thorium ion in the isomeric state is visualized as a blue sphere. **b**: Electron capture on the surface. The energy is dissipated in form of phonons (indicated as black circles). Electrons are visualized as yellow spheres. **c**: An IC electron is released by the isomeric decay. **d**: The IC electron triggers a secondary-electron cascade, which is accelerated towards the phosphor screen. **e**: The hole, left by the IC process, is closed by electron capture on the MCP surface. Again phonons are produced. With kind permission of Nature Research [1]

The successful direct observation of ^{229m}Th has opened the possibility for a multitude of measurements, in order to investigate the isomeric decay behavior under different conditions. Isomer detection measurements were successfully performed during extraction of $^{229}Th^{2+}$ and $^{229}Th^{3+}$, however, no isomeric decay was observed during the extraction of $^{229}Th^{1+}$. This observation was first attributed to a very low extraction efficiency for the 1+ charge state [1], but further investigations with higher extraction rates show that there is no detectable isomeric decay during the extraction of $^{229}Th^{1+}$ (Sect. 5.2.2). This is a clear hint towards an unexpectedly reduced lifetime of ^{229m}Th in Th^{1+} ions to below a few ms (corresponding to the time required for ion extraction). It is hard to explain such a lifetime reduction of six orders of magnitude compared to the photonic decay in Th^{1+}. One potential explanation might be an on-resonance electronic-bridge channel, which is, however, unlikely, as it would require a nearly exact resonance with an electronic shell state [4]. An alternative, but equally unlikely explanation, would be that IC is not energetically forbidden, due to an isomeric energy value exceeding the 2+ ionization potential of thorium of about 11.9 eV. This, however, would be in conflict with the currently best available energy value for ^{229m}Th of (7.8 ± 0.5) eV [5]. Alternatively, the experimental result could also be explained by buffer-gas quenching of the isomeric state in Th^{1+}.

The internal conversion decay channel was also investigated during on-the-fly molecule formation (Sect. 5.2.2). In this way $^{229}ThAr^{3+}$ and $^{229}ThO^{2+}$ could be produced. No isomeric decay was detectable for both cases. In combination with the non-observation of a photonic decay channel during the extraction of $^{229}ThO^{2+}$ (Sect. 5.2.1), this leads to the conclusion that the isomeric state is highly sensitive to molecule formation. This result can potentially be explained by means of the high density of rotational and vibrational modes of di-atomic molecules in the VUV region, which might be excited by the nuclear transition.

Further measurements were performed to probe the IC decay on different surface materials, of which the most important case is the collection on MgF_2 (Sect. 5.2.2), due to its large band gap of approximately 11.9 eV. As this band gap is significantly above the expected isomeric energy value, there is the hope that IC is suppressed [6], in this way opening the possibility for an enhanced photonic decay channel of ^{229m}Th. Experimentally, however, also on the MgF_2-coated surface an IC decay with a short lifetime was observed. This is in agreement with the non-observation of a photonic decay channel during the optical measurements, where ^{229m}Th was also collected on an MgF_2-coated surface (Sect. 5.2.1). A potential explanation can be found in surface effects, like remaining hydrocarbon contamination or unbound surface electrons [7]. A secure exclusion of surface effects would require significantly more experimental effort.

The half-life of ^{229m}Th was investigated for Th^{2+} ions (Sect. 5.2.3). By storing the thorium ions in the RFQ and subsequent detection of any remaining isomeric decay signal, the half-life was determined to be longer than 1 min [1]. It was not possible to probe for longer lifetimes, due to the limited storage time of Th^{2+}. This measurement, however, is sufficient to securely say that the internal conversion decay channel is energetically forbidden in Th^{2+}, resulting in an upper limit for the isomeric energy of 18.3 eV, corresponding to the Th^{2+} ionization potential [1]. Similarly, the IC decay channel is obviously allowed for neutral ^{229}Th, resulting in an estimated lower limit of the isomeric energy of 6.3 eV (corresponding to the ionization potential of neutral thorium). A more careful discussion would have to include the effective work function of the collection material. Most recently, the half-life of the isomeric decay during internal conversion was determined to about 7 μs [8].

While most of the isomeric properties behave exactly as theoretically predicted [9, 10], there are a few experimental surprises. One is the non-observation of an internal conversion decay channel during the extraction of Th^{1+}, which could potentially be explained by a significantly enhanced electronic-bridge process that can only occur in case of resonance with an electronic shell state [4]. If this interpretation turns out to be correct, there is hope for an enhanced population of the isomeric state making use of the reverse process, as envisaged by several groups (see Sect. 3.6.4). On the other hand, the isomeric state seems to be very sensitive to its chemical environment and strongly prefer a non-radiative decay with a short half-life during chemical reaction or surface contact. While significantly more experimental effort is required to provide clean surfaces and probe the isomeric decay under fully controlled chemical conditions, the first impression is that it might be very hard if not even impossible to suppress the non-radiative decay of the isomeric state in a way required for the development of a solid-state nuclear clock as proposed in Ref. [11].

6.2 Outlook

The direct detection of ^{229m}Th enables several new experiments and investigations required for the development of a nuclear frequency standard. Most importantly, it paves the way for direct nuclear laser excitation of ^{229m}Th. For this purpose, a new laser excitation scheme is presented in Sect. 2.3.3, which makes use of the isomer's internal conversion decay channel and the characteristic IC half-life of ^{229m}Th of about 7 μs [8]. Opposed to the general assumption that the isomer's direct nuclear laser excitation would require a better knowledge of the transition energy, the proposed excitation scheme circumvents this requirement, thereby allowing for nuclear laser spectroscopy of ^{229m}Th already today. This proposal can be interpreted as laser-based conversion-electron Mössbauer spectroscopy (CEMS) in the optical region and would also allow for the development of a proof-of-principle device of a CEMS-based solid-state nuclear clock.

Many experiments that are currently performed on ^{229m}Th depend on the availability of a more precise isomeric energy value. One of the main ongoing experimental objectives of our group is to determine the energy value of the excited state to better than 0.1 eV in precision, based on the direct isomeric decay signal. For this purpose, a magnetic-bottle electron spectrometer with a measured energy-resolving power of better than 50 meV [12] was developed and is currently tested.

One further objective is the isomeric lifetime determination in the 2+ and 3+ charge state. The availability of this value would lead to a more precise estimation of electronic-bridge processes and in this way allow to determine the precision of a thorium-ion-based nuclear clock. Currently, the determination of the ^{229m}Th$^{2+,3+}$ half-life is limited by the available ion storage time in the RFQ (Sect. 5.2.3). It is planned to develop a cryogenic linear Paul trap in close collaboration with the MPIK and PTB [13, 14], in order to increase the ion storage times to up to 10 hours. The cryogenic Paul trap will also provide full laser access, in this way providing the basis for future nuclear laser-spectroscopy experiments on thorium ions.

The failure of our optical experiments to observe any photonic decay signal of ^{229m}Th can clearly be interpreted in terms of internal conversion, which is dominating the radiative decay by 9 orders of magnitude [8]. It might be possible to suppress the internal conversion decay channel by implanting the thorium ions into completely clean surfaces of high band gap (e.g. MgF_2, CaF_2). Although thorium ion collection on MgF_2 was already performed (Sect. 5.2.2) and no suppression of the IC decay channel was observed, surface effects like hydrocarbon contamination or band gap reduction could have triggered the IC decay [7, 10]. For future experiments, it would be of interest if chemical conditions can be obtained which suppress the IC decay channel. Such an observation would be of high importance for the development of a solid-state nuclear clock as described in Ref. [11].

One could further think, if there is still the possibility to also observe the radiative isomeric decay with the help of a VUV-optical system as originally proposed by our group [15, 16]. To our best current knowledge, however, this would require long-time storage of thorium ions in the 2+ or 3+ charge state in a Paul trap. Opposed to surface collection, the number of stored ions is limited by the maximum achievable charge density in the trap, posing strong limitations for a corresponding experiment. At least in principle this problem could be solved by allowing for long detection times in a nearly background-free environment (e.g. in an underground laboratory) or by triggering the photon detector in accordance with the isomeric decay, which could be detected in parallel by applying the double-resonance method to a Coulomb crystal of $^{229}Th^{3+}$ [17]. Both proposals are, however, of significant experimental effort.

In conclusion, the direct detection of ^{229m}Th, besides providing new information on the isomeric properties already today, has opened the possibility for a multitude of new experimental investigations. These include direct nuclear laser-excitation, the precise determination of the isomeric energy value, isomeric lifetime measurements for different charge states and chemical conditions and laser-spectroscopy of the hyperfine structure of ^{229m}Th. All these efforts are important steps towards nuclear quantum optics with ^{229}Th and the development of a first nuclear-based frequency standard.

References

1. von der Wense L et al (2016) Direct detection of the ^{229}Th nuclear clock transition. Nature 533:47–51
2. Peko BL et al (2000) Absolute detection efficiencies of low energy H, H^-, H^+, H_2^+, H_3^+ incident on a multichannel plate detector. Nucl Instrum Methods Phys Res Sect B 171:597–604
3. Bay HL, Winters HF, Coufal HJ, Eckstein W (1992) Energy transfer to a copper surface by low energy noble gas ion bombardment. Appl Phys A Sol Surf 55:174–278
4. Porsev SG et al (2010) Excitation of the isomeric ^{229m}Th nuclear state via an electronic bridge process in $^{229}Th^+$. Phys Rev Lett 105:182501
5. Beck BR et al (2009) Improved value for the energy splitting of the ground-state doublet in the nucleus ^{229m}Th. LLNL-PROC-415170
6. Tkalya EV et al (2000) Decay of the low-energy nuclear isomer $^{229}Th^m(3/2^+, 3.5\pm1.0$ eV) in solids (dielectrics and metals): a new scheme of experimental research. Phys Rev C 61:064308
7. Borisyuk PV et al (2015) Band structure and decay channels of thorium-229 low lying isomeric state for ensemble of thorium atoms adsorbed on calcium fluoride. Phys Stat Solidi C 12:1333–1337
8. Seiferle B et al (2017) Lifetime measurement of the ^{229}Th nuclear isomer. Phys Rev Lett 118:042501
9. Karpeshin FF, Trzhaskovskaya MB (2007) Impact of the electron environment on the lifetime of the $^{229}Th^m$ low-lying isomer. Phys Rev C 76:054313
10. Tkalya EV et al (2015) Radiative lifetime and energy of the low-energy isomeric level in ^{229}Th. Phys Rev C 92:054324
11. Rellergert WG et al (2010) Constraining the evolution of the fundamental constants with a solid-state optical frequency reference based on the ^{229}Th nucleus. Phys Rev Lett 104:200802
12. Yamakita Y et al (2000) A highly sensitive electron spectrometer for crossed-beam collisional ionization: a retarding-type magnetic bottle analyzer and its application to collision-energy resolved Penning ionization electron spectroscopy. Rev Sci Instrum 71:3042–3049
13. Schwarz M et al (2012) Cryogenic linear paul trap for cold highly charged ion experiments. Rev Sci Instrum 83:083115
14. Schmöger L et al (2015) Coulomb crystallization of highly charged ions. Science 347:1233–1236

15. von der Wense L et al (2013) Towards a direct transition energy measurement of the lowest nuclear excitation in ^{229}Th. JINST 8:P03005
16. Seiferle B (2015) Setup of a VUV detection system for the direct identification of the fluorescence radiation of ^{229m}Th. Master Thesis, LMU Munich, Germany
17. Campbell CJ et al (2011) Wigner Crystals of ^{229}Th for optical excitation of the nuclear isomer. Phys Rev Lett 106:223001

Appendix A
^{233}U Source Investigation

Detailed investigations of the small-area ^{233}U source 1 and its source material were carried out. These investigations can be divided into the following steps:

1. The age of the ^{233}U source material as well as the fractional content of ^{232}U were investigated, based on ^{233}U γ and α spectroscopy (Sect. A.2).
2. These numbers, together with the activity of the ^{233}U source, were used as input parameters for the calculation of daughter-isotope activities for the decay chains of ^{233}U and ^{232}U (Sect. A.3).
3. The number of α-recoil isotopes leaving the source per second was measured for ^{229}Th and some daughter nuclides. These numbers allowed to estimate the α-recoil efficiencies of the ^{233}U source (Sect. A.4).
4. The measured α-recoil efficiencies were compared to two different theoretical predictions, based on different numerical models. These simulations allowed to estimate the α-recoil efficiencies and therefore also the absolute α-recoil rates for all isotopes in the ^{233}U and ^{232}U chains, respectively (Sect. A.5).
5. Alpha-recoil isotope activities caused by ion implantation into the source surrounding is estimated as an effect of second order and leads to a minor correction of the previously calcuated values (Sect. A.6).
6. The absolute number of α-recoil nuclides, extracted from the buffer-ga s stopping cell, was measured for the whole ^{233}U and ^{232}U decay chains. These values, combined with α-recoil rates of the ^{233}U source, allowed to estimate the charge-state dependent extraction efficiencies for all isotopes extracted from the ^{233}U and ^{232}U decay chain (Sect. A.7).

In the following, these investigations will be described, starting with the production process of ^{233}U (Sect. A.1). In the end, also the factor of chemical purification of ^{233}U material is investigated (Sect. A.8).

L. Von der Wense, *On the Direct Detection of ^{229m}Th*,
Springer Theses, https://doi.org/10.1007/978-3-319-70461-6

A.1 Production and Decay of ^{233}U

In order to allow for an unambiguous identification of the ^{229m}Th isomer, the ^{233}U source material, which is used for α-recoil isotope production, has to be carefully investigated. This allows for an exclusion of possible nuclear background caused by short-lived daughter isotopes (see Sect. 5.3.3).

^{233}U is a fissile material that is not naturally abundant. Large amounts of ^{233}U were artificially produced in nuclear reactors during the cold war [1]. It is mainly produced from highly enriched ^{232}Th by neutron irradiation via the reaction chain

$$^{232}\text{Th} \xrightarrow{(n,\gamma)} {}^{233}\text{Th} \xrightarrow[21.8\text{ min}]{(\beta-)} {}^{233}\text{Pa} \xrightarrow[26.9\text{ d}]{(\beta-)} {}^{233}\text{U}. \tag{A.1}$$

Besides ^{233}U, also ^{232}U is produced in this process, leading to a significant ingrowth of activity, as the half-lives involved in the ^{232}U decay chain are comparatively short. The reaction chains for the production of ^{232}U from ^{232}Th are

$$\begin{aligned} &^{232}\text{Th} \xrightarrow{(n,\gamma)} {}^{233}\text{Th} \xrightarrow[21.8\text{ min}]{(\beta-)} {}^{233}\text{Pa} \xrightarrow[26.9\text{ d}]{(\beta-)} {}^{233}\text{U} \xrightarrow{(n,2n)} {}^{232}\text{U} \\ &^{232}\text{Th} \xrightarrow{(n,\gamma)} {}^{233}\text{Th} \xrightarrow[21.8\text{ min}]{(\beta-)} {}^{233}\text{Pa} \xrightarrow{(n,2n)} {}^{232}\text{Pa} \xrightarrow[1.3\text{ d}]{(\beta-)} {}^{232}\text{U}. \end{aligned} \tag{A.2}$$

While the decay chains of ^{233}U and ^{232}U are the most prominent sources of α-recoil isotopes, a few other nuclides were shown to be present in the source material as trace amounts by γ spectroscopy (Fig. A.3) and α spectroscopy (Fig. A.5). These are ^{231}Pa, ^{238}Pu, ^{239}Pu and ^{241}Am. ^{231}Pa is a strong byproduct in the ^{232}Th fuel cycle, and is produced by the reaction chain

$$^{232}\text{Th} \xrightarrow{(n,2n)} {}^{231}\text{Th} \xrightarrow[25.5\text{ h}]{(\beta-)} {}^{231}\text{Pa}. \tag{A.3}$$

The presence of plutonium indicates that the material in use was in contact with the ^{238}U fuel cycle as well. The corresponding reaction chains are

$$\begin{aligned} &^{238}\text{U} \xrightarrow{(n,\gamma)} {}^{239}\text{U} \xrightarrow[23.5\text{ min}]{(\beta-)} {}^{239}\text{Np} \xrightarrow[2.36\text{ d}]{(\beta-)} {}^{239}\text{Pu} \\ &^{238}\text{U} \xrightarrow{(n,2n)} {}^{237}\text{U} \xrightarrow[6.75\text{ d}]{(\beta-)} {}^{237}\text{Np} \xrightarrow{(n,\gamma)} {}^{238}\text{Np} \xrightarrow[2.1\text{ d}]{(\beta-)} {}^{238}\text{Pu}. \end{aligned} \tag{A.4}$$

The production process of ^{241}Am is detailed in Fig. A.1. These are the only nuclides that were measured to be present in the ^{233}U source material, however, presumably further isotopes are contained as trace contaminations. When searching for the isomeric decay of ^{229}Th, even trace amounts of other isotopes (with an activity as low as 1/4000 of the ^{233}U activity, corresponding to 50 Bq) could potentially lead to a dilution of any obtained signal. A more systematic study is carried out in order to cover all isotopes that are potentially contained in the ^{233}U source material.

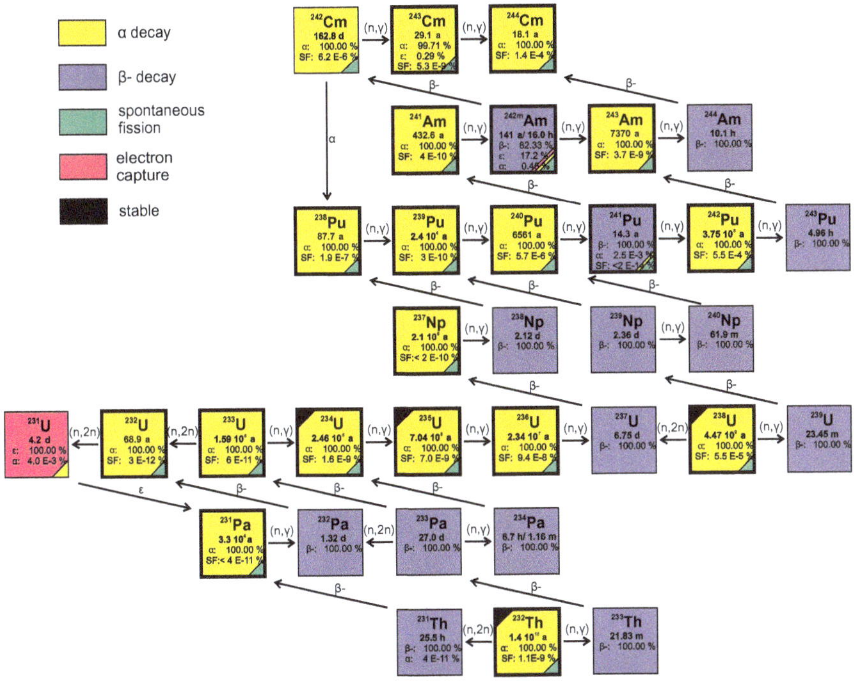

Fig. A.1 Transmutation chart of nuclides as obtained by neutron irradiation in nuclear reactors, starting with the isotopes ^{232}Th, ^{238}U, ^{235}U and ^{234}U (naturally abundant as indicated by a black corner). Only nuclides up to ^{244}Cm are shown. The nuclides framed by black boxes are long-lived and therefore potentially contained in the ^{233}U source material, together with their decay chains (see Fig. A.2)

In the process of nuclear-fuel production, naturally occurring ^{232}Th as well as ^{238}U (together with small amounts of ^{235}U and ^{234}U) is neutron irradiated in nuclear reactors. The nuclear transmutation processes triggered by this irradiation are shown for both transmutation cycles in Fig. A.1, together with the half-lives and dominant decay branches of the isotopes involved.

When the irradiation process is stopped, the short-lived isotopes will decay and only the long-lived nuclides are potentially included in the remaining nuclear-fuel material. Therefore, the starting material potentially contains a mixture of 19 different nuclides, as indicated by black boxes in Fig. A.1. No isotopes heavier than ^{244}Cm have to be considered. The reason is that the fractional abundance of ^{245}Cm compared to ^{244}Cm, as produced in the nuclear-fuel cycle, is small [2]. As no activity of ^{244}Cm could be detected within the α spectroscopy (Fig. A.5), it must already be below 1/250 of the ^{233}U activity of the source (otherwise it would have led to a detectable signal). The half-life ratio between ^{244}Cm and ^{245}Cm is 1/470. Therefore, even if the fractional abundance of ^{245}Cm was in the same range, the ^{245}Cm activity in the source must be below a fraction of 10^{-5} of the ^{233}U activity and thus below 2 Bq, which is too low to cause the detected signal attributed to ^{229m}Th (see Sect. 5.2.2,

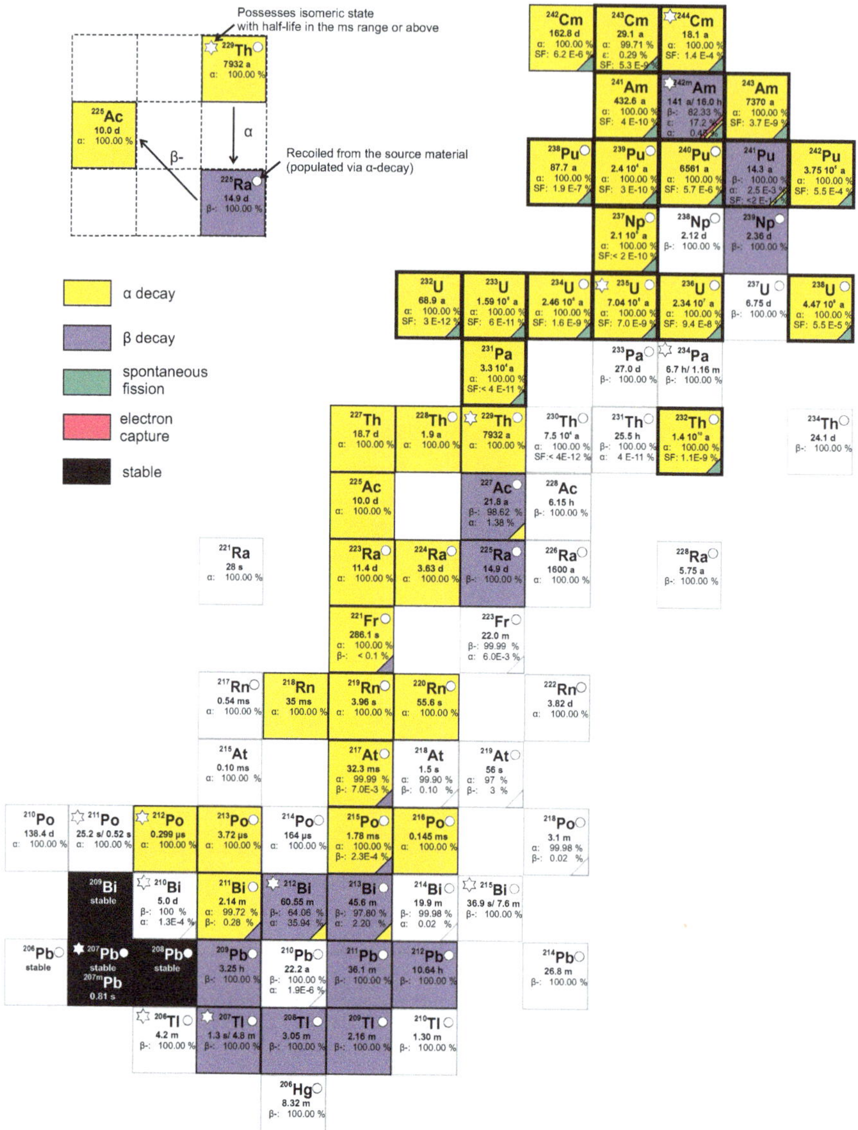

Fig. A.2 Chart of nuclides potentially contained in the ^{233}U source material. The starting isotopes are framed by thick black boxes. Nuclides that are expected to play a negligible role are shown without color. Nuclides that can potentially recoil from the source as being populated via α decay are labelled with a white circle. Nuclides with one or more isomeric states carry a white star. With kind permission of Nature Research [11]

^{233}U source 1). The fractional abundances of all isotopes of higher masses are even significantly below that of ^{245}Cm and are therefore not considered [2].

There is no information about the exact isotopic composition of the source material available, therefore the decay chains of all of these 19 starting isotopes will be taken into account in the following. Many of the heavy nuclides possess a spontaneous fission decay branch. However, even the most significant branching ratio is just $5.5 \cdot 10^{-6}$ (for ^{242}Pu). In order to explain the detected signal by spontaneous fission, an activity of about 9 MBq or above would be required, which is far above the total activity of the source. Even when taking into account that during the production process a significant amount of nuclides were produced due to neutron induced fission, the short-lived nuclides would already be decayed, while the long-lived nuclides are pure β^- emitters (^{137}Cs $t_{1/2} = 30.2$ a, ^{90}Sr $t_{1/2} = 28.8$ a, ^{85}Kr $t_{1/2} = 10.8$ a) and therefore cannot produce any recoil ions. Therefore fission-related effects will not play a role in this context.

The complete list of starting nuclides and successively populated daughter isotopes is shown in Fig. A.2 together with their half-lives and branching ratios. The starting isotopes are shown with a black frame. For completeness, all of the populated isotopes are shown, even if their activities can be assumed to play only a negligible role due to a small branching ratio, or due to a long half-life of the mother nuclide. These isotopes are shown with white boxes. Isotopes, which can potentially recoil

Table A.1 List of isomeric states of nuclides known to be potentially contained in the source material. Only few of the listed isomers are populated in the decay chain and could thus contribute to the background

Isotope	Isomer energy	Isomer half-life	Decay channel	Population
^{244}Cm	1.04 MeV	34 ms	IT: 100.00 %	Not populated
^{242}Am	48.6 keV	141 a	IT: 99.55 %, α: 0.45 %	100 % populated
	2.20 MeV	14.0 ms	SF: 100 %, $\alpha \leq 5.0 \cdot 10^{-3}$, IT	Not populated
^{235}U	76 eV	26 min	IT: 100 %	70 % from ^{239}Pu
^{234}Pa	73.9 keV	1.16 min	$\beta-$: 99.84 %, IT: 0.16 %	78 % from ^{234}Th
^{229}Th	$\sim$ 7.8 eV	$\sim$7 μs (IC)	IT	2 % from ^{233}U
^{212}Po	2.922 MeV	45.1 s	α: 99.93 %, IT: 0.07 %	Not populated
^{211}Po	1.462 MeV	25.2 s	α: 99.98 %, IT: 0.02 %	Not populated
^{215}Bi	1.348 MeV	36.9 s	IT: 76.2 %, $\beta-$: 23.8 %	Not populated
^{212}Bi	0.250 MeV	25.0 min	α: 67.0 %, $\beta-$: 33.0 %	Not populated
	1.91 MeV	7.0 min	$\beta-$: 100 %	Not populated
^{210}Bi	0.271 MeV	$3.04 \cdot 10^6$ a	α: 100 %	Not populated
^{207}Pb	1.633 MeV	0.806 s	IT: 100 %	$8.1 \cdot 10^{-4}$ % from ^{211}Po
^{207}Tl	1.348 MeV	1.33 s	IT: 100 %	$9 \cdot 10^{-4}$ % from ^{211}Bi
^{206}Tl	2.643 MeV	3.74 min	IT: 100 %	Not populated

out of the source as being populated via α decay, are labelled with a white circle, the isotopes which do possess one or more isomeric states carry a white star. A complete list of those isomers potentially playing a role in the context of background effects is listed in Table A.1 together with their corresponding energies and half-lives.

Nuclear excited states with half-lives in the μs range or below do not have to be considered, as the extraction time from the source to the point of detection is well within the ms range, so that prompt transitions already take place in the buffer-gas stopping cell. The isotopes shown in Fig. A.2, combined with the isomers contained in Table A.1, provide a complete list of radioactive nuclides which could potentially lead to background effects during the detection of the ^{229m}Th isomeric transition. This serves as a basis for a detailed discussion of possible background effects in Sect. 5.3.3.

A.2 Determination of ^{232}U Fraction and Material Age

A γ-ray energy spectrum of the ^{233}U-source material was measured in 2007 with the help of a germanium detector (where the resolution at that time suffered from neutron damage) in order to allow for a determination of the ^{229}Th to ^{228}Th activity ratio $R_1 = A_{^{229}\text{Th}}/A_{^{228}\text{Th}}$. The full spectrum is shown in Fig. A.3 (black). More recently also a new γ-ray energy spectrum (shown in red) became available. It was measured with improved detector resolution in 2015 at the Institute for Radiochemistry (IRC) of the University of Mainz, using the same ^{233}U material. This spectrum was, however, not available at the time of the ^{233}U material evaluation and does for this reason not provide the basis for the following quantitative investigation. Selected γ-ray lines are numbered and their corresponding energies, as well as transitions, line intensities (p_{int}) and branching ratios (p_{bra}) according to the ^{233}U and ^{232}U decay chains (see Fig. 4.5) are listed in Table A.2. Line integrals I_k were determined with the analysis code Origin and are detailed in Table A.3. Only the lines assigned with an asterisk were used for the evaluation of the source properties. Some of the lines correspond to two or more γ decays, which could not be distinguished due to the limited detector resolution.

Lines 27 and 29 correspond to single and double escape lines of the 2614 keV decay to the first excited state in ^{208}Pb. Their intensity values were added to the corresponding line. Line 26 cannot be used for the analysis, as this line also contains the 511 keV annihilation line.

The activity of a considered element is directly proportional to the integral over a corresponding γ-ray line

$$A_k \sim I_k. \tag{A.5}$$

The proportionality factor consists of 3 parts. One factor (c_{geo}) takes all external geometric circumstances and the length of the γ-ray measurement into account. This factor is constant for all lines. One further factor (c_{int}) corresponds to the line intensities and takes the ratio of decays of the mother nuclide into account, which lead to a γ-photon of the considered line. The third factor (c_{eff}) takes the energy-

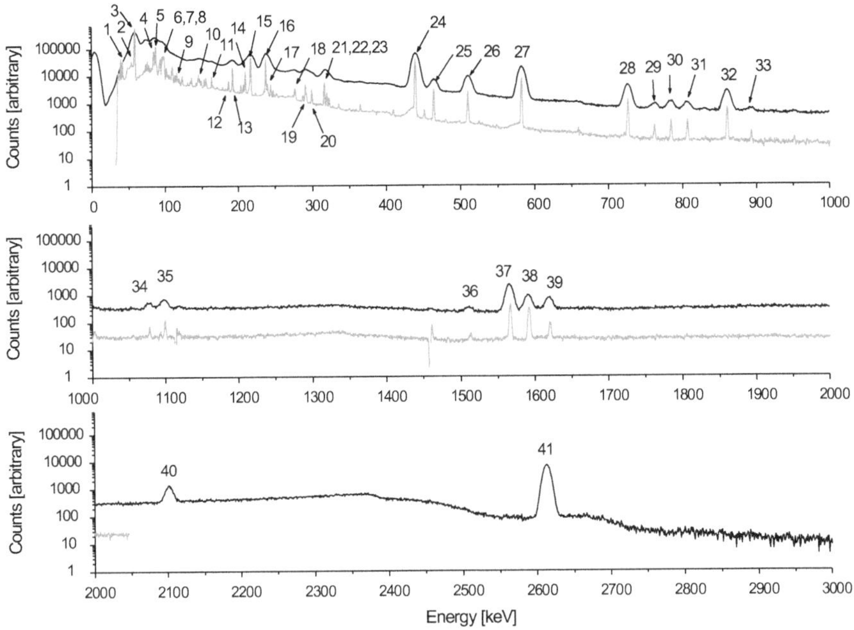

Fig. A.3 γ-ray energy spectra of the ^{233}U source material taken with germanium detectors. The upper spectrum was measured in 2007 and used to evaluate the ^{229}Th to ^{228}Th ratio [4]. The lower spectrum was measured with the same ^{233}U material in 2015 with a higher resolution. For peaks assigned with a number, the corresponding energies, transitions, intensities and branching ratios are listed in Table A.2 [3, 5]. See also Fig. A.11, where the low-energy spectral region below 500 keV is displayed for ^{233}U source material before and after chemical purification. With kind permission of The European Physical Journal (EPJ) [4]

dependent detection efficiency of the germanium detector into account. The resulting equation for the activity is

$$A_k = c_{\text{geo}} \cdot c_{\text{int}} \cdot c_{\text{eff}} \cdot I_k. \tag{A.6}$$

This equation holds for each line individually. The thorium (^{229}Th and ^{228}Th) activities A_{Th} are inferred from lines corresponding to the daughter nuclides. For this reason, the decay chains have to be calculated backwards, which is done by taking all decay branches (c_{bra}) and potential non-equilibrium cases (c_{equ}) into consideration. This results in

$$A_{\text{Th}} = c_{\text{bra}} \cdot c_{\text{equ}} \cdot c_{\text{geo}} \cdot c_{\text{int}} \cdot c_{\text{eff}} \cdot I_k. \tag{A.7}$$

Given the line intensities (p_{int}) as well as the branching ratios (p_{bra}) in percent as listed in Table A.2, the coefficients c_{int} and c_{bra} for a single decay-line are given as

$$c_{\text{int}} = \frac{100}{p_{\text{int}}}, \quad c_{\text{bra}} = \frac{100}{p_{\text{bra}}}. \tag{A.8}$$

Table A.2 Assignments of the transitions indicated in the γ-ray energy spectra shown in Figs. A.3 and A.11. Besides the energies of the transitions, their relative decay intensities and branching ratios are given [3, 5]. Lines assigned with an asterisk were used for the quantitative evaluation

No	E [keV]	Decay	Transition	p_{int} [%]	p_{bra} [%]
1	42.4	$^{233}U \rightarrow {}^{229}Th$	$7/2^+ \rightarrow 5/2^+$	0.0862	100
2	54.7	$^{233}U \rightarrow {}^{229}Th$	$9/2^+ \rightarrow 7/2^+$	0.0182	100
3	59.5	$^{241}Am \rightarrow {}^{237}Np$	$5/2^- \rightarrow 5/2^+$	35.9	
4	85.4	$^{229}Th \rightarrow {}^{225}Ra$	Kα 2	14.7	100
5	88.5	$^{229}Th \rightarrow {}^{225}Ra$	Kα 1	23.9	100
6	89.9	$^{233}U \rightarrow {}^{229}Th$	Kα 2	0.0059	100
7	93.4	$^{233}U \rightarrow {}^{229}Th$	Kα 1	0.0096	100
8	97.1	$^{233}U \rightarrow {}^{229}Th$	$9/2^+ \rightarrow 5/2^+$	0.0203	100
9	117.1	$^{233}U \rightarrow {}^{229}Th$	$5/2^- \rightarrow 5/2^+$	0.0023	100
	118.9	$^{233}U \rightarrow {}^{229}Th$	$7/2^- \rightarrow 5/2^+$	0.0041	100
	120.8	$^{233}U \rightarrow {}^{229}Th$	$11/2^+ \rightarrow 7/2^+$	0.0033	100
10	146	$^{233}U \rightarrow {}^{229}Th$	$5/2^- \rightarrow 5/2^+$	0.00657	100
11	164	$^{233}U \rightarrow {}^{229}Th$	$3/2^- \rightarrow 3/2^+$	0.00623	100
12	187	$^{233}U \rightarrow {}^{229}Th$	$5/2^- \rightarrow 5/2^+$	0.0019	100
	188	$^{225}Ac \rightarrow {}^{221}Fr$	$3/2^+ \rightarrow 3/2^-$	0.54	100
13	193	$^{229}Th \rightarrow {}^{225}Ra$	$7/2^+ \rightarrow 3/2^+$	4.4	100
14	210	$^{229}Th \rightarrow {}^{225}Ra$	$9/2^+ \rightarrow 7/2^+$	2.99	100
15	217	$^{233}U \rightarrow {}^{229}Th$	$5/2^- \rightarrow 5/2^+$	0.0032	100
	218	$^{221}Fr \rightarrow {}^{217}At$	$5/2^- \rightarrow 9/2^-$	11.6	100
16 *	238	$^{212}Pb \rightarrow {}^{212}Bi$	$0^- \rightarrow 1^-$	43.6	100
17 *	240	$^{224}Ra \rightarrow {}^{220}Rn$	$2^+ \rightarrow 0^+$	4.1	100
18	277	$^{208}Tl \rightarrow {}^{208}Pb$	$4^- \rightarrow 5^-$	6.3	35.9
19	291	$^{233}U \rightarrow {}^{229}Th$	$5/2^+ \rightarrow 5/2^+$	0.00537	100
20	300	$^{212}Pb \rightarrow {}^{212}Bi$	$1^- \rightarrow 2^-$	3.3	100
21	317	$^{233}U \rightarrow {}^{229}Th$	$5/2^+ \rightarrow 3/2^+$	0.00776	100
22	320	$^{233}U \rightarrow {}^{229}Th$	$5/2^+ \rightarrow 5/2^+$	0.00290	100
23	323	$^{213}Bi \rightarrow {}^{213}Po$	$3/2^+ \rightarrow 1/2^+$	0.165	100
24*	440	$^{213}Bi \rightarrow {}^{213}Po$	$7/2^+ \rightarrow 9/2^+$	26.1	100
25*	465	$^{209}Tl \rightarrow {}^{209}Pb$	$1/2^+ \rightarrow 5/2^+$	96.9	2.2
26	510	$^{208}Tl \rightarrow {}^{208}Pb$	$5^- \rightarrow 5^-$	22.6	35.9
	511	e^+ e^- annihil.			
27*	570	$^{212}Po \rightarrow {}^{208}Pb$	$5^- \rightarrow 3^-$	2.0	64.1
	583	$^{208}Tl \rightarrow {}^{208}Pb$	$5^- \rightarrow 3^-$	84.5	35.9

(continued)

Table A.2 (continued)

No	E [keV]	Decay	Transition	p_{int} [%]	p_{bra} [%]
28*	727	$^{212}\mathrm{Bi} \rightarrow {}^{212}\mathrm{Po}$	$2^+ \rightarrow 0^+$	6.7	100
29*	763	$^{208}\mathrm{Tl} \rightarrow {}^{208}\mathrm{Pb}$	$5^- \rightarrow 5^-$	1.8	35.9
30*	785	$^{212}\mathrm{Bi} \rightarrow {}^{212}\mathrm{Po}$	$2^+ \rightarrow 2^+$	1.1	100
31*	807	$^{213}\mathrm{Bi} \rightarrow {}^{213}\mathrm{Po}$	$7/2^+ \rightarrow 11/2^+$	0.29	100
32*	860	$^{208}\mathrm{Tl} \rightarrow {}^{208}\mathrm{Pb}$	$4^- \rightarrow 3^-$	12.4	35.9
33	893	$^{212}\mathrm{Bi} \rightarrow {}^{212}\mathrm{Po}$	$1^+ \rightarrow 2^+$	0.378	100
34*	1078	$^{212}\mathrm{Bi} \rightarrow {}^{212}\mathrm{Po}$	$2^+ \rightarrow 2^+$	0.564	100
35*	1100	$^{213}\mathrm{Bi} \rightarrow {}^{213}\mathrm{Po}$	$7/2^+ \rightarrow 9/2^+$	0.259	100
36	1512	$^{212}\mathrm{Bi} \rightarrow {}^{212}\mathrm{Po}$	$2^+ \rightarrow 0^+$	0.29	100
37*	1567	$^{209}\mathrm{Tl} \rightarrow {}^{209}\mathrm{Pb}$	$5/2^+ \rightarrow 9/2^+$	99.8	2.2
38	1592	double escape			
39*	1620	$^{212}\mathrm{Bi} \rightarrow {}^{212}\mathrm{Po}$	$1^+ \rightarrow 0^+$	1.47	100
40	2103	single escape			
41*	2610	$^{212}\mathrm{Po} \rightarrow {}^{208}\mathrm{Pb}$	$3^- \rightarrow 0^+$	2.6	64.1
	2614	$^{208}\mathrm{Tl} \rightarrow {}^{208}\mathrm{Pb}$	$3^- \rightarrow 0^+$	99.754	35.9

Table A.3 γ lines of the energy spectrum shown in Fig. A.3. Only lines used for the analysis of the ^{229}Th to ^{228}Th activity ratio are listed together with the corresponding decay chains, line integrals I_k as well as population branching ratios c_{pop} and activity values $\hat{A}_{\mathrm{Th}}$ as defined in Eq. (A.13)

Line	Energy [keV]	Decay chain	Integral (I_k)	c_{pop}	$\hat{A}_{\mathrm{Th}}$
16	238	^{228}Th	$(5.83 \pm 0.08) \cdot 10^5$	2.10 ± 0.02	$(1.22 \pm 0.03) \cdot 10^6$
17	240	^{228}Th			
24	440	^{229}Th	$(7.25 \pm 0.01) \cdot 10^5$	3.86 ± 0.02	$(2.80 \pm 0.02) \cdot 10^6$
25	465	^{229}Th	$(5.13 \pm 0.38) \cdot 10^4$	49.4 ± 1.0	$(2.52 \pm 0.3) \cdot 10^6$
27	570	^{228}Th	$(2.25 \pm 0.01) \cdot 10^5$	3.14 ± 0.02	$(7.07 \pm 0.07) \cdot 10^5$
	583	^{228}Th			
28	727	^{228}Th	$(4.34 \pm 0.04) \cdot 10^4$	15.0 ± 0.2	$(6.52 \pm 0.15) \cdot 10^5$
29	763	^{228}Th	$(3.40 \pm 0.21) \cdot 10^3$	155.4 ± 2.7	$(5.28 \pm 0.44) \cdot 10^5$
30	785	^{228}Th	$(5.87 \pm 0.30) \cdot 10^3$	90.7 ± 1.0	$(5.33 \pm 0.33) \cdot 10^5$
31	807	^{229}Th	$(4.94 \pm 0.26) \cdot 10^3$	343 ± 14	$(1.69 \pm 0.16) \cdot 10^6$
32	860	^{228}Th	$(2.58 \pm 0.03) \cdot 10^4$	22.3 ± 0.2	$(5.75 \pm 0.15) \cdot 10^5$
34	1078	^{228}Th	$(2.78 \pm 0.37) \cdot 10^3$	177 ± 6	$(4.93 \pm 0.87) \cdot 10^5$
35	1100	^{229}Th	$(4.45 \pm 0.42) \cdot 10^3$	386 ± 25	$(1.72 \pm 0.28) \cdot 10^6$
37	1567	^{229}Th	$(2.36 \pm 0.031) \cdot 10^4$	47.9 ± 1.0	$(1.13 \pm 0.04) \cdot 10^6$
39	1620	^{228}Th	$(5.35 \pm 0.16) \cdot 10^3$	68.0 ± 1.4	$(3.64 \pm 0.17) \cdot 10^5$
41	2610	^{228}Th	$(1.053 \pm 0.0075) \cdot 10^5$	2.665 ± 0.0014	$(2.81 \pm 0.02) \cdot 10^5$
	2614	^{228}Th			

This relation is only valid if the line is populated via just one decay branch. For the important case that one line is populated by two different decay channels (which is the case if the detector resolution does not allow for the separation of these two lines), only the total population coefficient c_{pop} can be evaluated as

$$c_{\text{pop}} = \frac{1 \cdot 10^4}{p_{\text{int }1} \cdot p_{\text{bra }1} + p_{\text{int }2} \cdot p_{\text{bra }2}} \tag{A.9}$$

and substitutes the product of c_{int} and c_{bra}. Therefore also defining

$$c_{\text{pop}} = \frac{1 \cdot 10^4}{p_{\text{int}} \cdot p_{\text{bra}}} \tag{A.10}$$

for a single-populated line, one obtains

$$A_{\text{Th}} = c_{\text{pop}} \cdot c_{\text{equ}} \cdot c_{\text{geo}} \cdot c_{\text{eff}} \cdot I_k. \tag{A.11}$$

For all later considerations, only activity ratios are of interest, therefore constant factors are divided out. We account for that already at this point by setting $c_{\text{geo}} = 1$ and defining the new variable $\tilde{A}_{\text{Th}}$ as

$$\tilde{A}_{\text{Th}} = c_{\text{pop}} \cdot c_{\text{equ}} \cdot c_{\text{eff}} \cdot I_k. \tag{A.12}$$

Further, from dividing by the so far unknown detection efficiency c_{eff} and the equilibrium factor c_{equ}, one obtains

$$\hat{A}_{\text{Th}} = \frac{\tilde{A}_{\text{Th}}}{c_{\text{eff}} \cdot c_{\text{equ}}} = c_{\text{pop}} \cdot I_k, \tag{A.13}$$

where the new variable $\hat{A}_{\text{Th}}$ was defined. The values for I_k, c_{pop}, and $\hat{A}_{\text{Th}}$ are given in Table A.3.

In a next step, the energy dependent detection efficiency coefficient c_{eff} is evaluated. Knowing that the time of the source material production was already several decades ago, the whole ^{232}U-decay chain can be assumed to be in equilibrium, which leads to $c_{\text{equ}} = 1$ (this is not the case for the ^{233}U-decay chain). The lines originating from the ^{232}U decay chain (namely lines 16, 17, 27, 28, 29, 30, 32, 34, 39 and 41) are then used to calculate c_{eff} for each line separately, using Eq. (A.13) and setting $c_{\text{equ}} = 1$

$$c_{\text{eff}} = \frac{\tilde{A}_{\text{Th}}}{\hat{A}_{\text{Th}}}. \tag{A.14}$$

The absolute value of $\tilde{A}_{\text{Th}}$ is unknown, but the function can be normalized with respect to any arbitrary line of the decay chain, as the normalization factor is independent of the energy and will therefore drop out, as soon as the ratio of two different lines will be evaluated. Dropping the normalization coefficient already now and normalizing with respect to line 27, one obtains

$$c_{\text{eff}}(E_\gamma) = \frac{\tilde{A}_{\text{Th}}(27)}{\hat{A}_{\text{Th}}}, \tag{A.15}$$

or, more commonly used, the detection efficiency of the germanium detector

$$d_{\text{eff}}(E_\gamma) = \frac{1}{c_{\text{eff}}(E_\gamma)} = \frac{\hat{A}_{\text{Th}}}{\tilde{A}_{\text{Th}}(27)}. \tag{A.16}$$

The values for d_{eff} are shown in Fig. A.4, together with an analytical fit corresponding to

$$d_{\text{eff}}(E_\gamma) = 51.29 \cdot E_\gamma^{-0.621}. \tag{A.17}$$

The values are normalized to line 27 and the arbitrary factor of normalization will cancel out in the end of the calculation.

Further, it is known that the ^{233}U decay chain is in secular equilibrium from ^{229}Th downwards. For this reason, the equilibrium coefficient can be set to unity for all considered lines ($c_{\text{equ}} = 1$). This allows to calculate $\tilde{A}_{\text{Th}}$ for all considered γ-ray lines as

$$\tilde{A}_{\text{Th}} = c_{\text{eff}} \cdot \hat{A}_{\text{Th}}. \tag{A.18}$$

The calculated values are listed in Table A.4.

From these values it is possible to determine the averages for $\tilde{A}_{^{229}\text{Th}}$ as well as $\tilde{A}_{^{228}\text{Th}}$ to $2.28 \cdot 10^6$ and $7.08 \cdot 10^5$, respectively. The ratio of these values equals the activity ratio of both isotopes in the year 2007:

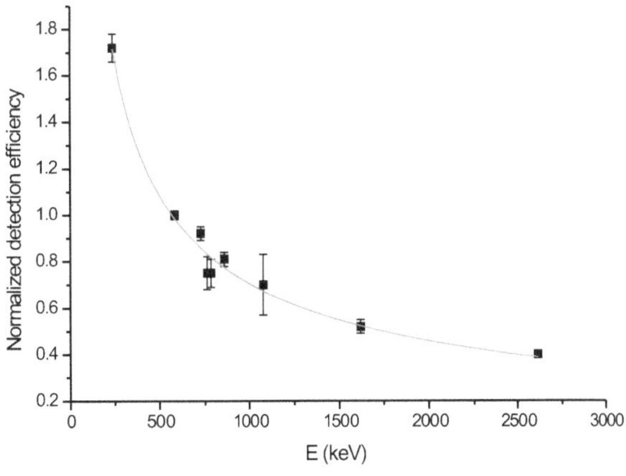

Fig. A.4 Energy dependent relative detection efficiency of the germanium detector used to acquire the γ-ray energy spectrum shown in Fig. A.3. The efficiency is normalized to the arbitrary line 27

Table A.4 List of γ lines used for the evaluation of the ^{233}U source material listed together with their energies, corresponding decay chains, activity values $\hat{A}_{Th}$ as defined in Eq. (A.13), detection efficiency factors c_{eff} and activity values $\tilde{A}_{Th}$, corrected for the detection efficiencies in accordance with Eq. (A.18)

Line	Energy [keV]	Decay chain	$\hat{A}_{Th}$	c_{eff}	$\tilde{A}_{Th}$
16	238	^{228}Th	$(1.22 \pm 0.03) \cdot 10^6$	0.58	$(7.08 \pm 0.17) \cdot 10^5$
17	240	^{228}Th			
24	440	^{229}Th	$(2.80 \pm 0.02) \cdot 10^6$	0.85	$(2.38 \pm 0.02) \cdot 10^6$
25	465	^{229}Th	$(2.52 \pm 0.3) \cdot 10^6$	0.88	$(2.22 \pm 0.26) \cdot 10^6$
27	570	^{228}Th	$(7.07 \pm 0.07) \cdot 10^5$	1.00	$(7.07 \pm 0.07) \cdot 10^5$
	583	^{228}Th			
28	727	^{228}Th	$(6.52 \pm 0.15) \cdot 10^5$	1.17	$(7.63 \pm 0.18) \cdot 10^5$
29	763	^{228}Th	$(5.28 \pm 0.44) \cdot 10^5$	1.20	$(6.34 \pm 0.53) \cdot 10^5$
30	785	^{228}Th	$(5.33 \pm 0.33) \cdot 10^5$	1.22	$(6.50 \pm 0.40) \cdot 10^5$
31	807	^{229}Th	$(1.69 \pm 0.16) \cdot 10^6$	1.24	$(2.10 \pm 0.20) \cdot 10^6$
32	860	^{228}Th	$(5.75 \pm 0.15) \cdot 10^5$	1.30	$(7.48 \pm 0.20) \cdot 10^5$
34	1078	^{228}Th	$(4.93 \pm 0.87) \cdot 10^5$	1.49	$(7.35 \pm 1.30) \cdot 10^5$
35	1100	^{229}Th	$(1.72 \pm 0.28) \cdot 10^6$	1.51	$(2.60 \pm 0.42) \cdot 10^6$
37	1567	^{229}Th	$(1.13 \pm 0.04) \cdot 10^6$	1.88	$(2.12 \pm 0.08) \cdot 10^6$
39	1620	^{228}Th	$(3.64 \pm 0.17) \cdot 10^5$	1.92	$(6.99 \pm 0.33) \cdot 10^5$
41	2610	^{228}Th	$(2.81 \pm 0.02) \cdot 10^5$	2.58	$(7.25 \pm 0.05) \cdot 10^5$
	2614	^{228}Th			

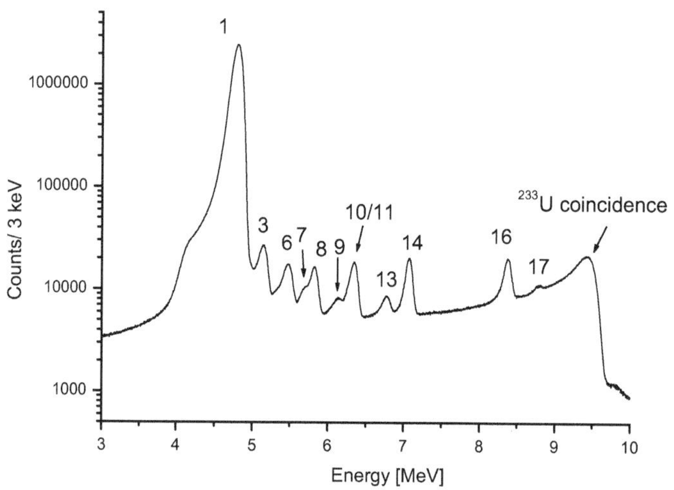

Fig. A.5 α-energy spectrum measured with a silicon detector (Ametek, BU-017-450-100) placed in 5 mm distance in front of the ^{233}U source. The acquisition time was 1 h. Line assignments are detailed in Table A.5. Besides ^{233}U and its daughter products, also ^{239}Pu and ^{238}Pu (^{241}Am) are inferred to be present in the source material. With kind permission of The European Physical Journal (EPJ) [4]

Table A.5 List of α-decay channels of the isotopes contained in the ^{232,233}U decay chains, together with their relative intensities taken from [3]. As they can be seen, also lines of the ^{239}Pu, ^{238}Pu, ^{241}Am and ^{231}Pa decay chains are listed. The lines are listed corresponding to the assignments given in the α-energy spectra shown in Figs. A.5, A.7 and A.9. Only lines with more than 5% intensity are listed

No	Isotope	Half life	Energy [keV]	Intensity [%]
1	^{233}U	$159.2 \cdot 10^3$ a	4783	13.2
			4824	84.3
2	^{229}Th	7932 a	4814	9.30
			4838	5.00
			4845	56.20
			4901	10.20
			4967	5.97
			5053	6.60
3	^{239}Pu	$2.41 \cdot 10^4$ a	5105	11.94
			5144	17.11
			5156	70.77
4	^{232}U	68.9 a	5263	31.55
			5320	68.15
5	^{228}Th	1.9 a	5340	27.20
			5423	72.20
6	^{238}Pu	87.7 a	5456	28.98
			5499	70.91
	^{241}Am	432.6 a	5442	13.1
			5485	84.8
7	^{224}Ra	3.6 d	5448	5.06
			5685	94.92
8	^{225}Ac	10.0 d	5732	8.00
			5790	8.60
			5792	18.10
			5830	50.70
9	^{212}Bi	60.55 min	6051	25.13
			6090	9.75
10	^{220}Rn	55.6 s	6288	99.89
11	^{221}Fr	286.1 s	6126	15.10
			6341	83.40
12	^{211}Bi	2.14 min	6278	16.19
			6623	83.54
13	^{216}Po	0.145 s	6778	99.99
14	^{217}At	32.3 ms	7067	99.89
15	^{215}Po	1.78 ms	7386	100
16	^{213}Po	3.72 μs	8376	100
17	^{212}Po	0.30 μs	8785	100

$$R_1 = \frac{A_{^{229}\mathrm{Th}}}{A_{^{228}\mathrm{Th}}} = 3.2 \pm 0.2. \quad (A.19)$$

The age of the source material was determined by evaluating the state of (non-)equilibrium of the ^{233}U decay chain. There is a constant ingrowth of ^{229}Th activity into the source and the ratio between the ^{233}U activity and the ^{229}Th activity allows to calculate the time elapesed since material production. This ratio was determined with an α-energy spectrum taken from the ^{233}U source, which is shown in Fig. A.5. Lines, as numbered in this spectrum, are detailed in Table A.5, together with further line assignments that will later be of interest. Also half-lives of the corresponding nuclides, α energies and relative intensities are listed.

The α spectrum was taken with a silicon detector (Ametek, BU-017-450-100, 100 μm thickness, 450 mm^2 active area) in 2014. The detector was placed in 5 mm distance to the small-area ^{233}U source 1 and α decays were detected for 1 h.

Based on the comparison of line 1 and line 16 of this spectrum, the ^{233}U to ^{229}Th activity ratio was determined (in 2014) to be

$$R_2 = \frac{A_{^{233}\mathrm{U}}}{A_{^{229}\mathrm{Th}}} = 250 \pm 10. \quad (A.20)$$

The age of the ^{233}U material was correspondingly determined to be $t = 1/(\lambda_{^{229}\mathrm{Th}} \cdot R_2) = 45 \pm 5$ years, where $\lambda_{^{229}\mathrm{Th}} = \ln(2)/t_{1/2}$ denotes the decay constant of ^{229}Th. The source material was thus produced around 1969. Also the absolute ^{233}U source activity was determined based on the α-energy spectrum shown in Fig. A.5 to be (200 ± 10) kBq. This value corresponds to an absolute number of $1.45 \cdot 10^{18}$ ^{233}U atoms in the source. Further, the ^{229}Th to ^{228}Th activity ratio of the source in 2007 (R_1) is used to determine the relative abundance of ^{232}U in the source material at the time of material production around 1969. For this purpose, the following equation was used:

$$\frac{N_{^{232}\mathrm{U}}}{N_{^{233}\mathrm{U}}} = \frac{\lambda_{^{229}\mathrm{Th}}\lambda_{^{233}\mathrm{U}}\left(\lambda_{^{228}\mathrm{Th}} - \lambda_{^{232}\mathrm{U}}\right)t}{R_1 \cdot \lambda_{^{228}\mathrm{Th}}\lambda_{^{232}\mathrm{U}}e^{-\lambda_{^{232}\mathrm{U}}t}} = (6.1 \pm 0.3) \cdot 10^{-7}. \quad (A.21)$$

This fraction leads to an absolute number of ^{232}U atoms of $8.85 \cdot 10^{11}$.

A.3 Calculating the Daughter Activities

The absolute numbers of ^{233}U and ^{232}U atoms in the source at the time of material production (around 1969) would have been $N_{^{233}\mathrm{U}} = 1.45 \cdot 10^{18}$ and $N_{^{232}\mathrm{U}} = 8.85 \cdot 10^{11}$, respectively, as determined in the previous section. These values serve as input parameters for the calculation of numbers and activities for all daughter nuclides of both decay chains, as they were present in the source in 2014. The year 2014 is chosen, as α-recoil implantation measurements were performed in this year and the calculated numbers will later be used in order to obtain recoil efficiencies of the source. The time evolution of the source is modeled with the Bateman equation [6]. This equation describes the population of an isotopic species as populated by one mother nuclide, it reads

$$N_k(t) = N_0\left(\prod_{i=1}^{k-1} b_i\lambda_i\right)\sum_{j=1}^{k}\frac{e^{-\lambda_j t}}{\prod_{i=1,i\neq j}^{k}(\lambda_i-\lambda_j)}. \tag{A.22}$$

Here N_k is the number of nuclei of the k-th element in the decay chain, λ_k is the corresponding decay constant (listed in Table A.6), b_k is the corresponding branching ratio and N_0 the starting number of nuclei. Calculations were performed numerically (see Appendix B.1.1), the results are shown for 2014 in Table A.6.

Table A.6 Calculated absolute numbers of nuclei and corresponding activities for all isotopes of the decay chains of ^{233}U and ^{232}U, as contained in α-recoil ion source 1 [4]. Also half-lives ($t_{1/2}$) and decay constants (λ) are listed [3]

Isotope	$t_{1/2}$	λ [s^{-1}]	Number	Activity [s^{-1}]
^{233}U	$1.59 \cdot 10^5$ a	$1.38 \cdot 10^{-13}$	$1.45 \cdot 10^{18}$	$2.00 \cdot 10^5$
^{229}Th	7932 a	$2.79 \cdot 10^{-12}$	$2.83 \cdot 10^{14}$	$7.91 \cdot 10^2$
^{225}Ra	14.9 d	$5.42 \cdot 10^{-7}$	$1.46 \cdot 10^9$	$7.90 \cdot 10^2$
^{225}Ac	10.0 d	$8.02 \cdot 10^{-7}$	$9.84 \cdot 10^8$	$7.89 \cdot 10^2$
^{221}Fr	286.1 s	$2.36 \cdot 10^{-3}$	$3.34 \cdot 10^5$	$7.89 \cdot 10^2$
^{217}At	32.3 ms	21.46	36.76	$7.89 \cdot 10^2$
^{213}Bi	45.6 m	$2.53 \cdot 10^{-4}$	$3.12 \cdot 10^6$	$7.89 \cdot 10^2$
^{213}Po	3.72 μs	$1.65 \cdot 10^5$	$4.67 \cdot 10^{-3}$	$7.72 \cdot 10^2$
^{209}Tl	2.16 m	$5.35 \cdot 10^{-3}$	$3.24 \cdot 10^3$	17.4
^{209}Pb	3.25 h	$5.92 \cdot 10^{-5}$	$1.33 \cdot 10^7$	$7.89 \cdot 10^2$
^{209}Bi	stable	0	$5.59 \cdot 10^{11}$	0
^{232}U	68.9 a	$3.19 \cdot 10^{-10}$	$5.63 \cdot 10^{11}$	$1.80 \cdot 10^2$
^{228}Th	1.9 a	$1.15 \cdot 10^{-8}$	$1.61 \cdot 10^{10}$	$1.85 \cdot 10^2$
^{224}Ra	3.63 d	$2.19 \cdot 10^{-6}$	$8.43 \cdot 10^7$	$1.85 \cdot 10^2$
^{220}Rn	55.6 s	$1.25 \cdot 10^{-2}$	$1.48 \cdot 10^4$	$1.85 \cdot 10^2$
^{216}Po	0.145 ms	4.62	40.0	$1.85 \cdot 10^2$
^{212}Pb	10.64 h	$1.81 \cdot 10^{-5}$	$1.02 \cdot 10^7$	$1.85 \cdot 10^2$
^{212}Bi	60.55 m	$1.91 \cdot 10^{-4}$	$9.67 \cdot 10^5$	$1.85 \cdot 10^2$
^{212}Po	0.299 μs	$2.31 \cdot 10^6$	$5.12 \cdot 10^{-5}$	$1.19 \cdot 10^2$
^{208}Tl	3.05 m	$3.78 \cdot 10^{-3}$	$1.75 \cdot 10^4$	66.4
^{208}Pb	stable	0	$3.06 \cdot 10^{11}$	0

A.4 Measurement of the α-Recoil Isotope Activities

A set of measurements was performed in order to experimentally determine the α-recoil activities of the small-area ^{233}U source 1 for ^{221}Fr, ^{224,225}Ra and ^{228,229}Th. In these measurements, α-recoil isotopes, emerging from the ^{233}U source, were directly implanted under vacuum conditions into a silicon detector surface (Ametek, type BU-017-450-100) and important use was made of the different half-lives of the recoil nuclides. The silicon detector was mounted movable under vacuum in 5 mm distance in front of the ^{233}U source. In case of ^{221}Fr ($t_{1/2} = 286$ s), an implantation time of 10 min was chosen. After implantation, the detector was moved (within 2 min) out of sight of the source with the help of a linear translation stage, and α-recoil isotope implantation was stopped. Subsequently, the α decays occurring on the silicon detector were measured for further 15 min, which allowed for the detection of ^{221}Fr α decays. None of the long-lived (^{228,229}Th and ^{224,225}Ra) nuclides had to be considered in this measurement, as the corresponding activities are too low to lead to a significant contribution on this short time scale.

In order to also investigate the ^{224}Ra ($t_{1/2} = 3.6$ d) and ^{225}Ra ($t_{1/2} = 14.9$ d) implantation rates, respectively, a long implantation time of 5 days was chosen, and also the times for α-decay detection were accordingly increased, as detailed in the following. For the detection of the ^{228}Th ($t_{1/2} = 1.9$ a) and ^{229}Th ($t_{1/2} = 7932$ a) recoil activities, the ^{224,225}Ra had to first fully decay (leading to a required waiting time of 200 days) and the time for α-decay detection was increased to 220 days.

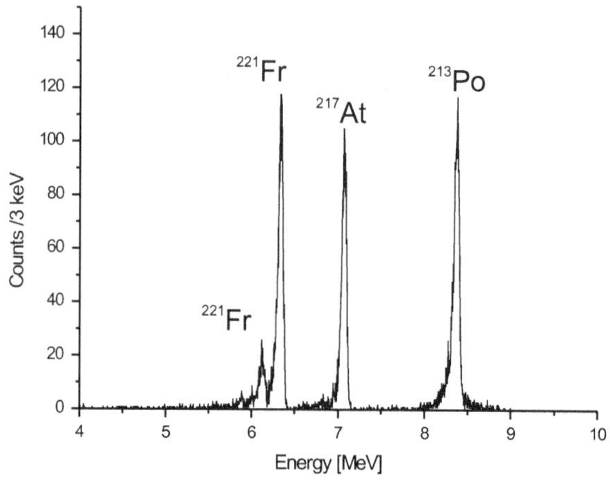

Fig. A.6 α-energy spectrum obtained after 10 min of α-recoil isotope implantation from ^{233}U source 1 into the silicon detector surface. After implantation, the detector was moved away from the source and occurring α decays were measured for 15 min. The spectrum is dominated by the α decay of ^{221}Fr ($t_{1/2} = 286$ s). ^{217}At ($t_{1/2} = 32.2$ ms) and ^{213}Po ($t_{1/2} = 3.72$ μs) are successively occurring daughter decays

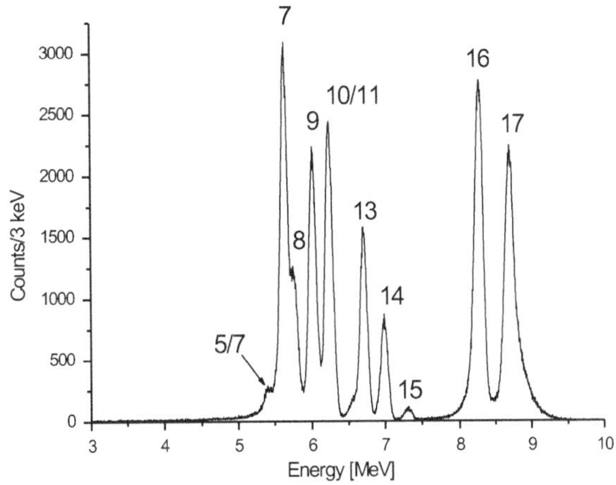

Fig. A.7 α-energy spectrum obtained after 5 days of α-recoil implantation from ^{233}U source 1 and further 1 day of detection. The spectrum is dominated by the α decay of ^{224}Ra ($t_{1/2} = 3.6$ d, line 7)

The α-energy spectrum obtained after 10 min of implantation time for the investigation of the ^{221}Fr decay is shown in Fig. A.6. The number of ^{221}Fr α decays measured by the detector N_{count} is related to the α-recoil activity of the ^{233}U source $A_{\text{recoil}}(^{221}\text{Fr})$ via

$$A_{\text{recoil}}(^{221}\text{Fr}) = \frac{2 \cdot N_{\text{count}} \cdot g \cdot \lambda}{\left(1 - e^{-\lambda t_{\text{acc}}}\right)\left(e^{-\lambda t_1} - e^{-\lambda t_2}\right)} = 45.3\ \text{s}^{-1}. \tag{A.23}$$

Here λ denotes the ^{221}Fr decay constant ($2.42 \cdot 10^{-3}\ \text{s}^{-1}$), t_{acc} is the accumulation time on the detector surface and t_1 and t_2 denote the times when detection was started and terminated, respectively, measured with respect to the point when α-recoil isotope implantation was stopped. The factor of 2 corrects for the fact that only half of the ^{221}Fr α decays are detected by the silicon detector as being emitted into the detector's hemisphere and g is a geometry factor, taking the size of the detection surface and the distance between detector and source into account. g was determined numerically to be 1.49. This value does also account for the fact that α-recoil isotope emission does not occur isotropically, but instead self-collimation occurs in the source material, as the recoil isotopes, which are emitted in the direction perpendicular to the surface, have to overcome the smallest amount of surface material (Appendix A.5). Such effects were also confirmed experimentally [4].

A similar evaluation was performed in order to determine the ^{224}Ra α-recoil isotope activity. For this purpose, isotope implantation into the silicon detector surface was carried out for 5 days. Subsequently, the detector was shifted away from the source and 1 day of α-decay detection was started. The corresponding measurement is shown in Fig. A.7 (α decay of ^{224}Ra is marked as line no. 7).

The number of detected ^{224}Ra isotopes ($t_{1/2} = 3.66$ d) was inferred to be $N_{\text{count}} = 1.66 \cdot 10^5$. Equation A.23 was used to calculate the α-recoil isotope activity for ^{224}Ra, with $\lambda = 2.21 \cdot 10^{-6}\ \text{s}^{-1}$. The result is $A_{\text{recoil}}(^{224}\text{Ra}) = 10.2\ \text{s}^{-1}$.

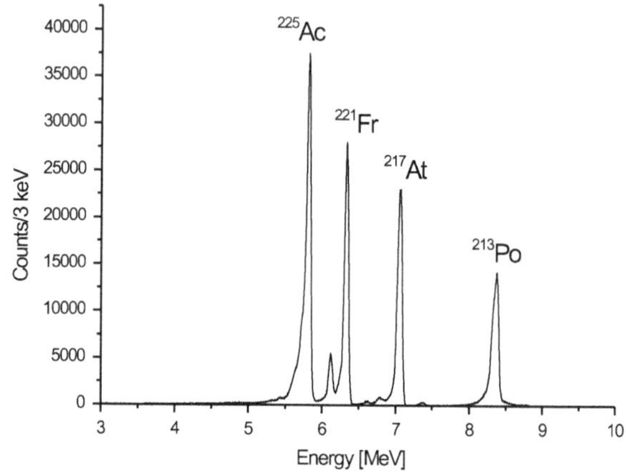

Fig. A.8 α-energy spectrum obtained after 5 days of α-recoil implantation and 20 days of decay followed by 10 days of data acquisition. The spectrum is dominated by the α decay of ^{225}Ac ($t_{1/2} = 10.0$ d), as ^{224}Ra has already decayed

The case of ^{225}Ra ($t_{1/2} = 14.8$ d) is slightly complicated by the fact that it is a β^- emitter and its decays are therefore not directly registered by the silicon detector. Instead, the α decays of its daughter isotope ^{225}Ac ($t_{1/2} = 10.0$ d) are detected and can be calculated back in order to infer the α-recoil isotope efficiency for ^{225}Ra. For this purpose, the silicon-detector measurement (of 5 days of accumulation) was read out after about 20 days ($1.81 \cdot 10^6$ s) and read out again after about 30 days ($2.7 \cdot 10^6$ s) of data acquisition time. The difference of both spectra is shown in Fig. A.8. In this spectrum, ^{224}Ra is not dominating anymore due to its shorter half-life of 3.66 days and instead the ^{225}Ac α decay becomes visible.

The number of detected ^{225}Ac decays is inferred to be $N_{\text{count}} = 1.14 \cdot 10^6$. The equation required to calculate the corresponding ^{225}Ra α-recoil isotope activity $A_{\text{recoil}}(^{225}\text{Ra})$ reads

$$\begin{aligned} A_{\text{recoil}}(^{225}\text{Ra}) =& \frac{2N_{\text{count}} \cdot g \cdot \lambda_{\text{Ac}}}{\lambda_{\text{Ac}} - \lambda_{\text{Ra}}} \left(1 - e^{-\lambda_{\text{Ra}} t_{\text{acc}}}\right) \left[\frac{1}{\lambda_{\text{Ac}}} \left(e^{-\lambda_{\text{Ac}} t_2} - e^{-\lambda_{\text{Ac}} t_1}\right) + \frac{1}{\lambda_{\text{Ra}}} \left(e^{-\lambda_{\text{Ra}} t_1} - e^{-\lambda_{\text{Ra}} t_2}\right)\right] \\ &+ \frac{2N_{\text{count}} \cdot g \cdot \lambda_{\text{Ra}}}{\lambda_{\text{Ac}} - \lambda_{\text{Ra}}} \left(e^{-\lambda_{\text{Ac}} t_1} - e^{-\lambda_{\text{Ac}} t_2}\right) \left[\frac{1}{\lambda_{\text{Ra}}} \left(1 - e^{-\lambda_{\text{Ra}} t_{\text{acc}}}\right) - \frac{1}{\lambda_{\text{Ac}}} \left(1 - e^{-\lambda_{\text{Ac}} t_{\text{acc}}}\right)\right]. \end{aligned}$$

Using the ^{225}Ac and ^{225}Ra decay rates of $\lambda_{\text{Ac}} = 8.02 \cdot 10^{-7}$ s^{-1} and $\lambda_{\text{Ra}} = 5.42 \cdot 10^{-7}$ s^{-1}, respectively, the α-recoil isotope activity is inferred to be $A_{\text{recoil}}(^{225}\text{Ra}) = 42.8$ s^{-1}.

Due to the comparably long half-lives of ^{225}Ra ($t_{1/2} = 14.8$ d) and ^{225}Ac ($t_{1/2} = 10.0$ d), their activities are dominating the α-energy spectrum for longer than 100 days. However, after that time, the activities of ^{229}Th and ^{228}Th start to be measurable. A corresponding energy spectrum is shown in Fig. A.9. After 5 days of accumulation, the short-lived isotopes had 200 days time to decay before the measurement was started. Subsequently, α decays were recorded for 220 days.

The background corrected number of detected ^{229}Th decays amounts to $N_{\text{count}} = 7.58 \cdot 10^4$. This can be used to calculate the α-recoil activity of the source via

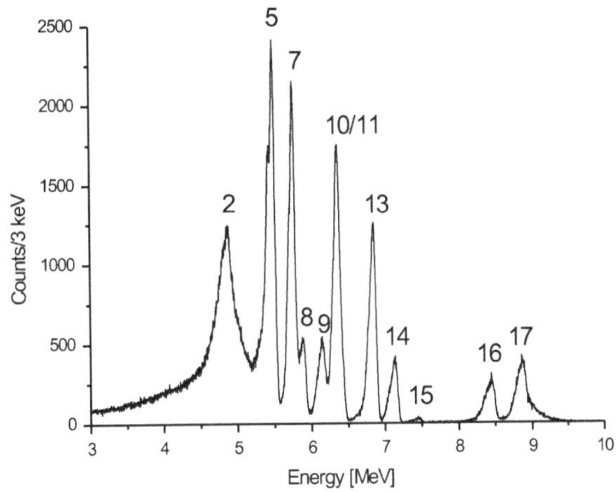

Fig. A.9 α-energy spectrum obtained after 5 days of α-recoil implantation from ^{233}U source 1, followed by 200 days of decay and further 220 days of detection. The spectrum is dominated by the α decay of ^{228}Th ($t_{1/2} = 1.9$ a), as the activities of ^{225}Ra ($t_{1/2} = 14.9$ d) and ^{225}Ac $t_{1/2} = 10.0$ d) have already faded away

Table A.7 Measured absolute α-recoil activities and recoil efficiencies of the ^{233}U source

Nuclide	Rec. act. [s^{-1}]	Rec. eff. [%]
^{229}Th	$(9.9 \pm 1.0) \cdot 10^3$	5.0 ± 0.5
^{228}Th	8.1 ± 2.0	4.5 ± 1.0
^{225}Ra	42.8 ± 4.3	5.4 ± 0.5
^{224}Ra	10.2 ± 1.0	5.5 ± 0.6
^{221}Fr	45.3 ± 4.5	5.7 ± 0.6

$$A_{\text{recoil}}(^{229}\text{Th}) = \frac{2N_{\text{count}} \cdot g}{\lambda \cdot t_{\text{acc}} \cdot (t_2 - t_1)} = 9930 \text{ s}^{-1}. \tag{A.24}$$

Here $\lambda = 2.77 \cdot 10^{-12}$ s^{-1} denotes the ^{229}Th decay constant. The same energy spectrum is also used to evaluate the ^{228}Th α-recoil activity. For this purpose, Eq. (A.23) is applied using $\lambda = 1.16 \cdot 10^{-8}$ s^{-1} and the number of ^{228}Th α decays is inferred to be about $N_{\text{count}} \approx 1.9 \cdot 10^5$. The corresponding result for the α-recoil activity is $A_{\text{recoil}}(^{228}\text{Th}) \approx 8.1$ s^{-1}.

An overview over the measured absolute α-recoil isotope activities is given in Table A.7. Also the corresponding α-recoil efficiencies are tabulated, based on the comparison with the ^{233}U-source α activities as listed in Table A.6.

A.5 Calculation of α-Recoil Efficiencies

The measured α-recoil isotope efficiencies, presented in Sect. A.4, can be compared to theoretical calculations. The following model was applied for this purpose. In a first step, it is assumed that all α-recoil isotopes emerge from a defined point of decay and

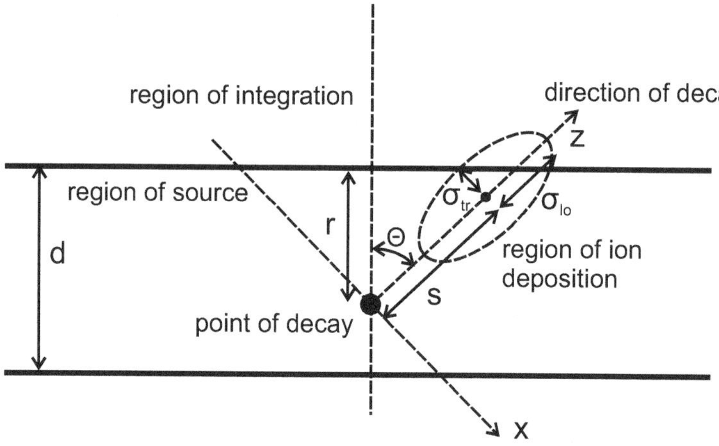

Fig. A.10 Schematic drawing of the α-recoil process. Variables are used for modeling the α-recoil isotope activity of the ^{233}U source. See text for explanation. With kind permission of The European Physical Journal (EPJ) [4]

are emitted into the same direction. These isotopes will be stopped in a region with a distance of the stopping length s away from the point of decay. Due to scattering processes in the source material, not all particles will terminate in the same point, but instead they will be distributed with standard deviations in longitudinal σ_{lo} and transversal σ_{tr} direction around the center position. The corresponding probability density of isotopes, stopped in ^{233}U material of predefined chemical composition, is modeled by a 3-dimensional Gaussian function. In case that the material surface intersects this Gaussian region, there will be a part of the Gaussian distribution outside of the surface material. It is assumed that this is the fraction of ions which will leave the source material, due to their kinetic α-recoil energy for a well defined point and direction of decay (see Fig. A.10).

This fraction of ions is calculated by integrating the normalized Gaussian function over the hemisphere with no source material. This integration can be analytically performed for a defined distance r between the material surface and the point of decay, as well as a given polar angle θ. It reads

$$
\begin{aligned}
T(r,\theta) &= \frac{1}{(2\pi)^{3/2}\sigma_{\text{tr}}^2\sigma_{\text{lo}}} \int_{-\infty}^{\infty} dx \int_{-\infty}^{\infty} dy \int_{\frac{r+x\sin(\theta)}{\cos(\theta)}}^{\infty} e^{-\left(\frac{x^2+y^2}{2\sigma_{\text{tr}}^2}+\frac{(z-s)^2}{2\sigma_{\text{lo}}^2}\right)}\, dz \\
&= \frac{1}{2}\left[1+\text{erf}\left(\frac{s\cos(\theta)-r}{\sqrt{2}\sigma_{\text{tr}}\sin(\theta)}\cdot\left[\left(\frac{\sigma_{\text{lo}}\cos(\theta)}{\sigma_{\text{tr}}\sin(\theta)}\right)^2+1\right]^{-1/2}\right)\right].
\end{aligned}
\tag{A.25}
$$

Here erf denotes the error function. The transmission function $T(r,\theta)$ can then be used to calculate the total fraction of α-recoil isotopes leaving the source material.

Table A.8 Measured and simulated α-recoil efficiencies of the ^{233}U source. The SRIM simulations systematically underestimate the actual recoil activity. The measured values are well reproduced by the MDrange (MDH) simulations, when a polycrystalline source with about 80 nm crystal size is assumed

Isotope	F_{meas} (%)	F_{SRIM} (%)	F_{MDH} (%)
^{229}Th	5.0	1.6	5.3
^{228}Th	4.5	1.7	5.5
^{225}Ra	5.4	1.9	5.5
^{224}Ra	5.5	2.0	5.5
^{221}Fr	5.7	2.1	5.8

This is done by averaging $T(r, \theta)$ over the depth r and direction θ of decay. The corresponding integral is

$$F = \frac{1}{4\pi d} \int_0^d dr \int_0^{2\pi} d\phi \int_0^{\pi} \sin(\theta) d\theta \; T(r, \theta). \tag{A.26}$$

This integral was solved numerically (Appendix B.1.2).

In a first approach, SRIM simulations [7] were performed in order to calculate the stopping length s, as well as the standard deviations σ_{lo} and σ_{tr} for nuclides in the ^{233}U and ^{232}U decay chains individually (Appendix B.1.3). The individual calculation is required due to different α-recoil energies E_{rec}, which are determined by the α-decay energies E_α via $E_{rec} = E_\alpha \cdot m_\alpha / m_{rec}$. However, the obtained α-recoil efficiencies were significantly below the measured values (Table A.8). Also results of MDrange (MDH) simulations [8] are tabulated and will be introduced below.

A comparison between the theoretical model and measured α-recoil isotope efficiencies of different uranium sources (obtained by Hashimoto et al. [8]) revealed the opposite result of a systematic overestimation (see Table A.9) [4]. In this study, eight ^{238}U sources (UO_2 and U_3O_8 with different ^{235}U contents) were electro-deposited

Table A.9 Measured and simulated α-recoil efficiencies for the 8 different ^{238}U sources investigated in Ref. [8]. The SRIM simulations systematically overestimate the actual recoil activities. The deposited material of all sources is UO_2, except for EU-3, for which UO_2 was ignited under air to produce U_3O_8

Source	d [nm]	F_{meas} (%)	F_{SRIM} (%)
EU-5	11.8	20	30
NU-D	15.5	17	25
EU-6	16.4	13	24
EU-2	53.8	5.8	7.8
EU-3	78.7	6.3	6.5
EU-4	92.7	3.6	4.5
NU-C	167.9	1.7	2.5
NU-B	256.4	1.1	1.6

with different thicknesses onto 20 mm diameter mirror-polished stainless-steel plates and probed for their ^{234}Th-recoil efficiencies.

The reason for the observed underestimation of the α-recoil efficiency in the SRIM calculations might be the different production technique via evaporation applied for the small-area ^{233}U source 1. This production procedure may have led to the formation of a micro-crystalline structure of the source, in this way opening the possibility of channeling. In order to further investigate this possibility, a different type of simulation was performed. This time the MDrange program code [8] was employed, which allows to take channeling-effects into account by taking respect for the explicit crystal lattice structure (Appendix B.1.4). It was found that this effect allows to explain the observed α-recoil efficiencies, when assuming a polycrystalline source structure with 80 nm crystal sizes. The α-recoil efficiencies F, as calculated in accordance with Eq. (A.26) and based on MDrange simulations, are listed in Table A.10. In combination with the isotopic activities listed in Table A.6, these values allow to estimate the α-recoil activities for all nuclides of the ^{233}U and ^{232}U decay chains, respectively. The corresponding values are listed in Table A.11.

Table A.10 List of α-recoil ions with corresponding dominant α-decay energies, intensities and α-recoil ion energies, together with stopping lengths s and standard deviations $\sigma_{\rm lo}$ and $\sigma_{\rm tr}$ as obtained by MDrange simulations [8]. Note, that the α-decay energies, as listed in the table, correspond to the mother nuclide (e.g. ^{233}U in case of ^{229}Th). The fractions of α-recoil ions, which are able to leave the ^{233}U source F, were calculated in accordance with Eq. (A.26)

Isotope	E_α [keV]	Int. [%]	$E_{\rm rec.}$ [keV]	s [nm]	$\sigma_{\rm lo}$ [nm]	$\sigma_{\rm tr}$ [nm]	F
^{229}Th	4824	84.3	84.3	75.9	35.1	5.2	$5.34\cdot10^{-2}$
^{225}Ra	4845	56.2	86.1	77.5	35.9	5.3	$5.45\cdot10^{-2}$
^{221}Fr	5830	50.7	105.5	82.4	33.5	6.4	$5.78\cdot10^{-2}$
^{217}At	6341	83.4	116.9	84.6	34.7	6.2	$5.93\cdot10^{-2}$
^{213}Bi	7067	99.9	132.7	92.8	37.2	7.4	$6.51\cdot10^{-2}$
^{209}Tl	5869	1.9	112.3	85.8	35.7	6.5	$6.02\cdot10^{-2}$
^{209}Pb	8376	100	160.3	97.1	37.6	8.4	$6.82\cdot10^{-2}$
^{228}Th	5320	68.2	93.3	78.4	33.6	5.4	$5.50\cdot10^{-2}$
^{224}Ra	5423	72.2	96.8	78.6	34.4	5.6	$5.52\cdot10^{-2}$
^{220}Rn	5685	94.9	103.4	81.0	34.1	6.3	$5.69\cdot10^{-2}$
^{216}Po	6288	99.9	116.4	85.3	34.7	6.3	$5.98\cdot10^{-2}$
^{212}Pb	6778	100	127.9	87.8	35.4	7.0	$6.16\cdot10^{-2}$
^{208}Tl	6051	25.1	116.4	87.4	34.6	6.5	$6.13\cdot10^{-2}$
^{208}Pb	8785	100	168.9	99.3	39.2	8.9	$6.98\cdot10^{-2}$

Table A.11 Calculated α activities as well as α-recoil efficiencies and absolute α-recoil activities for all isotopes of the decay chains of ^{233}U and ^{232}U, as contained in the α-recoil ion source 1 used

Isotope	Activity [s^{-1}]	Recoil efficiency F	Recoil activity A_{recoil} [s^{-1}]
^{233}U	$2.00 \cdot 10^5$	0	0
^{229}Th	$7.91 \cdot 10^2$	$5.34 \cdot 10^{-2}$	$(10.7 \pm 1.1) \cdot 10^3$
^{225}Ra	$7.90 \cdot 10^2$	$5.45 \cdot 10^{-2}$	43.1 ± 4.3
^{225}Ac	$7.89 \cdot 10^2$	0	(Produced by β decay)
^{221}Fr	$7.89 \cdot 10^2$	$5.78 \cdot 10^{-2}$	45.6 ± 4.6
^{217}At	$7.89 \cdot 10^2$	$5.93 \cdot 10^{-2}$	46.8 ± 4.7
^{213}Bi	$7.89 \cdot 10^2$	$6.51 \cdot 10^{-2}$	51.4 ± 5.1
^{213}Po	$7.72 \cdot 10^2$	0	(Produced by β decay)
^{209}Tl	17.4	$6.02 \cdot 10^{-2}$	1.0 ± 0.1
^{209}Pb	$7.89 \cdot 10^2$	$6.82 \cdot 10^{-2}$	52.7 ± 5.3
^{209}Bi	0	0	(Produced by β decay)
^{232}U	$1.80 \cdot 10^2$	0	0
^{228}Th	$1.85 \cdot 10^2$	$5.50 \cdot 10^{-2}$	9.9 ± 1.0
^{224}Ra	$1.85 \cdot 10^2$	$5.52 \cdot 10^{-2}$	10.2 ± 1.0
^{220}Rn	$1.85 \cdot 10^2$	$5.69 \cdot 10^{-2}$	10.5 ± 1.1
^{216}Po	$1.85 \cdot 10^2$	$5.98 \cdot 10^{-2}$	11.1 ± 1.1
^{212}Pb	$1.85 \cdot 10^2$	$6.16 \cdot 10^{-2}$	11.4 ± 1.1
^{212}Bi	$1.85 \cdot 10^2$	0	(Produced by β decay)
^{212}Po	$1.19 \cdot 10^2$	0	(Produced by β decay)
^{208}Tl	66.4	$6.13 \cdot 10^{-2}$	4.1 ± 0.4
^{208}Pb	0	$6.98 \cdot 10^{-2}$	8.3 ± 0.8

A.6 Implanted α-Recoil Isotope Activities

Calculations were performed in order to also estimate the α-recoil isotope activities, which are accumulated in the surrounding of the ^{233}U source as implanted into the surfaces of the buffer-gas stopping cell (i.e. the interior of the buffer-gas stopping cell). As it turns out, in most cases this effect is negligible, however, for some of the short-lived daughter isotopes, such effects can become significant. The following estimates assume that the intrinsic α-recoil activities of the ^{233}U source are not significantly reduced by the nuclides leaving the source (e.g. that the recoil efficiency is small, which is valid in case of a thick active layer) and that the nuclides are homogeneously distributed within the source material (e.g. no reduction of the isotope activity occurs near the surface).

α-recoil nuclides of all isotopic species contained in the ^{233}U source material and emitted due to α decay are implanted into the surrounding of the ^{233}U source. Denoting the implantation rate of the k-th isotope of the decay chain ϵ_k, the system of differential equations to be solved in order to estimate the implanted α-recoil activity reads

$$\begin{aligned}\dot{N}_1(t) &= -\lambda_1 N_1(t) + \epsilon_1\\ \dot{N}_2(t) &= b_1\lambda_1 N_1(t) - \lambda_2 N_2(t) + \epsilon_2\\ &\vdots\\ \dot{N}_k(t) &= b_{k-1}\lambda_{k-1}N_{k-1}(t) - \lambda_k N_k(t) + \epsilon_k.\end{aligned} \tag{A.27}$$

This system is solved by the Bateman equation with source terms for $N_k(0) = 0$ for all k as

$$N_k^{\text{acc}}(t) = \sum_{i=1}^{k}\left(\prod_{l=i}^{k-1} b_l\lambda_l\right)\sum_{j=i}^{k}\frac{\epsilon_i\left(1-e^{-\lambda_j t}\right)}{\lambda_j\prod_{l=i,l\neq j}^{k}(\lambda_l-\lambda_j)}. \tag{A.28}$$

When buffer-gas enters the system, the implantation into the metal surfaces stops. At that time the isotopes start to decay with non-zero starting values $N_k(0) \neq 0$ for all isotopes in the decay chain. This decay is modeled by summing the Bateman equations (Eq. A.22) for individual isotopes over the whole decay chain. The solution reads

$$N_k^{\text{dec}}(t) = \sum_{i=1}^{k}\left(\prod_{l=i}^{k-1} b_l\lambda_l\right)\sum_{j=i}^{k}\frac{N_i(0)e^{-\lambda_j t}}{\prod_{l=i,l\neq j}^{k}(\lambda_l-\lambda_j)}. \tag{A.29}$$

Therefore, the complete solution of the system of differential equations (A.27) is given as the sum of Eqs. (A.28) and (A.29) [9][1]:

$$N_k(t) = \sum_{i=1}^{k}\left(\prod_{l=i}^{k-1} b_l\lambda_l\right)\sum_{j=i}^{k}\left[\frac{N_i(0)e^{-\lambda_j t}}{\prod_{l=i,l\neq j}(\lambda_l-\lambda_j)} + \frac{\epsilon_i\left(1-e^{-\lambda_j t}\right)}{\lambda_j\prod_{l=i,l\neq j}^{k}(\lambda_l-\lambda_j)}\right]. \tag{A.30}$$

The corresponding numbers N_0 and activities A_0 are calculated numerically (see Appendix B.1.5) for all nuclides of both decay chains. The activity levels in secular equilibrium are listed in Table A.12.

For the calculation of ϵ_0 it is assumed that 1/4 of the α-recoil isotopes is able to leave the surface. As these isotopes are back implanted into the vacuum chamber, they still contribute to the population of the daughter nuclides. After about 150 days a secular equilibrium is reached. When buffer-gas is introduced, the implantation stops and the activity starts to decay in accordance with Eq. (A.29). Therefore, a new equilibrium situation will emerge. The corresponding activities A_{surf}, as well as α-recoil activities values ϵ_{surf} as obtained after one hour of free decay, are also listed in Table A.12. The branching ratios b_k were used to account for the fact that 1/4 of the isotopes will leave the surface and therefore will not populate the next decay level. The total α-recoil activity ϵ_{total} is then defined as the source activity ϵ listed in Table A.11 plus ϵ_{surf}. The values for ϵ_{total} are also listed in Table A.12 and will be used for the estimation of the extraction efficiency in the following section.

[1] Note the typo in Ref. [9]: the first product should go from i to $k-1$.

Table A.12 Calculated numbers (N_0) of isotopes contained in the metal surfaces of the surrounding of the ^{233}U source due to α-recoil ion implantation. The values are given together with the corresponding activities (A_0) and estimated α-recoil ion rates (ϵ_0) after 150 days of continuous implantation, when secular equilibrium has been reached. The values A_{surf} and ϵ_{surf} give the calculated activities and α-recoil activities of the surface 1 h after the accumulation was stopped due to buffer gas inlet. The values ϵ_{surf} were added to the calculated ^{233}U source α-recoil ion activities A_{recoil} as listed in Table A.11, in order to yield the total estimated α-recoil ion activity ϵ_{total} of the system

Isotope	N_0	A_0 [s^{-1}]	ϵ_0 [s^{-1}]	A_{surf} [s^{-1}]	ϵ_{surf} [s^{-1}]	ϵ_{total} [s^{-1}]
^{233}U	0	0	0	0	0	0
^{229}Th	$1.39 \cdot 10^{11}$	0.39	0	0.39	0	$(10.7 \pm 1.1) \cdot 10^3$
^{225}Ra	$8.00 \cdot 10^7$	43.4	0.098	43.3	0.098	43.1 ± 4.3
^{225}Ac	$5.40 \cdot 10^7$	43.3	0	43.3	0	0
^{221}Fr	$3.77 \cdot 10^4$	88.9	10.8	32.5	10.8	56.4 ± 5.6
^{217}At	6.32	136	22.2	24.4	8.1	54.9 ± 5.5
^{213}Bi	$7.39 \cdot 10^5$	187	34.0	87.7	6.1	57.5 ± 5.8
^{213}Po	$1.11 \cdot 10^{-3}$	183	0	85.7	0	0
^{209}Tl	$7.73 \cdot 10^2$	4.13	1.03	1.50	0.48	1.5 ± 0.2
^{209}Pb	$4.07 \cdot 10^6$	241	46.0	213	21.4	74.1 ± 7.4
^{209}Bi	$2.98 \cdot 10^9$	0	0	0	0	0
^{232}U	0	0	0	0	0	0
^{228}Th	$1.19 \cdot 10^8$	1.37	0	1.37	0	9.9 ± 1.0
^{224}Ra	$5.26 \cdot 10^6$	11.5	0.34	11.4	0.34	10.5 ± 1.1
^{220}Rn	$1.76 \cdot 10^3$	22.0	2.88	8.58	2.85	13.4 ± 1.3
^{216}Po	7.17	33.1	5.50	6.43	2.15	13.3 ± 1.3
^{212}Pb	$2.46 \cdot 10^6$	44.5	8.28	42.0	1.61	13.0 ± 1.3
^{212}Bi	$2.33 \cdot 10^5$	44.5	0	43.8	0	0
^{212}Po	$1.24 \cdot 10^{-5}$	28.5	0	28.4	0	0
^{208}Tl	$4.62 \cdot 10^3$	17.4	3.99	11.8	3.93	8.0 ± 0.8
^{208}Pb	$7.22 \cdot 10^8$	0	7.8	0	7.1	15.4 ± 1.5

A.7 Determination of Ion Extraction Efficiencies

The previous investigations give an estimate of the number of α-recoil isotopes that could potentially be extracted from the buffer-gas stopping cell. The actual numbers of extracted ions are below these values, according to the extraction efficiencies. Measurements were performed in order to infer the absolute number of extracted ions for individual charge states. The measurements performed for ^{229}Th were already discussed in Sect. 5.1.1. Besides ^{229}Th, also most of the daughter isotopes contained in the ^{233}U and ^{232}U decay chains were investigated, as will be detailed in the following.

A silicon detector (Ametek, BU-016-300-100) was placed behind the triode extraction system (see Fig. 5.1) and set onto a voltage offset of −1250 V, in order to implant a desired ion species directly into the detector surface. The ion species was chosen by calibrating the QMS with the help of mass scans, which were performed using an MCP detector (Hamamatsu, type F2223) that was placed off-axis. Detection was carried out in parallel to ion accumulation for the nuclides of short half-lives. Different accumulation and detection times were chosen for the investigation of the different nuclides.

For the investigation of ^{229}Th, Th^{3+} ions were accumulated on the detector surface for 5 days, following 100 days of continuous α-decay detection. The resulting α-energy spectrum is shown in Fig. 5.3. The number of detected ^{229}Th α-decays is $N_{count} = 5575$. This number is related to the extraction rate R via

$$R(^{229}\mathrm{Th}) = \frac{2N_{\mathrm{count}}}{\lambda \cdot t_{\mathrm{acc}} \cdot t_{\mathrm{det}}} = 1070\ \mathrm{s}^{-1}, \tag{A.31}$$

which corresponds to about 10% extraction efficiency, given the $(10.7 \pm 1.1) \cdot 10^3\ \mathrm{s}^{-1}$ total ^{229}Th α-recoil isotope activity as listed in Table A.12. Here t_{acc} and t_{det} denote the accumulation and detection time, respectively. In Table A.13 the numbers of detected decays N_{count}, the estimated numbers of real decays N_{real} as well as t_{acc} and t_{det} are listed, together with the calculated extraction rates and the corresponding extraction efficiencies for all ion species.

In order to calculate the extraction rates for ^{229}Th and ^{228}Th, Eq. (A.31) was used. As ^{225}Ra is a pure β^- emitter, the extraction rate was calculated based on the ^{225}Ac α decay with the help the following equation

$$R(^{225}\mathrm{Ra}) = \frac{2N_{\mathrm{count}}\left(\lambda_{\mathrm{Ac}} - \lambda_{\mathrm{Ra}}\right)}{t_{\mathrm{acc}}\left[\lambda_{\mathrm{Ac}}\left(e^{-\lambda_{\mathrm{Ra}} t_1} - e^{-\lambda_{\mathrm{Ra}} t_2}\right) - \lambda_{\mathrm{Ra}}\left(e^{-\lambda_{\mathrm{Ac}} t_1} - e^{-\lambda_{\mathrm{Ac}} t_2}\right)\right]}, \tag{A.32}$$

where t_1 corresponds to the start time of the measurement (2.6 days after 2 h of accumulation) and t_2 is the time when the measurement was terminated (25.7 days after accumulation). For ^{224}Ra the equation

$$R(^{224}\mathrm{Ra}) = \frac{2\,\lambda\,N_{\mathrm{count}}}{\left(1 - e^{-\lambda t_{\mathrm{acc}}}\right)\left(1 - e^{-\lambda t_{\mathrm{det}}}\right)} \tag{A.33}$$

was used for evaluation, as ^{224}Ra is an α emitter and the decays can be directly detected with the silicon detector. For ^{221}Fr, ^{217}At, ^{213}Bi, ^{220}Rn and ^{216}Po, accumulation and detection were performed in parallel and the extraction rates were determined via

$$R = \frac{N_{\mathrm{real}}}{t_{\mathrm{det}} + 1/\lambda\left(e^{-\lambda t_{\mathrm{det}}} - 1\right)}. \tag{A.34}$$

Here N_{real} denotes the expected amount of α decays. This value is two times the detected number of α decays, as only α particles emitted into the hemisphere of the

Table A.13 Determination of the extraction efficiencies for nuclides contained in the ^{233}U and ^{232}U decay chains, respectively. Besides the ion species and the obtained extraction efficiencies, also the times for accumulation t_{acc} and detection t_{det} are listed, together with the number of registered α-decay events N_{count}, the corresponding number of real α decays N_{real} and the extraction rate R. The calculation of the extraction rates is detailed in the text

Isotope	t_{acc}	t_{det}	N_{count}	N_{real}	R [s^{-1}]	Efficiency [%]
$^{229}Th^{3+}$	5 d	100 d	5575	11150	1070	10 ± 2.0
$^{225}Ra^{2+}$	2 h	(25.7–2.6) d	26178	52356	17.7	41 ± 8
$^{221}Fr^{1+}$	2 h	2 h	41168	82336	12.1	21 ± 4.2
$^{221}Fr^{2+}$	2 h	2 h	31248	62496	9.2	16 ± 3.2
$^{221}Fr^{3+}$	2 h	2 h	3	6	$\leq 8.8 \cdot 10^{-4}$	$\leq 1.6 \cdot 10^{-3}$
$^{217}At^{1+}$	2 h	2 h	17041	34082	4.7	8.6 ± 1.7
$^{217}At^{2+}$	2 h	2 h	25741	51482	7.2	13 ± 2.6
$^{217}At^{3+}$	2 h	2 h	66	132	$1.8 \cdot 10^{-2}$	$(3.3 \pm 0.7) \cdot 10^{-2}$
$^{213}Bi^{1+}$	2 h	2 h	4700	9607	2.5	4.3 ± 0.9
$^{213}Bi^{2+}$	2 h	2 h	23098	47212	12.1	21 ± 4.2
$^{213}Bi^{3+}$	2 h	2 h	91	186	$4.8 \cdot 10^{-2}$	$(8.3 \pm 1.6) \cdot 10^{-2}$
$^{228}Th^{3+}$	5 d	100 d	19453	38906	0.9	9.1 ± 1.8
$^{224}Ra^{2+}$	2 h	63 h	3607	7214	2.6	24.8 ± 5
$^{220}Rn^{1+}$	2 h	2 h	2091	5576	0.8	5.8 ± 1.2
$^{220}Rn^{2+}$	2 h	2 h	3336	8896	1.25	9.3 ± 1.9
$^{220}Rn^{3+}$	2 h	2 h	19	51	$7.1 \cdot 10^{-3}$	$(5.3 \pm 1.1) \cdot 10^{-2}$
$^{216}Po^{1+}$	2 h	2 h	3494	6988	1.0	7.3 ± 1.5
$^{216}Po^{2+}$	2 h	2 h	3870	7740	1.1	8.1 ± 1.6
$^{216}Po^{3+}$	2 h	2 h	1	2	$2.8 \cdot 10^{-4}$	$\leq 2.1 \cdot 10^{-3}$
$^{212}Pb^{1+}$	2 h	(20-5) h	302	942	0.28	2.2 ± 0.4
$^{212}Pb^{2+}$	2 h	(20-5) h	1544	4817	1.4	11 ± 2.2
$^{212}Pb^{3+}$	2 h	(20-4) h	4	12.5	$1.3 \cdot 10^{-3}$	$\leq 1.2 \cdot 10^{-2}$

silicon detector are registered. However, in case of ^{213}Bi it takes also the branching ratio of 97.8% to ^{213}Po into account, because ^{213}Bi is a β^- emitter and thus not directly detected. The extraction efficiency of ^{220}Rn was estimated based on the successively occurring ^{216}Po decays. The energy of the ^{220}Rn α decay is close to that of ^{221}Fr and can therefore not be properly distinguished. A reduction of 25% of ^{216}Po decays compared to ^{220}Rn, occurring on the detector surface, was assumed, taking the α-recoil loss into account.

The evaluation for ^{212}Pb is more complicated, as it involves two β^- decays and its extraction efficiency was calculated based on the successively occurring α decay of ^{212}Po. The equation used for the evaluation reads

$$R(^{212}\text{Pb}) = \frac{N_{\text{real}}}{\lambda_{\text{Po}}\left[f_1(t_{\text{acc}})\int_{t_1}^{t_2} f_2(t)dt + f_3(t_{\text{acc}})\int_{t_1}^{t_2} f_4(t)dt\right]}, \quad (A.35)$$

where

$$f_1(t_{\text{acc}}) = \frac{\lambda_{\text{Pb}}e^{-\lambda_{\text{Bi}}t_{\text{acc}}} - \lambda_{\text{Bi}}e^{-\lambda_{\text{Pb}}t_{\text{acc}}} + \lambda_{\text{Bi}} - \lambda_{\text{Pb}}}{\lambda_{\text{Bi}}(\lambda_{\text{Bi}} - \lambda_{\text{Pb}})},$$

$$f_2(t) = \lambda_{\text{Bi}}\left(\frac{e^{-\lambda_{\text{Bi}}t}}{\lambda_{\text{Po}} - \lambda_{\text{Bi}}} + \frac{e^{-\lambda_{\text{Po}}t}}{\lambda_{\text{Bi}} - \lambda_{\text{Po}}}\right),$$

$$f_3(t_{\text{acc}}) = \frac{1 - e^{-\lambda_{\text{Pb}}t_{\text{acc}}}}{\lambda_{\text{Pb}}}$$

and

$$f_4(t) = \lambda_{\text{Bi}}\lambda_{\text{Pb}}\left(\frac{e^{-\lambda_{\text{Pb}}t}}{(\lambda_{\text{Bi}} - \lambda_{\text{Pb}})(\lambda_{\text{Po}} - \lambda_{\text{Pb}})} + \frac{e^{-\lambda_{\text{Bi}}t}}{(\lambda_{\text{Pb}} - \lambda_{\text{Bi}})(\lambda_{\text{Po}} - \lambda_{\text{Bi}})} + \frac{e^{-\lambda_{\text{Po}}t}}{(\lambda_{\text{Pb}} - \lambda_{\text{Po}})(\lambda_{\text{Bi}} - \lambda_{\text{Po}})}\right).$$

Here t_1 and t_2 are detailed in Table A.13 and N_{real} includes for the geometrical efficiency of the detector and the 64.1% decay branch from ^{212}Bi to ^{212}Po. Accumulation and detection were performed separately.

The important result of these measurements is that only ^{229}Th can be extracted to a significant amount in the 3+ charge state. The physical reason is the low 3rd ionization potential of thorium of just 18.3 eV, which is significantly below the first ionization potential of helium of 24.6 eV. For this reason, it is energetically favorable for the electrons to stay attached to the helium atom instead of reducing the thorium charge state during stopping in the buffer gas. This was the first time that a strong extraction efficiency of ions in the 3+ charge state from a buffer-gas stopping cell was reported [4].

A.8 Chemical Purification

Before the large-area ^{233}U source (source 2) was produced by electro-deposition, the radioactive material was chemically purified (23.4.2015) in order to remove most of the daughter products of the ^{233}U and ^{232}U decay chains. Ion-exchange chromatography (Dowex AG 1-X8) was used for this purpose. γ-ray energy spectra were taken with a germanium detector (2 h acquisition time) before and after chemical purification of the ^{233}U source material. The corresponding spectra are shown in Fig. A.11, the peaks are numbered in accordance with Table A.2.

The lines 11 (169 keV, ^{233}U) and 13 (193 keV, ^{229}Th) of both spectra were compared, in order to infer a lower limit for the factor of purification of the ^{233}U material. When defining

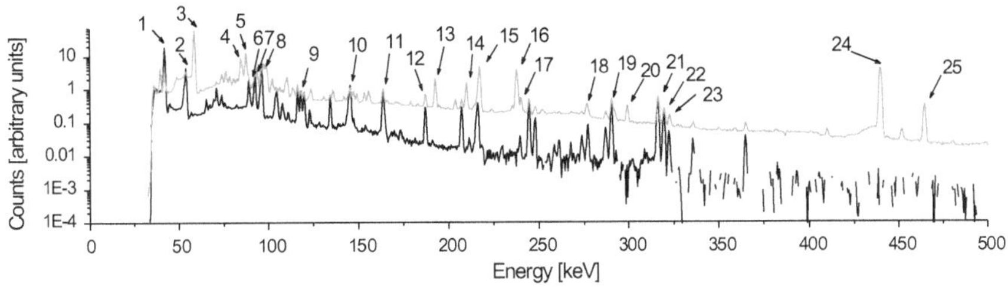

Fig. A.11 γ-ray energy spectra for the low-energy range (up to 500 keV) of the ^{233}U source material, measured with a germanium detector after 2 h of acquisition time. The upper spectrum was measured before chemical purification of the ^{233}U source material, it corresponds to the material composition of source 1. The lower spectrum was measured after chemical purification of the source material, it corresponds to the composition of ^{233}U source 2. Corresponding line numbers are detailed in Table A.2 [3, 5]

$$A = \frac{I\,(^{229}\text{Th, unpurified})}{I\,(^{233}\text{U, unpurified})} \quad \text{and} \quad B = \frac{I\,(^{229}\text{Th, purified})}{I\,(^{233}\text{U, purified})},$$

the factor of purification P is obtained as $P = A/B$. By integration of the γ-ray energy spectra one obtains $A \approx 2.95$ and $B \approx 9.7 \cdot 10^{-3}$, the factor of purification is thus inferred to be larger than $P \approx 300$.

Bibliography

1. Orth DA (1978) SRP Thorium processing experience. ANS annual meeting, San Diego, CA, USA
2. Stevens CM et al (1954) Curium isotopes 246 and 247 from pile-irradiated plutonium. Phys Rev 94:974
3. NNDC interactive chart of nuclides. http://www.nndc.bnl.gov/chart [2016, March 22nd], Brookhaven National Laboratory, Brookhaven
4. von der Wense L et al (2015) Determination of the extraction efficiency for ^{233} U source α -recoil ions from the MLL buffer-gas stopping cell. Eur Phys J A 51:29
5. Data base of the Radiochemistry Society. http://www.radiochemistry.org/periodictable/gamma_spectra [2016, September 15th], Richland, Washington
6. Bateman H (1910) Solution of a system of differential equations occuring in the theory of radioactive transformations. In: Proceedings of the Cambridge Philosophical Society IS, pp 423–427
7. Ziegler JF, Biersack JP, Littmark U (1985) The stopping and range of ions in matter. Pergamon, New York
8. Nordlund K (1995) Molecular dynamics simulation of ion ranges in the 1–100 keV energy range. Comput Mater Sci 3:448–456
9. Hashimoto T et al (1981) Ranges of α-recoil ^{234}Th atoms in uranium oxides. J inorg nucl Chem 43(10):2233
10. Magill J, Galy J (2005) Radioactivity radionuclides radiation. Springer, Berlin (2005), p 47
11. von der Wense L et al (2016) Direct detection of the ^{229}Th nuclear clock transition. Nature 533:47–51

Appendix B
Numerical Calculations and Code Implementation

Many of the calculations required for this thesis were performed numerically. Self-developed Matlab (version 7.0) codes were used for evaluation, except for a few simulations, for which existing programs were used (SIMION, SRIM, MDrange). In the following, the numerical calculations and corresponding Matlab source codes will be provided. The chapter is divided into three sections. In the first section, numerical investigations related to the ^{233}U source and ion extraction will be described. In the second section, the program code related to the optical system will be given and in the third section the programs used for image evaluation will be detailed. All Matlab-based program codes can be made available on request.

B.1 Ion Extraction and ^{233}U Source Investigation

In this section, Matlab source code related to the investigation of the ^{233}U source will be provided and the parameters used for the SIMION, SRIM and MDrange simulations will be detailed.

B.1.1 The Bateman Equation

The time-dependent population of daughter nuclides from a single mother nuclide is described by the Bateman equation [1]. It reads

$$N_k(t) = N_0\left(\prod_{i=1}^{k-1} b_i\lambda_i\right)\sum_{j=1}^{k}\frac{e^{-\lambda_j t}}{\prod_{i=1,i\neq j}^{k}\left(\lambda_i-\lambda_j\right)}, \tag{B.1}$$

L. Von der Wense, *On the Direct Detection of* ^{229m}*Th*,
Springer Theses, https://doi.org/10.1007/978-3-319-70461-6

where N_k is the number of nuclei of the k-th daughter isotope, λ_k the corresponding decay constant, b_k the branching ratio, N_0 the starting number of mother nuclei and t the time since start of the decay.

The Matlab program that calculates the number of nuclei and activity values for the ^{233}U decay chain with a starting number of ^{233}U nuclei of $1.45 \cdot 10^{18}$ and a time of decay of 45 years is given in the following. The calculated values are listed in Table A.6. The decay constants have to be adapted in order to also calculate the ^{232}U decay chain.

```
function Bateman
%Implementation of Bateman equation
clc;
format long

No=1.45e18; %U233 starting number of nuclei
tdecaytime=1.419e9; %time to decay

%Decay constants U233 decay chain
L(1)=1.38e−13; %U233
L(2)=2.79e−12; %Th229
L(3)=5.42e−7; %225Ra
L(4)=8.02e−7; %225Ac
L(5)=2.36e−3; %221Fr
L(6)=21.46; %217At
L(7)=2.53e−4; %213Bi
L(8)=1.65e5; %213Po
L(9)=5.92e−5; %209Pb


%Branching ratios U233 decay chain
X(1)=1; %U233
X(2)=1; %Th229
X(3)=1; %225Ra
X(4)=1; %225Ac betazerfall
X(5)=1; %221Fr
X(6)=1; %217At
X(7)=0.978; %213Bi
X(8)=1; %213Po betazerfall
X(9)=1; %209Pb

Ndecay=sourcesolution_decay(No,L,X,tdecaytime); %calculates number of nuclei
A=activity(Ndecay,L); %calculates activities

%results
Ndecay(:)
A(:)
end

function [A]=activity(N,L)
for k=1:length(N)
    A(k)=N(k)*L(k);
end
end

function [N]=sourcesolution_decay(No,L,X,t)
for k=1:length(L)
    N(k)=0;
    p=1;
    N(k)=singlesum(No,L,X,t,k,p);
end
end

function Np=singlesum(No,L,X,t,k,p)
kpsum=0;
for j=1:k−p+1
    kpsum=kpsum+exp(−L(j+p−1)*t)/kpproduct_2(L,k,p,j);
end
Np=No(p)*kpproduct_1(L,X,k,p)*kpsum;
end

function prod1=kpproduct_1(L,X,k,p)
prod1=1;
for i=1:k−p
    prod1=prod1*L(i+p−1)*X(i+p−1);
end
end

function prod2=kpproduct_2(L,k,p,j)
prod2=1;
```

```
for l=1:k-p+1
    if l>j | l<j
        prod2=prod2*(L(l+p-1)-L(j+p-1));
    end
end
end
```

B.1.2 α-Recoil Efficiencies

In Sect. A.5, an analytic model was developed to infer the α-recoil efficiencies for the ^{233}U source under investigation. The result is the following integral (Eq. A.26)

$$F = \frac{1}{4\pi d}\int_0^d dr \int_0^{2\pi} d\phi \int_0^{\pi} \sin(\theta) d\theta \, T(r,\theta), \tag{B.2}$$

where $T(r, \theta)$ is defined as

$$T(r,\theta) = \frac{1}{2}\left[1 + \mathrm{erf}\left(\frac{s\cos(\theta) - r}{\sqrt{2}\sigma_{\mathrm{tr}}\sin(\theta)} \cdot \left[\left(\frac{\sigma_{\mathrm{lo}}\cos(\theta)}{\sigma_{\mathrm{tr}}\sin(\theta)}\right)^2 + 1\right]^{-1/2}\right)\right]. \tag{B.3}$$

Here d is the thickness of the source material, s is the stopping length of the α-recoil ion in the source material, σ_{lo} is the longitudinal straggling and σ_{tr} is the transversal straggling. These are input parameters, which were obtained by MDrange or SRIM simulations, respectively (see Appendix B.1.3). In the following, the source code used to solve the integral (B.2) is provided. Parameters are inserted as obtained for the large-area ^{233}U source 2, the result is an efficiency of 34.7%.

```
function transmitanalytic
% Calculates the recoil efficiency of a given source material
clc;
l=10.7; %stopping length in nm
sigz=5.8; %longitudinal straggle in nm
sigx=5.1; %vertical straggle in nm
dmax=6.9; %material thickness in nm

Dint(l,sigz,sigx,dmax) %result
end

function intd=Dint(l,sigzmax,sigxmax,dmax)
m=100;
dstart=0.00001;
dstep=(dmax-dstart)/m;
dvalue=[dstart:dstep:dmax];
intd=0;
for i=1:m
    intd=intd+(Thetaint(dvalue(i), l,sigzmax,sigxmax)+Thetaint(dvalue(i+1),l,sigzmax,sigxmax))/2*dstep;
end
intd=intd/dmax;
end

function inttheta=Thetaint(d,l,sigzmax,sigxmax)
n=100;
thetastart=pi/1000;
thetaend=pi;
thetastep=(thetaend-thetastart)/n;
thetavalue=[thetastart:thetastep:thetaend];
inttheta=0;
for i=1:n
```

```
    inttheta=inttheta+(Transmit(d,thetavalue(i),l,sigzmax,sigxmax)+Transmit(d,thetavalue(i+1),l,sigzmax,sigxmax))/2*thetastep*sin((thetavalue(i)+
            thetavalue(i+1))/2);
end
inttheta=inttheta/2;
end

function T=Transmit(d,theta,l,sigz,sigx)
if cos(theta)>0
    T=0.5*(1+erf((l*cos(theta)-d)/(sqrt(2)*sigx*sin(theta))/sqrt((sigz*cos(theta)/(sigx*sin(theta)))^2+1)));
else
    T=0.5*(1+erf((l*cos(theta)-d)/(sqrt(2)*sigx*sin(theta))/sqrt((sigz*cos(theta)/(sigx*sin(theta)))^2+1)));
end
end
```

B.1.3 SRIM Simulations

A program for simulation of the stopping range of ions in matter (SRIM) was developed by J.F. Ziegler and J.P. Biersack. It is based on the ZBL (Ziegler-Biersack-Littmark) stopping formalism [2] and online available at www.srim.org. SRIM version 2008 was used for the calculations. For investigation of the large-area ^{233}U source 2, 10 000 ^{229}Th ions with a kinetic energy of 84.3 keV were simulated in metallic ^{233}U material with a density of 18.75 g/cm^3. The simulated material area was 100 nm × 100 nm. The results are a stopping range s of 10.7 nm, a longitudinal straggling $\sigma_{\rm lo}$ of 5.8 nm and a lateral projected straggling $\sigma_{\rm tr}$ of 5.1 nm. These numbers are used together with the source thickness as starting parameters for the calculation described in Sect. B.1.2.

The SRIM-based simulations gave systematically too small results in case of ^{233}U source 1. The presumable reason is the different production process. Source 2 was produced via electrodeposition, in this way providing a completely amorphous surface layer as correctly described by the SRIM simulations. Source 1, however, was produced via evaporation in vacuum, which may have led to the formation of a microcrystalline surface layer. Such a surface structure allows for channeling effects along the crystal channels, in this way increasing the α-recoil efficiency of the source. These effects are not taken into account by the SRIM simulations.

B.1.4 MDrange Simulations

The MDrange package was developed by K. Nordlund to take channeling effects into account during the stopping process of ions in matter [3]. The program code is available for download at http://beam.helsinki.fi/~knordlund/mdrange.tar.gz. This program was used to calculate the stopping range of α-recoil isotopes in the UF_4 material of the small-area ^{233}U source 1. Before the start of the program, a file (atoms.in) has to be provided, where the relative positions of the 60 uranium and fluorine atoms in the UF_4 unit cell are defined.

When starting the program, several input parameters will be requested. These include the different atom types (uranium and fluorine), the number of atoms in the unit cell (60), the lattice length ($a = 12.73$ nm, $b = 10.75$ nm, $c = 6.67$ nm). Note that MDrange only works with rectangular lattices, therefore the monoclinic crystal structure of UF_4 is not fully supported. As an approximation, the actual crystal lattice parameter $c = 8.43$ nm was projected onto the vertical axis, leading to $c = 6.67$ nm. Further requested parameters are the projectile (e.g. Th), the mass (e.g. 229 u) the recoil energy (e.g. 84.3 keV), the size of the domains of the polycrystalline target (80 nm), the standard deviation (10 nm), the maximum angle (90°, corresponding to random orientation), the number of projectiles (1000), the temperature of the initial state (300 K) and the Debye temperature (160 K).

Following the parameter input, the uranium mass has to be changed from 238 u to 233 u in the file param.in (type[1].m, line 4). Running the program results in the calculation of the stopping range s as well as the longitudinal straggling σ_{lo}.

In order to also calculate the transversal straggling, a change in the file mdrange/md-h/recoil.c is required. In line 388 of this file, the code which calculates the radial projection of the ion distribution

```
recoil->r_proj=sqrtf((recoil->sx*recoil->sx+recoil->sy*recoil->sy+recoil->sz*recoil->sz)-(recoil->vx0*recoil->sx+recoil->vy0*recoil->sy+recoil->vz0*recoil->sz)/recoil->vel0*(recoil->vx0*recoil->sx+recoil->vy0*recoil->sy+recoil->vz0*recoil->sz)/recoil->vel0);
```

has to be exchanged by a code which calculates the transversal projection of the ion distribution

```
recoil->r_proj=sqrtf(((recoil->x0-recoil->sx)*(recoil->x0-recoil->sx)+(recoil->y0-recoil->sy)*(recoil->y0-recoil->sy)+(recoil->z0-recoil->sz)*(recoil->z0-recoil->sz))-(recoil->vx0*(recoil->sx-recoil->x0)+recoil->vy0*(recoil->sy-recoil->y0)+recoil->vz0*(recoil->sz-recoil->z0))/recoil->vel0*(recoil->vx0*(recoil->sx-recoil->x0)+recoil->vy0*(recoil->sy-recoil->y0)+recoil->vz0*(recoil->sz-recoil->z0))/recoil->vel0);
```

When the program is compiled and rerun, the expectation value of the transversal projection $E(|t|)$, as well as its standard deviation $\sigma(|t|)$, will be the outputs of the computation. From these values the standard deviation of the gaussian distribution can be calculated via

$$\sigma_{\text{tr}} = \sqrt{\frac{\sigma(|t|)^2 + E(|t|)^2}{2}}.$$

The results for ^{229}Th are $s = 75.9$ nm, $\sigma_{\text{lo}} = 35.1$ nm, $\sigma_{\text{tr}} = 5.4$ nm. Numerical results for all other nuclides in the ^{233}U and ^{232}U decay chains are listed in Table A.10, together with the recoil efficiency F of the source as obtained from the calculation described in Sect. B.1.2.

B.1.5 The Bateman Equation with Source Terms

In Sect. B.1.1 the Bateman equation, describing the population of daughter isotopes from a single mother nuclide, was already discussed. At several points, e.g. during continuous radioactive ion extraction from the buffer-gas stopping cell, a constant source term has to be considered. Further, the source can be stopped when the ion accumulation is terminated, thus leading to the free decay as described by the Bateman equation. This time, however, the decay chain is not only populated by a single starting isotope, but instead from various nuclides that have build up during the ion extraction. The extended Bateman equation, which represents the solution of the complete system reads (see Eq. A.30)

$$N_k(t) = \sum_{i=1}^{k} \left(\prod_{l=i}^{k-1} b_l \lambda_l \right) \sum_{j=i}^{k} \left[\frac{N_i(0) e^{-\lambda_j t}}{\prod_{l=i, l \neq j} (\lambda_l - \lambda_j)} + \frac{\epsilon_i \left(1 - e^{-\lambda_j t}\right)}{\lambda_j \prod_{l=i, l \neq j}^{k} (\lambda_l - \lambda_j)} \right]. \tag{B.4}$$

In this equation, $N_k(0)$ denotes the starting number of nuclei of the k-th isotope (already existing contamination) and ϵ_k denotes the population rates for each nuclide. In case of the calculation of implanted α-recoil activities, the values of A_{recoil} as listed in Table A.11 for ^{233}U source 1 have to be used for ϵ_k.

The program code applied to calculate the numbers and activity values of implanted α-recoil isotopes as listed in Table A.12 (N_0 and A_0) is provided in the following.

```
function Bateman_extended
clc;
format long

%Starting values (contamination)
No(1)=0; %U233
No(2)=0; %Th229
No(3)=0; %225Ra
No(4)=0; %225Ac
No(5)=0; %221Fr
No(6)=0; %217At
No(7)=0; %213Bi
No(8)=0; %213Po
No(9)=0; %209Pb
No(10)=0; %209Bi

%Decay constants U233 decay chain
L(1)=1.38e−13; %U233
L(2)=2.79e−12; %Th229
L(3)=5.42e−7; %225Ra
L(4)=8.02e−7; %225Ac
L(5)=2.36e−3; %221Fr
L(6)=21.46; %217At
L(7)=2.53e−4; %213Bi
L(8)=1.65e5; %213Po
L(9)=5.92e−5; %209Pb
L(10)=1e−15; %209Bi

%branching ratios U233
X(1)=1; %U233
X(2)=1; %Th229
X(3)=1; %225Ra
X(4)=1; %225Ac
X(5)=1; %221Fr
X(6)=1; %217At
X(7)=1; %213Bi beta
X(8)=1; %213Po
X(9)=1; %209Pb
X(10)=1;

```

```
%Population rates U233 decay chain
E(1)=0; %U233
E(2)=10700; %Th229
E(3)=43.1; %225Ra
E(4)=0; %225Ac betazerfall
E(5)=45.6; %221Fr
E(6)=46.8; %217At
E(7)=51.4; %213Bi
E(8)=0; %213Po betazerfall
E(9)=52.7; %209Pb
E(10)=0; %209Bi betazerfall

taccumulation=1.296e7; %time of accumulation
tdecaytime=1e−20; %time of free decay

%calculation of time dependent values (allowing for time−dependent plot)
Npoints=100;
tacc=[0:taccumulation/Npoints:taccumulation];
tdecay=[0:tdecaytime/Npoints:tdecaytime];
for i=1:length(tacc)
    [Nacc(i,:)]=sourcesolution_accumulation(L,X,E,tacc(i))+sourcesolution_decay(No,L,X,tacc(i));
end
for i=1:length(tdecay)
    [Ndecay(i,:)]=sourcesolution_decay(Nacc(Npoints+1,:),L,X,tdecay(i));
end
for i=1:2*Npoints+2
    if i<=Npoints+1
        N(i,:)=Nacc(i,:);
        t(i)=tacc(i);
    else
        N(i,:)=Ndecay(i−(Npoints+1),:);
        t(i)=tacc(Npoints+1)+tdecay(i−(Npoints+1));
    end
    [A(i,:)]=activity(N(i,:),L);
end

%results
N1=N(2*Npoints+2,1)
N2=N(2*Npoints+2,2)
N3=N(2*Npoints+2,3)
N4=N(2*Npoints+2,4)
N5=N(2*Npoints+2,5)
N6=N(2*Npoints+2,6)
N7=N(2*Npoints+2,7)
N8=N(2*Npoints+2,8)
N9=N(2*Npoints+2,9)
N10=N(2*Npoints+2,10)

A1=A(2*Npoints+2,1)
A2=A(2*Npoints+2,2)
A3=A(2*Npoints+2,3)
A4=A(2*Npoints+2,4)
A5=A(2*Npoints+2,5)
A6=A(2*Npoints+2,6)
A7=A(2*Npoints+2,7)
A8=A(2*Npoints+2,8)
A9=A(2*Npoints+2,9)
A10=A(2*Npoints+2,10)
end

function [N]=sourcesolution_accumulation(L,X,E,t)
for k=1:length(L)
    N(k)=0;
    for j=1:k
        N(k)=N(k)+sumnuclide(L,X,E,t,k,j);
    end
end
end

function s=sumnuclide(L,X,E,t,k,j)
s1=0;
for i=1:j
    s1=s1+productnuclide(L,X,E,k,j,i);
end
s=−s1*exp(−L(j)*t)+productnuclide3(X,k,j)*E(j)/L(k);
end

function p=productnuclide(L,X,E,k,j,i)
p1=1;
for l=i:k−1
    p1=p1*X(l)*L(l);
end
p2=1;
for l=i:k
    if l>j | l<j
        p2=p2*(L(l)−L(j));
    end
end
p=E(i)*p1/(L(j)*p2);
```

```
end

function p3=productnuclide3(X,k,j)
p3=1;
for l=j:k-1
    p3=p3*X(l);
end
end

function [A]=activity(N,L)
for k=1:length(N)
    A(k)=N(k)*L(k);
end
end

function I=t_integral(t,A)
n=length(t);
for k=1:length(A(1,:))
    I(k)=0;
    for i=1:n-1
        I(k)=I(k)+((A(i,k)+A(i+1,k))/2*(t(i+1)-t(i)));
    end
end
end

function [N]=sourcesolution_decay(No,L,X,t)
for k=1:length(L)
    N(k)=0;
    for p=1:k
        N(k)=N(k)+singlesum(No,L,X,t,k,p);
    end
end
end

function Np=singlesum(No,L,X,t,k,p)
kpsum=0;
for j=1:k-p+1
    kpsum=kpsum+exp(-L(j+p-1)*t)/kpproduct_2(L,k,p,j);
end
Np=No(p)*kpproduct_1(L,X,k,p)*kpsum;
end

function prod1=kpproduct_1(L,X,k,p)
prod1=1;
for i=1:k-p
    prod1=prod1*L(i+p-1)*X(i+p-1);
end
end

function prod2=kpproduct_2(L,k,p,j)
prod2=1;
for l=1:k-p+1
    if l>j | l<j
        prod2=prod2*(L(l+p-1)-L(j+p-1));
    end
end
end
```

B.1.6 SIMION Simulations

Ion trajectories were simulated with the help of SIMION [4]. SIMION is a commercial code package, developed for ion trajectory calculation in arbitrary electric potentials, that is available at www.simion.com. For the presented work, SIMION version 8.1 was used in combination with 32 GB RAM. This allowed to simulate the complete setup, from the ^{233}U source to the ion collection on the micro-electrode. For this purpose, the setup was split into potential arrays of different resolutions, that were loaded into the same work-bench file. In this way, the important part of ion collection, on the micro electrode could be simulated with a high resolution in the μm-range. Simulations were mainly used to verify the functionality of the quadrupole-mass separator and for design and development of the triode extraction system (Sect. 4.2.8), as well as the collection surface (Sect. 4.3.1).

SIMION allows to simulate ion trajectories in buffer-gas backgrounds. For this purpose, important use was made of the SDS (statistical diffusion simulation) model, which was used to calculate ion trajectories in the high-pressure region of the buffer-gas stopping cell, as well as the HS1 (hard-sphere) collision model, used to calculate the collisions with the buffer-gas background in the low-pressure region of the RFQ. Implementation of both types of collision models is provided together with the commercial SIMION code package.

B.2 Optical Simulations

A Matlab program code was developed to simulate the VUV optical system. The program code was used for three purposes: (i) to numerically investigate a variety of different optical systems during the process of development, (ii) to provide values of magnification and transmission as well as focusing efficiency and (iii) to numerically estimate the tolerances for the process of optical alignment.

The complete program code is lengthy and can be made available on request. It allows to test a large variety of optical systems, including spherical and aspherical lenses (for a given set of asphere-parameters) as well as spherical and parabolic mirrors. In the following, a shortened version of the program code is presented, which only allows to investigate the properties of the optical system consisting of two parabolic mirrors, as applied in the measurements described in Sect. 5.2.1.

The program is based on analytical ray-tracing. Photons are emitted from a round surface (of radius x_{max}), which is positioned at the point (x_{center}, y_{center}). The surface is segmented in radial (n_f) and angular (n_k) direction and the center of each segment is used as a starting point to propagate photons in polar (n_c) and azimuthal (n_w) direction. The total number of propagated light rays therefore amounts to $n_f \cdot n_k \cdot n_c \cdot n_w$. Each of the propagated rays is provided with a weight, which corresponds to the surface area of its source segment, multiplied by the surface area of its spherical surface segment, normalized to one. In this way, an equal photon density distribution is simulated.

All light rays are numerically propagated through the optical system, in this case consisting of two parabolic mirrors. The distance between the light source and the first optical element (z_{start}) corresponds to the focal length of the first parabolic mirror. The distance between the two optical elements has to be inserted as parameter d. The distance between the point of detection and the second optical element (z_{end}) is predefined as the focal length of the second parabolic mirror.

The detection plane is defined by its outer dimensions (max_x and max_y) as well as the pixel size ($boxsize$). A further input parameter is the number of emitted photons per second from the light source ($startnum$), as required to calculate physical quantities in the end.

```
 function lensesystemadv
% 3D simulation of optical system consisting of 2 parabolic mirrors. Allows to break axial symmetry. Mirrors can be tilted
format long
clc;

xmax=0.025; %radius of light source
xcenter=0; %x−shift of light source
ycenter=0; %y−shift of light source
nf=10; %number of radial surface segments
nk=10; %number of angular surface segments (even)
nc=10; %number of polar propagated photons
nw=10; %number of azimuthal propagated photons (even)
zstart=10; %defines distance between light source and first optical element
d=100; %distance between two optical elements in mm (from entrance plane to entrance plane)
zend=2; %defines distance between 2nd optical element and detector
max_x=0.2; %x−length of detection plane (in mm)
max_y=0.2; %y−length of detecion plane (in mm)
boxsize=0.005; %Pixelsize of detection plane (in mm)
startnum=1; %number of photons to be emitted per second from light source

%calculate result
mirrorareadistributionorigin_3DS(d,zstart,zend,nc,nw,nf,nk,max_x,max_y,boxsize,xmax,startnum,xcenter,ycenter);
end

function mirrorareadistributionorigin_3DS(d,zstart,zend,nc,nw,nf,nk,max_x,max_y,boxsize,xmax,startnum,xcenter,ycenter)
%propagate photons
[xend,yend,gewicht,loss,ntotal]=sourceextension_3DS(d,zstart,zend,nc,nw,nf,nk,xmax,xcenter,ycenter);
%evaluate the resulting light distribution
[X,Y,znorm2(:,:,1),zges,counter(:,:,1),gfactor,Rnorm2(:,1),yg,max_r]=new_distribution(nf,nk,ntotal,max_x,max_y,boxsize,xend(:,:,:,1),yend(:,:,:,1),···
        gewicht,loss,startnum);
nys=length(Rnorm2(:,1));
Rnormdouble(1:nys)=flipud(Rnorm2);
Rnormdouble(nys+1:2*nys)=Rnorm2;
ygdouble(1:length(yg))=−flipud(yg(:));
ygdouble(length(yg)+1:2*length(yg))=yg;

%calculate FWHM numbers
[peaksizeFWHM,probabFWHM]=evaluation(yg,Rnorm2,max_r,gfactor,startnum);
FWHM=peaksizeFWHM
Transmission_efficiency=gfactor
Imaging_efficiency=probabFWHM

%plot results
nxc=round(max_x/boxsize);
nyc=round(max_y/boxsize);
scrsz = get(0,'ScreenSize');
figure('Position',[1 scrsz(4)/4 scrsz(3)/2 scrsz(4)/2]);
image(znorm2);
imagesc(znorm2,[0,max(max(znorm2))]);
axis image
scale=max_x/4;
set(gca,'XTick',mod(nxc,nxc/max_x*scale)/2:nxc/max_x*scale:nxc)
set(gca,'XTickLabel',−max_x/2:scale:max_x/2)
set(gca,'YTick',mod(nyc,nyc/max_y*scale)/2:nyc/max_y*scale:nyc)
set(gca,'YTickLabel',max_x/2:−scale:−max_x/2)
title('Area Density Distribution');
xlabel('x [mm]');
ylabel('y [mm]');
zlabel('photon density [1/(mm^2 s)]');
scrsz = get(0,'ScreenSize');
figure('Position',[1 scrsz(4)/4 scrsz(3)/2 scrsz(4)/2]);
plot(ygdouble,Rnormdouble);
end

function [peaksizeFWHM,probabFWHM]=evaluation(yg,Rnorm2,max_r,gfactor,startnum)
%evaluation of results
%calculation of values for FWHM
nys=length(yg);
peaksizeFWHM=0;
peaksizenote=0;
normheight=max(Rnorm2)/2; %Full Width Half Maximum
for i=1:nys−1
       if Rnorm2(nys−i) >normheight & peaksizenote==0
         y1=yg(nys−i+1);
         y2=yg(nys−i);
         z1=Rnorm2(nys−i+1);
         z2=Rnorm2(nys−i);
         peaksizeFWHM=y1−abs((y2−y1)/(z2−z1))*(normheight−z1);
         peaksizenote=1;
       elseif Rnorm2(nys−i) >normheight & peaksizenote==0 & i==1
         peaksizeFWHM=100;
         peaksizenote=2;
       else
       end
end
peaksizeFWHM=2*peaksizeFWHM; %calculate diameter from radius
ycenter=[0:max_r/nys:max_r];
%fraction of photons in FWHM relative to all photons on detector
```

```
probabFWHM=0;
probabnote=0;
for i=1:nys
    if ycenter(i+1)<peaksizeFWHM/2 & probabnote==0
        probabFWHM=probabFWHM+Rnorm2(i)*(pi*(ycenter(i+1)^2−ycenter(i)^2))/(startnum*gfactor);
    elseif ycenter(i+1)>peaksizeFWHM/2 & probabnote==0
        probabFWHM=probabFWHM+Rnorm2(i)*(pi*((peaksizeFWHM/2)^2−ycenter(i)^2))/(startnum*gfactor);
        probabnote=1;
    else
    end
end
end

function [xendout,yendout,gewichtout,lossout,ntotalout]=sourceextension_3DS(d,zstart,zend,nc,nw,nf,nk,xmax,xcenter,ycenter)
%calculates propagation of light rays for extended source
xbound=[0:xmax/nf:xmax]; %nf+1 punkte
%run through source segments
for i=1:nf
    for j=1:nk
        xstart(i,j)=(xbound(i)+xbound(i+1))/2*cos(j*2*pi/nk)+xcenter;
        ystart(i,j)=(xbound(i)+xbound(i+1))/2*sin(j*2*pi/nk)+ycenter;
        %calculate endpoints for segments individually
        [xend(i,j,:,:),yend(i,j,:,:),gewicht(i,j,:),loss(i,j,:),ntotalin(i,j)]=singlepoint_3DS(d,zstart,zend,nc,nw,xstart(i,j),ystart(i,j));
    end
end
for i=1:nf
    densityvalue(i)=xbound(i+1)^2−xbound(i)^2;
end
for i=1:nf
    gewichtx(i)=densityvalue(i)/sum(densityvalue)/nk;
    for j=1:nk
        totalgewicht(i,j,:)=gewichtx(i)*gewicht(i,j,:);
    end
end
xendout=xend;
yendout=yend;
gewichtout=totalgewicht;
lossout=loss;
ntotalout=ntotalin;
end

function [xendout,yendout,gewichtout,lossout,ntotalout]=singlepoint_3DS(d,zstart,zend,nc,nw,xstart,ystart);
%calculates propagation of light rays for point source
betamax=pi/2;
betag=[−betamax:2*betamax/nw:betamax]; %nw+1 points
for i=1:nw
    beta(i)=(betag(i)+betag(i+1))/2;
    %propagate light through optical system
    [xend(i,:,:),yend(i,:,:),gewicht(i,:),loss(i,:),ncnew(i)]=opticcalc_final_3DS(d,zstart,zend,nc,nw,xstart,ystart,beta(i));
end
counter=1;
for i=1:nw
    for j=1:ncnew(i)
        for k=1:length(zend)
            xendout(counter,k)=xend(i,j,k);
            yendout(counter,k)=yend(i,j,k);
        end
        gewichtout(counter)=gewicht(i,j);
        lossout(counter)=loss(i,j);
        counter=counter+1;
    end
end
ntotalout=length(xendout(:,1));
end


function [xend,yend,gewicht,lossout2,ncnew]=opticcalc_final_3DS(d,zstart,zend,nc,nw,xstart,ystart,beta)
%propagates light rays through optical system

%optical parameters of first mirror
D1=39;
delta1=0;
R1=−20;
kappa1=−1;
zeta1=0;
xi1=1;
R2=0;
kappa2=0;
zeta2=0;
xi2=0;
theta1=0; %mirror tilt
phi1=0.00; %mirror tilt
n=1;
opticelem1=1;
p1=[0 0 0 0 0 0 0 0 0 0 0 0 0 0 0 0 0 0 0 0];
p2=[0 0 0 0 0 0 0 0 0 0 0 0 0 0 0 0 0 0 0 0];

%optical parameters of second mirror
D2=39;
```

```
delta2=0;
R3=−4;
kappa3=−1;
zeta3=0;
xi3=1;
R4=0;
kappa4=0;
zeta4=0;
xi4=0;
theta2=0; %mirror tilt
phi2=0; %mirror tilt
n=1;
opticelem2=2;
p3=[0 0 0 0 0 0 0 0 0 0 0 0 0 0 0 0 0 0 0 0];
p4=[0 0 0 0 0 0 0 0 0 0 0 0 0 0 0 0 0 0 0 0];

almax=pi/2;
q1=parabcurve_3D(D1/2,0,abs(R1));
alg=[0:almax/nc:almax];%nc+1 Punkte
for i=1:nc
    gewichth(i)=1/(2∗nw)∗abs(cos(alg(i))−cos(alg(i+1)))/(1−cos(almax));%automatisch korrekte Normierung durch angepasste Gr\"{o}{\SS}e der Fl···
          \"{a}chenelemente
    alintermed(i)=(alg(i)+alg(i+1))/2;
end

for i=1:nc
    al(i)=−alintermed(nc+1−i);
    al(nc+i)=alintermed(i);
    gewicht(i)=gewichth(nc+1−i);
    gewicht(nc+i)=gewichth(i);
end
ncnew=2∗nc;
%propagate light rays
for i=1:2∗nc
    xin(i)=xstart+(zstart−q1)∗tan(al(i))∗cos(beta);
    yin(i)=ystart+(zstart−q1)∗tan(al(i))∗sin(beta);
    lossin1(i)=0;
    attenuationin1(i)=1;
    %first optical element is shallow parabolic mirror
    [alout1(i),betaout1(i),xout1(i),yout1(i),lossout1(i),attenuationout1(i)]=asphericmirrorray_3DS(al(i),beta,xin(i),yin(i),R1,D1,theta1,phi1,lossin1(i),···
          attenuationin1(i));
    xint(i)=xout1(i)+d∗tan(alout1(i))∗cos(betaout1(i));
    yint(i)=yout1(i)+d∗tan(alout1(i))∗sin(betaout1(i));
    if opticelem2==2 %deep parabolic mirror
        q2=parabcurve_3D(D2/2,0,abs(R3));
        [alout2(i),betaout2(i),xout2(i),yout2(i),lossout2(i),attenuationout2(i)]=asphericmirrorray2_3DS(alout1(i),−betaout1(i),xint(i),−yint(i),R3,D2,···
              theta2,phi2,lossout1(i),zend,attenuationout1(i));
        if zend>=q2
            xend(i,:)=xout2(i)+(zend−q2)∗tan(alout2(i))∗cos(−betaout2(i));
            yend(i,:)=−yout2(i)+(zend−q2)∗tan(alout2(i))∗sin(−betaout2(i));
        else
            xend(i,:)=xout2(i);
            yend(i,:)=yout2(i);
        end
    end
    gewicht(i)=gewicht(i)∗attenuationout2(i); %takes losses of transmission and reflection into account
end
end

function [X,Y,Znorm2,zges,counter,gfactor,Rnorm2,yg,max_r]=new_distribution(nf,nk,ntotal,max_x,max_y,boxsize,xend,yend,gewicht,loss,startnum)
%calculates light distribution from terminated light rays
nxc=round(max_x/boxsize);
nyc=round(max_y/boxsize);
[X,Y] = meshgrid(−max_x/2:max_x/nxc:max_x/2,−max_y/2:max_y/nyc:max_y/2);
Z=zeros(nyc+1,nxc+1);
counter=zeros(nyc+1,nxc+1);
for mi=1:nf
    for mj=1:nk
        for l=1:ntotal
            if loss(mi,mj,l)==0 & abs(xend(mi,mj,l))<max_x/2 & abs(yend(mi,mj,l))<max_y/2
                i=floor((xend(mi,mj,l)+max_x/2)/(max_x/nxc))+1;
                j=floor((yend(mi,mj,l)+max_y/2)/(max_y/nyc))+1;
                Z(j,i)=Z(j,i)+gewicht(mi,mj,l);
                counter(j,i)=counter(j,i)+1;
            end
        end
    end
end
for i=1:nxc+1
    for j=1:nyc+1
        Zrela2(j,i)=Z(j,i)∗1/(max_x/nxc∗max_y/nyc); %area density
    end
end

%calculate the weight of photons that arrive at detector
zges=0;
for i=1:nf
    for j=1:nk
        for l=1:ntotal
```

```
                if loss(i,j,l)==0
                    zges=zges+gewicht(i,j,l);
                end
            end
        end
    end
%fraction of photons that arrive at detector compared to all emitted photons
gfactor=0.5*zges; %factor 0.5 as photons are only emitted in one hemisphere
Znorm2=Zrela2/zges *startnum*gfactor; %normalisation to photons per s and mm^2

%radial symmetric distribution for calculation of FWHM
max_r=max_x/2;
nys=round(max_r/boxsize); %number of segments used to investigate spot size
ycenter=[0:max_r/nys:max_r];%nys+1 points equally distributed around the center
yg=[max_r/(2*nys):max_r/nys:max_r-max_r/(2*nys)];
R=zeros(1,nys);
for mi=1:nf
    for mj=1:nk
        for l=1:ntotal
            rend(mi,mj,l)=sqrt(xend(mi,mj,l)^2+yend(mi,mj,l)^2);
            if loss(mi,mj,l)==0 & rend<max_r
                i=floor((rend(mi,mj,l))/(max_r/nys))+1;
                R(i)=R(i)+gewicht(mi,mj,l); %radial weight distribution
            end
        end
    end
end
for i=1:nys
    Rrela2(i)=R(i)/(pi*abs(ycenter(i+1)^2-ycenter(i)^2)); %area density
end
Rnorm2=Rrela2/zges *startnum*gfactor; %normalisation to photons per s and mm^2
end

function [x,y] = findzerooutshiftedparab(R,theta,phi,xin,yin,al,beta,q)
f=abs(R)/2;
if abs(phi)>0
rem=-sin(phi)^2*cos(phi)/(4*f)+2*tan(al)*cos(phi)^2/(4*f)*sin(phi)*(cos(beta)*sin(theta)-sin(beta)*cos(theta))-tan(al)^2*cos(phi)/(4*f)*(cos(beta)...
        ^2*(cos(theta)^2+sin(theta)^2*cos(phi)^2)+sin(beta)^2*(sin(theta)^2+cos(theta)^2*cos(phi)^2)+2*cos(beta)*sin(beta)*sin(theta)*cos(theta)*...
        sin(phi)^2);
pc=(-1-sin(phi)^2+2*cos(phi)^2/(4*f)*sin(phi)*(xin*sin(theta)-yin*cos(theta))+tan(al)*cos(phi)*sin(phi)*(cos(beta)*sin(theta)-sin(beta)*cos(theta...
        ))-2*tan(al)*cos(phi)/(4*f)*(xin*cos(beta)*(cos(theta)^2+sin(theta)^2*cos(phi)^2)+yin*sin(beta)*(sin(theta)^2+cos(theta)^2*cos(phi)^2)+sin...
        (theta)*cos(theta)*sin(phi)^2*(xin*sin(beta)+yin*cos(beta))))/rem;
qc=(q+cos(phi)*sin(phi)*(xin*sin(theta)-yin*cos(theta))-cos(phi)/(4*f)*(xin^2*(cos(theta)^2+sin(theta)^2*cos(phi)^2)+yin^2*(sin(theta)^2+cos(theta...
        )^2*cos(phi)^2)+2*xin*yin*sin(theta)*cos(theta)*sin(phi)^2))/rem;
zc=-pc/2+sqrt(pc^2/4-qc);
x=xin+zc*tan(al)*cos(beta);
y=yin+zc*tan(al)*sin(beta);
else
pc=(2*tan(al)*(xin*cos(beta)+yin*sin(beta))+4*f)/tan(al)^2;
qc=(xin^2+yin^2-4*f*q)/tan(al)^2;
zc=-pc/2+sqrt(pc^2/4-qc);
x=xin+zc*tan(al)*cos(beta);
y=yin+zc*tan(al)*sin(beta);
end
end

function [mx,my]=parabdiff(R,theta,phi,x,y)
f=abs(R)/2;
if abs(phi)>0
    p=4*f/cos(phi)-2*x.*cos(phi)*sin(theta)/(sin(phi))+2*y.*cos(phi)*cos(theta)/(sin(phi))-4*f/(sin(phi)^2*cos(phi));
    q=-x.*4*f*sin(theta)/sin(phi)+y.*4*f*cos(theta)/sin(phi)+x.^2/(sin(phi)^2)*(cos(theta)^2+sin(theta)^2*cos(phi)^2)+y.^2/(sin(phi)^2)*(sin(theta)^2+...
        cos(theta)^2*cos(phi)^2)+2*x.*y.*sin(theta)*cos(theta);
    dpx=-2*cos(phi)*sin(theta)/sin(phi);
    dpy=2*cos(phi)*cos(theta)/sin(phi);
    dqx=-4*f*sin(theta)/sin(phi)+2*x/sin(phi)^2*(cos(theta)^2+sin(theta)^2*cos(phi)^2)+2*y*sin(theta)*cos(theta);
    dqy=4*f*cos(theta)/sin(phi)+2*y/sin(phi)^2*(sin(theta)^2+cos(theta)^2*cos(phi)^2)+2*x*sin(theta)*cos(theta);
    mx=-(-1/2*dpx-1/4*dpx*p*1/sqrt(p^2/4-q)+1/2*dqx*1/sqrt(p^2/4-q));
    my=-(-1/2*dpy-1/4*dpy*p*1/sqrt(p^2/4-q)+1/2*dqy*1/sqrt(p^2/4-q));
else
    mx=-1/abs(R)*x;
    my=-1/abs(R)*y;
end
end

function z=parabshape(x,y,theta,phi,R)
if abs(phi)>0
    f=abs(R)/2;
    p=4*f/cos(phi)-2*x.*cos(phi)*sin(theta)/(sin(phi))+2*y.*cos(phi)*cos(theta)/(sin(phi))-4*f/(sin(phi)^2*cos(phi));
    q=-x.*4*f*sin(theta)/sin(phi)+y.*4*f*cos(theta)/sin(phi)+x.^2/(sin(phi)^2)*(cos(theta)^2+sin(theta)^2*cos(phi)^2)+y.^2/(sin(phi)^2)*(sin(theta)^2+...
        cos(theta)^2*cos(phi)^2)+2*x.*y.*sin(theta)*cos(theta);
    z=-p./2-sqrt(p.^2/4-q);
else
    z=parabcurve_3D(x,y,R);
end
end

function z=parabcurve_3D(x,y,R)
z=1/(2*abs(R))*(x^2+y^2);
end
```

```
function [alout,betaout,xout,yout,lossout,attenuationout]=asphericmirrorray_3DS(alin,betain,xin,yin,R,D,theta,phi,lossin,attenuationin)
% propagates light through parabolic entrance mirror
if lossin==1
    lossout=1;
    alout=0;
    betaout=0;
    xout=0;
    yout=0;
    attenuationout=0;
else
    qo=parabcurve_3D(D/2,0,abs(R));
    [hx,hy]=findzerooutshiftedparab(R,theta,phi,xin,yin,alin,betain,qo);
    hz=parabshape(hx,hy,theta,phi,R);
    h=sqrt((hx*cos(theta)+hy*sin(theta))^2+(cos(phi)*(-hx*sin(theta)+hy*cos(theta))+hz*sin(phi))^2);
    if (h==NaN) | abs(imag(h))>0 | h>D/2 | h<6 % mirror center + hole
        lossout=1;
        alout=0;
        betaout=0;
        xout=0;
        yout=0;
        attenuationout=0;
    else
        l=qo-hz;
        [mx,my]=parabdiff(R,theta,phi,hx,hy);
        nx=mx/sqrt(mx^2+my^2+1);
        ny=my/sqrt(mx^2+my^2+1);
        nz=-1/sqrt(mx^2+my^2+1);
        vx=(xin-hx)/sqrt((xin-hx)^2+(yin-hy)^2+l^2);
        vy=(yin-hy)/sqrt((xin-hx)^2+(yin-hy)^2+l^2);
        vz=-l/sqrt((xin-hx)^2+(yin-hy)^2+l^2);
        thetain=acos(nx*vx+ny*vy+nz*vz);
        attenuationout=attenuationin*fresnelcalc_mirror(thetain); % reflection losses
        voutx=(2*nx^2-1)*vx+2*nx*ny*vy+2*nx*nz*vz;
        vouty=2*nx*ny*vx+(2*ny^2-1)*vy+2*ny*nz*vz;
        voutz=2*nx*nz*vx+2*ny*nz*vy+(2*nz^2-1)*vz;
        norm=-l/voutz;
        voutxnorm=norm*voutx;
        voutynorm=norm*vouty;
        xout=voutxnorm+hx;
        yout=voutynorm+hy;
        rout=sqrt(xout^2+yout^2);
        if rout>D/2
            xout=0;
            yout=0;
            alout=0;
            betaout=0;
            lossout=1;
            attenuationout=0;
        else
        if voutx>=0
            s=sqrt(voutx^2+vouty^2);
        else
            s=-sqrt(voutx^2+vouty^2);
        end
        alout=atan(s/abs(voutz));
        if voutx==0
            betaout=pi/2;
        else
            betaout=atan(vouty/voutx);
        end
        lossout=0;
        end
    end
end
end

function [alout,betaout,xout,yout,lossout,attenuationout]=asphericmirrorray2_3DS(alin,betain,xin,yin,R,D,theta,phi,lossin,zend,attenuationin)
% propagates light through parabolic exit mirror
if lossin==1
    lossout=1;
    alout=0;
    betaout=0;
    xout=0;
    yout=0;
    attenuationout=0;
else
    qo=parabcurve_3D(D/2,0,abs(R));
    [hx,hy]=findzerooutshiftedparab(R,theta,phi,xin,yin,alin,betain,qo);
    hz=parabshape(hx,hy,theta,phi,R);
    h=sqrt((hx*cos(theta)+hy*sin(theta))^2+(cos(phi)*(-hx*sin(theta)+hy*cos(theta))+hz*sin(phi))^2);
    if (h==NaN) | abs(imag(h))>0 | h>D/2 | h<6 % center hole
        lossout=1;
        alout=0;
        betaout=0;
        xout=0;
        yout=0;
        attenuationout=0;
    else
```

```
        l=qo-hz;
        [mx,my]=parabdiff(R,theta,phi,hx,hy);
        nx=mx/sqrt(mx^2+my^2+1);
        ny=my/sqrt(mx^2+my^2+1);
        nz=-1/sqrt(mx^2+my^2+1);
        vx=(xin-hx)/sqrt((xin-hx)^2+(yin-hy)^2+l^2);
        vy=(yin-hy)/sqrt((xin-hx)^2+(yin-hy)^2+l^2);
        vz=-l/sqrt((xin-hx)^2+(yin-hy)^2+l^2);
        thetain=acos(nx*vx+ny*vy+nz*vz);
        attenuationout=attenuationin*fresnelcalc_mirror(thetain);
        voutx=-((2*nx^2-1)*vx+2*nx*ny*vy+2*nx*nz*vz);
        vouty=-(2*nx*ny*vx+(2*ny^2-1)*vy+2*ny*nz*vz);
        voutz=-(2*nx*nz*vx+2*ny*nz*vy+(2*nz^2-1)*vz);
          if zend<qo
            lout=parabcurve_3D(hx,hy,abs(R))-zend;
          else
            lout=parabcurve_3D(hx,hy,abs(R))-qo;
          end
          norm=lout/voutz;
          voutxnorm=norm*voutx;
          voutynorm=norm*vouty;
          xout=voutxnorm+hx;
          yout=voutynorm+hy;
          rout=sqrt(xout^2+yout^2);
          if rout>D/2
            xout=0;
            yout=0;
            alout=0;
            betaout=0;
            lossout=1;
            attenuationout=0;
          else
            if voutx>=0
              s=sqrt(voutx^2+vouty^2);
            else
              s=-sqrt(voutx^2+vouty^2);
            end
            alout=atan(s/abs(voutz));
            if voutx==0
              betaout=pi/2;
            else
              betaout=atan(vouty/voutx);
            end
            lossout=0;
          end
      end
    end
    end

    function reflectivity=fresnelcalc_mirror(refangle)
    %calculates reflectivity losses of Al-mirror in dependence of angle
    reflectivity=0.598+0.349*cos(refangle)-0.123*cos(refangle)^2; %angular dependence of reflectivity MgF2 coated Al @ 157 nm
    %source: Nucl inst meth A 480 (2002) 65-70
    end
```

B.3 Image Evaluation

The decay signal of ^{229m}Th was detected with the help of an MCP detector combined with a phosphor screen for spatially resolved read out (see Sect. 4.4). The phosphor screen was imaged onto a CCD chip and individual image frames with 4 s exposure time were acquired. Typically, between 500 and 18 000 image frames were saved for one measurement (see Sect. 5.2.2). These image frames had to be numerically evaluated, in order infer the event positions of single MCP events on the phosphor screen and subsequently add these events to obtain an integrated image.

Three Matlab programs were used for image-frame evaluation. The first program (image_process) screens the image frames (saved in the form IMG.00001.tif, with continuous numbering) in order to infer the x and y positions of individual MCP events. The evaluation is started by running the program and opening the first image file. The events are distinguished from CCD noise and hot pixels by size and intensity.

These events are saved by the program as separate .dat files (validx and validy). The .dat files are post-processed by a different Matlab program (count_process), which integrates the events and produces a visualization as shown in Fig. 5.17. A third program (count_evaluation) was used to infer the background corrected number of detected events for the measurement. Parts of a Matlab program package developed for digital image correlation and tracking by C. Eberl were used as a basis and modified for our purposes [5].

```
function image_process
%generates x and y coordinate files from CCD camera frames (.tif).
%parts of the program code were developed for a digital image
%correlation tracking program by Christopher Eberl
%(publicly available) and adapted for our purposes
clc;

[Firstimagename,ImageFolder,tif_filenamelist]=create_tif_filenamelist; %create filenamelist for tif-files
peak_labelling(tif_filenamelist); %generates x and y coordinate files
%add_pictures(tif_filenamelist); %simple adding of image frames (optional)
end

function [Firstimagename,ImageFolder,tif_filenamelist]=create_tif_filenamelist
%creates .tif-filename list of frames
[Firstimagename ImageFolder]=uigetfile('*.tif','Open First Image');
if Firstimagename~~[];
    cd(ImageFolder);
end
if Firstimagename~~[];
    % Get the number of image name
    letters=isletter(Firstimagename);
    Pointposition=findstr(Firstimagename,'.');
    Firstimagenamesize=size(Firstimagename);
    counter=Pointposition-1;
    counterpos=1;
    letterstest=0;
    while letterstest==0
        letterstest=letters(counter);
        if letterstest==1
            break
        end
        Numberpos(counterpos)=counter;
        counter=counter-1;
        counterpos=counterpos+1;
        if counter==0
            break
        end
    end
    Filename_first = Firstimagename(1:min(Numberpos)-1);
    Firstfilenumber=Firstimagename(min(Numberpos):max(Numberpos));
    Lastname_first = Firstimagename(max(Numberpos)+1:Firstimagenamesize(1,2));
    Firstfilenumbersize=size(Firstfilenumber);
    onemore=10^(Firstfilenumbersize(1,2));
    tif_filenamelist(1,:)=Firstimagename;
    Firstfilenumber=str2num(Firstfilenumber);
    u=1+onemore+Firstfilenumber;
    ustr=num2str(u);
    tif_filenamelist(2,:)=[Filename_first ustr(2:Firstfilenumbersize(1,2)+1) Lastname_first];
    numberofimages=2;
    counter=1;
    while exist(tif_filenamelist((counter+1),:),'file') ==2;
        counter=counter+1;
        u=1+u;
        ustr=num2str(u);
        tif_filenamelist(counter+1,:)=[Filename_first ustr(2:Firstfilenumbersize(1,2)+1) Lastname_first];
        if exist(tif_filenamelist((counter+1),:),'file') ==0;
            tif_filenamelist(counter+1,:)=[];
            break
        end
        counter
    end
end
save('tif_filenamelist.mat','tif_filenamelist');
end

function peak_labelling(tif_filenamelist);
%evaluates image frames to obtain events from intensity (roi) and peak size
%(Amin and Amax)
filenumber=length(tif_filenamelist(:,1));
nonzero=0;
for n=1:filenumber
n
```

```
I2=imread(tif_filenamelist(n,:));
peak=max(max(I2));
mittel=mean(mean(I2));
roi=(I2>10); % subtract greyvalues to work only with real peaks
[labeled,numObjects] = bwlabel(roi,8); %label all peaks
powderdata=regionprops(labeled,'basic'); % get peak properties from bwlabel
powderarea=[powderdata.Area]; %define area variable
powdercentroid=[powderdata.Centroid]; %define position variable
powderboundingbox=[powderdata.BoundingBox]; %define bounding box variable
counter=0;
countermax=length(powdercentroid)/2;
powderxy=zeros(countermax,4);
for i=1:countermax; % get all data from the bwlabel (position, bounding box and area of peaks)
    counter=counter+1; %
    powderxy(i,1)=i; % number of the detected event
    powderxy(i,2)=powdercentroid(1, (i*2−1)); % x coordinate of event position
    powderxy(i,3)=powdercentroid(1, (i*2)); % y coordinate of event position
    powderxy(i,4)=powderarea(1, i); % area of bounding box
end

% crop away peaks which are too small or too big
Amin=3;%minimum peaksize
Amax=1000;%maximum peaksize
counter=0;
i=0;
% throw away useless peaks (defined by size)
while (counter<505) & (i<countermax)
    i=i+1;
    if Amin<powderxy(i,4) % crop all points with a small peak area
        if powderxy(i,4)<Amax % crop all points with a too big area
            counter=counter+1;
            cropxy(counter,1)=counter; % peaks get a new number
            cropxy(counter,2)=powderxy(i,2); % x
            cropxy(counter,3)=powderxy(i,3); % y
            cropxy(counter,4)=powderxy(i,4); % area bounding box
        end
    end
end

if counter>0 & length(cropxy(:,2))<500 %only images with less than 500 events are considered
    nonzero=nonzero+1;
    validnumber(nonzero)=length(cropxy(:,2));
    validx(1:validnumber(nonzero),nonzero)=cropxy(:,2);
    validy(1:validnumber(nonzero),nonzero)=cropxy(:,3);
end

% clear variables
clear powderxy
clear powderdata
clear powderarea
clear powdercentroid
clear powderboundingbox
clear counter
clear countermax
clear Amin
clear Amax
clear roi
clear cropxy
end

imageheight=length(I2(:,1));
imagelength=length(I2(1,:));
% generate single image from events
A=imageevaluation(imageheight,imagelength,validx,validy,validnumber);
% plot image on screen
figure;
imagesc(A,[0,30]);
colorbar('vert');

% save the coordinates
save('validx');
save('validy');
save validx.dat validx −ascii −tabs;
save validy.dat validy −ascii −tabs;
end

function array=imageevaluation(imageheight,imagelength,validx,validy,validnumber)
% Adds the detected events to build a single image
m=206;% number of pixels in y direction (original .tif files have 1032 pixels)
n=278;% number of pixels in x direction (original .tif files have 1392 pixels)
array=zeros(m+1,n+1);
for k=1:length(validx(1,:)) %run through pictures
    for l=1:validnumber(k)
        i=floor((validx(l,k))/(imagelength/n))+1;
        j=floor((validy(l,k))/(imageheight/m))+1;
        if array(j,i)<500
            array(j,i)=array(j,i)+1;
        end
        %array=circlefill(array,i,j); %optional, detected events are shown as filled circles
```

```
        end
    end
    end

    function arrayout=circlefill(arrayin,centerx,centery)
    %generates filled circles for detected events (optional)
    radius=7; %radius of the filled circles (in pixel numbers of image frame)
    for j=1:2*radius+1
        path(j)=floor(sqrt(radius^2-(radius-j+1)^2));
        for i=1:2*path(j)+1
            if centery-radius+j-1<length(arrayin(:,1)) & centery-radius+j-1>0 & centerx-path(j)+i-1<length(arrayin(1,:)) & centerx-path(j)+i-1>0
            if arrayin(centery-radius+j-1,centerx-path(j)+i-1)<500
                arrayin(centery-radius+j-1,centerx-path(j)+i-1)=arrayin(centery-radius+j-1,centerx-path(j)+i-1)+1;
            end
            end
        end
    end
    arrayout=arrayin;
    end

    function add_pictures(tif_filenamelist);
    % simple adding of image frames (optional)
    filenumber=length(tif_filenamelist(:,1));
    Iadd=imread(tif_filenamelist(1,:));
    Iadd=double(Iadd);
    for n=2:filenumber
        n
        toadd=imread(tif_filenamelist(n,:));
        toadd=double(toadd);
        Iadd=Iadd+toadd;
    end
    Iadd=Iadd/filenumber;
    figure, image(Iadd);
    imagesc(Iadd,[0,12]);
    colorbar('vert');
    end
```

```
function count_process
%angepasst an extreme Z\"{a}hlraten von neuer Quelle (12.2015)
clc;
%open the validx and validy files
[Image,PathImage] = uigetfile('*.tif','Open Image');
cd(PathImage)
load('validx')
load('validy')

imageheight=1032;%pixel size of image frames
imagelength=1392;
reallength=85.7; %real x-length of the image frames in mm
centerx=780; %define the center of the image (pixels)
centery=450;

binningsize=0.2; %binningsize in mm
smoothradius=0.3; %smoothradius in mm (for circlefill, smoothradius= n*binnningsize)
realfilterradius=10; %define the radius of the region of interest shown (in mm)
realitywidth=30; %size of the shown image in mm (squared)
measurementtime=2000; %time of the image integration in s

realityscale=imagelength/reallength; %wieviele Bildpixel im Ausgangsbild entsprechen 1 mm
scale=1/(binningsize*realityscale); %binningscale
fillradius=double(int8(smoothradius/binningsize)); %filled radius for each single count; (the larger, the smoother)

xlength=round(scale*imagelength);
ylength=round(scale*imageheight);
width=realitywidth*realityscale;
imagecenterx=round(scale*centerx);
imagecentery=round(scale*centery);
imagewidth=round(realitywidth/binningsize);
filtercenterx=imagecenterx;
filtercentery=imagecentery;
filterradius=round(realfilterradius/binningsize);

validnumber=validnumbertest(validx); %get number of events per frame
%use x and y coordinates of events to generate image
A=imageevaluation(imageheight,imagelength,validx,validy,validnumber,scale,fillradius,filtercenterx,filtercentery,filterradius);
%generate submatrix (outer squared dimension of image)
B1=submatrix(imagecenterx-(round(imagewidth/2)+1),imagecenterx+round(imagewidth/2),imagecentery-(round(imagewidth/2)+1),imagecentery+...
        round(imagewidth/2),A);
%convert to physical parameters
B=1/binningsize^2*1/zscaleparameter(fillradius)*1/measurementtime*B1;

%plot image
figure;
imagesc(B,[0,0.5]);
axis equal
```

```
axis([1 round(scale*width)+1 1 round(scale*width)+1]);
set(gca,'XTick',1:(imagewidth)/4:imagewidth+1);
set(gca,'XTickLabel',-realitywidth/2:realitywidth/4:realitywidth/2,'FontSize',14);
set(gca,'YTick',1:(imagewidth)/4:imagewidth+1);
set(gca,'YTickLabel',-realitywidth/2:realitywidth/4:realitywidth/2,'FontSize',14);
xlabel('x [mm]','FontSize',14);
ylabel('y [mm]','FontSize',14);
colorbar('vert');
pos=get(gca,'pos');
hc=colorbar('position',[pos(1)+0.7 pos(2) pos(3)-0.7 pos(4)]);
set(get(hc,'ylabel'),'String', 'signal intensity [cts/s mm^2]','FontSize',14);
set(hc,'Fontsize',14);
%save image
print -dtiff -r200 2D_image_2.tif

end

function zscale=zscaleparameter(fillradius)
%get numbers for conversion to physical parameters if circlefill is used
if fillradius==1
    zscale=5;
elseif fillradius==2
    zscale=13;
elseif fillradius==3
    zscale=29;
elseif fillradius==4
    zscale=49;
elseif fillradius==5
    zscale=81;
elseif fillradius==6
    zscale=113;
elseif fillradius==7
    zscale=149;
else fillradius==8
    zscale=197;
end
end

function array=imageevaluation(imageheight,imagelength,validx,validy,validnumber,scale,fillradius,filtercenterx,filtercentery,filterradius)
% generates integrated image from x and y coordinates of events
m=round(scale*imageheight);
n=round(scale*imagelength);
array=zeros(m+1,n+1);
for k=1:length(validx(1,:)) %run through pictures
    for l=1:validnumber(k)
        if hotpixeltest(validx(l,k),validy(l,k))==1 %removes hot pixels of MCP
        i=floor((validx(l,k))/(imagelength/n))+1;
        j=floor((validy(l,k))/(imageheight/m))+1;
        if filtertest(i,j,filtercenterx,filtercentery,filterradius)==1 % only region of interest is shown
            array=circlefill(array,i,j,fillradius);
        end
        end
    end
end
end

function a=hotpixeltest(x,y)
%removes hot pixels of MCP (coordinates manually included)
  a=1;
if x>=746 & x<=749& y>=519 & y<=522
    a=0;
end
end

function b=filtertest(i,j,filtercenterx,filtercentery,filterradius)
%generates filter to show only region of interest
b=0;
if i>filtercenterx-filterradius & i<filtercenterx+filterradius
   filterpath=floor(sqrt(filterradius^2-(filtercenterx-i)^2));
      if j>filtercentery-filterpath & j<filtercentery+filterpath
         b=1;
      end
end
end

function arrayout=circlefill(arrayin,centerx,centery,fillradius)
%presents single events as filled circles
radius=fillradius;
for j=1:2*radius+1
    path(j)=floor(sqrt(radius^2-(radius-j+1)^2));
    for i=1:2*path(j)+1
        if centery-radius+j-1<length(arrayin(:,1)) & centery-radius+j-1>0 & centerx-path(j)+i-1<length(arrayin(1,:)) & centerx-path(j)+i-1>0
        if arrayin(centery-radius+j-1,centerx-path(j)+i-1)<500
            arrayin(centery-radius+j-1,centerx-path(j)+i-1)=arrayin(centery-radius+j-1,centerx-path(j)+i-1)+1;
        else
            disp('error');
        end
        end
    end
```

```
end
arrayout=arrayin;
end

function validnumber=validnumbertest(validx);
%obtains number of events for each frame
for i=1:length(validx(1,:)) %number of frames
    validnumber(i)=length(validx(:,1));
    for j=1:length(validx(:,1))
        if validx(validnumber(i),i)==0
            validnumber(i)=validnumber(i)−1;
        end
    end
end
end

function B=submatrix(x1,x2,y1,y2,A)
B=A(y1:y2,x1:x2);
end
```

```
function count_evaluation
%obtains the number of events in a defined circular region of interest
clc;
%load the x and y coordinates of events
[Image,PathImage] = uigetfile('*.tif','Open Image');
cd(PathImage)
load('validx')
load('validy')

imageheight=1032;%pixel size of image frames
imagelength=1392;
validnumber=validnumbertest(validx); %get the number of events for each frame
array=imageevaluation(imageheight,imagelength,validx,validy,validnumber);
centerx=156;% center of the circle used for count evalation
centery=91;
radius=7; %radius of region for event count
total=countsinroi(array,centerx,centery,radius) %number of events in roi
centerx=145;% center of circle used for dark count evaluation
centery=120;
radius=7; %radius of region for event count
darks=countsinroi(array,centerx,centery,radius) %number of events in roi
counts=total−darks
end

function number=countsinroi(array,centerx,centery,radius)
%counts the number of events in roi
number=0;
for j=1:2*radius+1
    path(j)=floor(sqrt(radius^2−(radius−j+1)^2));
    for i=1:2*path(j)+1
        number=number+array(centery−radius+j−1,centerx−path(j)+i−1);
        array(centery−radius+j−1,centerx−path(j)+i−1)=1;
    end
end
%plots the roi
figure;
imagesc(array,[0,10]);
colorbar('vert');
end

function array=imageevaluation(imageheight,imagelength,validx,validy,validnumber)
% generates integrated image from x and y coordinates of events
m=206;%new pixelsize of the image
n=278;
array=zeros(m+1,n+1);
for k=1:length(validx(1,:)) %run through pictures
    for l=2:validnumber(k)
        i=floor((validx(l,k))/(imagelength/n))+1;
        j=floor((validy(l,k))/(imageheight/m))+1;
        if array(j,i)<255
            array(j,i)=array(j,i)+1;
        end
    end
end
%plots the image
figure;
imagesc(array,[0,10]);
colorbar('vert');
end

function validnumber=validnumbertest(validx);
%obtains the number of events for each image frame
for i=1:length(validx(1,:)) %number of images
    validnumber(i)=length(validx(:,1));
    for j=1:length(validx(:,1))
```

```
            if validx(validnumber(i),i)==0
                validnumber(i)=validnumber(i)-1;
            end
        end
    end
end
```

Bibliography

1. Bateman H (1910) Solution of a system of differential equations occuring in the theory of radioactive transformations. In: Proceedings of the Cambridge Philosophical Society IS (1910) pp 423–427
2. Ziegler JF, Biersack JP, Littmark U (1985) The stopping and range of ions in matter. Pergamon, New York
3. Nordlund K (1995) Molecular dynamics simulation of ion ranges in the 1–100 keV energy range. Comput Mater Sci 3:448–456
4. Dahl DA et al (1990) SIMION PC/PS2 electrostatic lens design program. Rev Sci Imstrum 61:607–609
5. Eberl C (2016) Digital image correlation and tracking. In: Matlab source code available at Mathworks file exchange. http://de.mathworks.com/matlabcentral/fileexchange/12413-digital-image-correlation-and-tracking [2016, September 21st]

Curriculum Vitae of the Author

Lars von der Wense was born in 1984 in Elmshorn, Germany. He received his secondary education in his hometown, already having a strong interest in physics. During his school days he received a Lion's club scholarship for a scientific stay at the Max-Planck-Institute for Extraterrestrial Physics in Garching. Moreover, he was a finalist in the German young scientists contest "Jugend Forscht" and his Abitur (German A levels) was awarded with a Prize of the German Physical Society. In 2003 Lars began to study Physics and Mathematics at the University of Hamburg, receiving his intermediate diploma for both subjects in 2006 with best grades. He continued his graduate studies in Physics and Mathematics at the University of Hamburg, receiving his diploma in theoretical physics in 2009 for a thesis titled "Classical Strings in Area Metric Backgrounds". During the time of his studies he was exempted from the study fees for performance reasons (being among the 5% best of his year) and was awarded a Gude-scholarship within the "Year of Mathematics 2008".

After finalizing his Physics studies in Hamburg, Lars proceeded to the University of Cambridge, UK, to participate in the Part III master course of Applied Mathematics in 2010. Having successfully passed the master course in Cambridge he moved to Garching in 2011 for his PhD thesis at the Ludwig-Maximilians-University Munich dealing with the development of a nuclear clock based on the ^{229}Th isotope. During the work on his PhD thesis he received a scholarship of the Max-Planck-Institute for

L. Von der Wense, *On the Direct Detection of ^{229m}Th*,
Springer Theses, https://doi.org/10.1007/978-3-319-70461-6

Quantum Optics. He finalized his PhD thesis in 2016 graded "summa cum laude" in which he describes the first direct detection of the ^{229}Th nuclear isomer - a nuclear excited state which had not been detected within 40 years of research and will allow for the first development of an atomic clock based on a nuclear transition. This scientific achievement was elected one of the top 10 physics breakthroughs of the year 2016 by Physics World. Furthermore, Lars was awarded the "Dissertation Award" for the best physics PhD thesis 2016 by the Ludwig-Maximilians-University Munich. After he received his PhD he stayed in the same group, where he is continuing his work as a PostDoc.

List of Publications

1. von der Wense, L., Thirolf, P.G., Kalb, D., Laatiaoui, M., *Towards a direct transition energy measurement of the lowest nuclear excitation in* ^{229}Th, JINST **8** (2013) P03005.
2. Thirolf, P.G., von der Wense, L., Kalb, D., Laatiaoui, M., *Towards an all-optical access to the lowest nuclear excitation in* ^{229m}Th, Acta Physica Polonica B **44** (2013) 391-394.
3. von der Wense, L., Seiferle, B., Laatiaoui, M., Thirolf, P.G., *Determination of the extraction efficiency for* ^{233}U *source* α*-recoil ions from the MLL buffer-gas stopping cell*, Eur. Phys. J. A **51** (2015) 29.
4. von der Wense, L., Seiferle, B., Laatiaoui, M., Thirolf, P.G., *The extraction of* $^{229}Th^{3+}$ *from a buffer-gas stopping cell*, Nucl. Instrum. Methods B **376** (2016) 260-264.
5. Seiferle, B., von der Wense, L., Laatiaoui, M., Thirolf, P.G., *A VUV detection system for the direct photonic identification of the first excited isomeric state of* ^{229}Th, Eur. Phys. J. D **70** (2016) 58.
6. von der Wense, L., Seiferle, B., Laatiaoui, M., Neumayr, J.B., Maier, H.J., Wirth, H.F., Mokry, C., Runke, J., Eberhardt, K., Düllmann, C.E., Trautmann, N.G., Thirolf, P.G., *Direct detection of the* ^{229} *Th nuclear clock transition*, Nature **533** (2016) 47-51.
7. Seiferle, B., von der Wense, L., Thirolf, P.G., *Lifetime measurement of the* ^{229} *Th nuclear isomer*, Phys. Rev. Lett. **118** (2017) 042501.
8. Seiferle, B., von der Wense, L., Thirolf, P.G., *Feasibility study of internal conversion electron spectroscopy of* ^{229m} *Th*, Eur. Phys. J. A **53** (2017) 108.
9. Thielking, J., Okhapkin, M.V., Glowacki, P., Meier, D.M., von der Wense, L., Seiferle, B., Düllmann, C.E., Thirolf, P.G., Peik. E., Laser spectroscopic characterization of the nuclear clock isomer ^{229m}Th, submitted for publication (2017) arXiv: 1709.05325[nucl-ex].
10. von der Wense, L., Seiferle, B., Stellmer, S., Weitenberg, J., Kazakov, G., Pálffy, A., Thirolf, P.G., A laser excitation scheme for ^{229m}Th, Phys. Rev. Lett. **119** (2017) 132503.

List of conference talks

23.03.2012 *Preparations for an optical access to the lowest excited nuclear state in ^{229}Th*, DPG Conference Hadrons and Nuclei HK 59.7, Mainz.

25.09.2012 *Towards an all-optical access to the lowest nuclear excitation in ^{229}Th*, EMMI Workshop on ^{229}Th, GSI Darmstadt.

05.03.2013 *Preparations for an optical access to the lowest excited nuclear state in ^{229}Th*, DPG Conference Hadrons and Nuclei HK 31.6, Dresden.

19.03.2014 *Preparations for an optical access to the lowest nuclear excitation in ^{229}Th*, DPG Conference Hadrons and Nuclei HK 32.6, Frankfurt.

05.05.2014 *The Munich concept: Preparations for an optical access to the lowest nuclear excitation in ^{229}Th*, "nuClock" preparatory meeting, TU Vienna.

26.03.2015 *Preparations for an optical access to the lowest nuclear excitation in ^{229}Th*, DPG Conference Hadrons and Nuclei HK 66.3, Heidelberg.

14.05.2015 *Towards a nuclear clock: On the extraction of $^{229}Th^{3+}$ from a buffer-gas stopping cell*, EMIS conference 2015, Grand Rapids, USA.

14.09.2015 *On the first time unambiguous direct identification of the isomeric ground-state decay of ^{229}Th*, "nuClock" kick-off meeting, LMU Munich, Garching.

02.11.2015 *Towards a nuclear clock: On the direct detection of the thorium-229 isomer*, Seminar für Kern- und Radiochemie, Institute for Radiochemistry, University of Mainz (invited).

05.11.2015 von der Wense, L., Seiferle, B., Thirolf, P.G., Laatiaoui, M., *Towards a nuclear clock: On the direct detection of the thorium-229 isomer*, Maier-Leibnitz-Laboratory Kolloquium, LMU and TU Munich, Garching (invited).

13.01.2016 *Towards a nuclear clock: On the direct detection of the thorium-229 isomer*, Special Seminar at MPIK, Heidelberg (invited).

29.02.2016 *Towards a nuclear clock: On the direct detection of the thorium-229 isomer*, DPG Conference mass spectrometry MS 1.1, Hannover (invited).

14.03.2016 *Direct detection of the thorium-229 isomer: Milestone towards a nuclear clock*, DPG Conference Hadrons and Nuclei HK 6.2, Darmstadt.

08.06.2016 *The extraction of $^{229}Th^{3+}$ from a buffer-gas stopping cell and the direct detection of ^{229m}Th*, SMI Conference, Lanzhou, China (invited).

18.07.2016 *A first half-life determination of ^{229m}Th*, "nuClock" check meeting, Brussels, Belgium.

24.10.2016 *^{229}Th activities at Munich: Application to gas cell extraction efficiencies and chemistry*, RESIST Conference, Paris, France (invited).

11.01.2017 *Towards a nuclear clock: On the direct detection of the thorium-229 isomer*, CQD colloquium at Kirchhoff Institute for Physics, Heidelberg (invited).

30.03.2017 *A new nuclear laser excitation scheme for* ^{229m}*Th*, DPG conference Hadrons and Nuclei HK 41.7, Münster.

18.05.2017 *On the direct detection of the* ^{229}*Th nuclear isomer*, 9th Workshop on Precision Physics and Fundamental Physical Constants (FFK), Warsaw, Poland (invited).

14.06.2017 *Towards a* ^{229}*Th-based nuclear clock*, 175 years of the D.I.Mendeleyev Institute for Metrology (VNIIM) and National Measurement System, St. Petersburg, Russia (invited).

19.09.2017 *A direct laser excitation scheme for* ^{229m}*Th*, NuClock Consortium Meeting, Heidelberg.

FSC
www.fsc.org
MIX
Papier | Fördert gute Waldnutzung
FSC® C083411

Kolibri 360 GmbH
Ferdinand-Jühlke-Straße 7
99095 Erfurt, Deutschland
produktsicherheit@kolibri360.de